GENETICS, GENOMICS AND BREEDING OF POTATO

Genetics, Genomics and Breeding of Crop Plants

Series Editor

Chittaranjan Kole
Department of Genetics and Biochemistry
Clemson University
Clemson, SC
USA

Books in this Series:

Published or in Press:

- Jinguo Hu, Gerald Seiler & Chittaranjan Kole: *Sunflower*
- Kristin D. Bilyeu, Milind B. Ratnaparkhe & Chittaranjan Kole: *Soybean*
- Robert Henry & Chittaranjan Kole: *Sugarcane*
- Kevin Folta & Chittaranjan Kole: *Berries*
- Jan Sadowsky & Chittaranjan Kole: *Vegetable Brassicas*
- James M. Bradeen & Chittaranjan Kole: *Potato*
- C.P. Joshi, Stephen DiFazio & Chittaranjan Kole: *Poplar*
- Anne-Françoise Adam-Blondon, José M. Martínez-Zapater & Chittaranjan Kole: *Grapes*
- Christophe Plomion, Jean Bousquet & Chittaranjan Kole: *Conifers*

Books under preparation:

- Dave Edwards, Jacqueline Batley, Isobel Parkin & Chittaranjan Kole: *Oilseed Brassicas*
- Marcelino Pérez de la Vega, Ana María Torres, José Ignacio Cubero & Chittaranjan Kole: *Cool Season Grain Legumes*

GENETICS, GENOMICS AND BREEDING OF POTATO

Editors

James M. Bradeen
Department of Plant Pathology
University of Minnesota
St. Paul, Minnesota
USA

Chittaranjan Kole
Department of Genetics and Biochemistry
Clemson University
Clemson, SC
USA

CRC Press is an imprint of the
Taylor & Francis Group, an **informa** business
A SCIENCE PUBLISHERS BOOK

CRC Press
Taylor & Francis Group
6000 Broken Sound Parkway NW, Suite 300
Boca Raton, FL 33487-2742

First issued in paperback 2017

CRC Press is an imprint of Taylor & Francis Group, an Informa business

ISBN 13: 978-1-138-11530-9 (pbk)
ISBN 13: 978-1-57808-715-0 (hbk)

Cover Illustrations:
Left: An 'Inca Gold' potato plant in full bloom in a research field in Minnesota. *Upper Right:* Tubers of 'French Fingerling' potato. *Lower Right:* Tubers of 'Lumpers' potato. *Inset:* A close up of a potato flower revealing fused petals and a rotate or wheel-like composition. Photos courtesy of Dimitre Mollov and James Bradeen (University of Minnesota, Department of Plant Pathology).

Library of Congress Cataloging-in-Publication Data

Genetics, genomics and breeding of potato / editors, James M. Bradeen,Chittaranjan Kole.
p. cm. -- (Genetics, genomics and breeding of crop plants)
Includes bibliographical references and index.
ISBN 978-1-57808-715-0 (hardcover : alk. paper)
1. Potatoes--Genetics. 2. Potatoes--Genome mapping. 3. Potatoes--Breeding. I. Bradeen, James M. (James Mathew) II. Kole, Chittaranjan. III. Series: Genetics, genomics and breeding of crop plants.
SB211.P8.G385 2011
635'.21--dc22

2011006585

For G.L.K.

Preface to the Series

Genetics, genomics and breeding has emerged as three overlapping and complimentary disciplines for comprehensive and fine-scale analysis of plant genomes and their precise and rapid improvement. While genetics and plant breeding have contributed enormously towards several new concepts and strategies for elucidation of plant genes and genomes as well as development of a huge number of crop varieties with desirable traits, genomics has depicted the chemical nature of genes, gene products and genomes and also provided additional resources for crop improvement.

In today's world, teaching, research, funding, regulation and utilization of plant genetics, genomics and breeding essentially require thorough understanding of their components including classical, biochemical, cytological and molecular genetics; and traditional, molecular, transgenic and genomics-assisted breeding. There are several book volumes and reviews available that cover individually or in combination of a few of these components for the major plants or plant groups; and also on the concepts and strategies for these individual components with examples drawn mainly from the major plants. Therefore, we planned to fill an existing gap with individual book volumes dedicated to the leading crop and model plants with comprehensive deliberations on all the classical, advanced and modern concepts of depiction and improvement of genomes. The success stories and limitations in the different plant species, crop or model, must vary; however, we have tried to include a more or less general outline of the contents of the chapters of the volumes to maintain uniformity as far as possible.

Often genetics, genomics and plant breeding and particularly their complimentary and supplementary disciplines are studied and practiced by people who do not have, and reasonably so, the basic understanding of biology of the plants for which they are contributing. A general description of the plants and their botany would surely instill more interest among them on the plant species they are working for and therefore we presented lucid details on the economic and/or academic importance of the plant(s); historical information on geographical origin and distribution; botanical origin and evolution; available germplasms and gene pools, and genetic and cytogenetic stocks as genetic, genomic and breeding resources; and basic information on taxonomy, habit, habitat, morphology, karyotype, ploidy level and genome size, etc.

Classical genetics and traditional breeding have contributed enormously even by employing the phenotype-to-genotype approach. We included detailed descriptions on these classical efforts such as genetic mapping using morphological, cytological and isozyme markers; and achievements of conventional breeding for desirable and against undesirable traits. Employment of the in vitro culture techniques such as micro- and megaspore culture, and somatic mutation and hybridization, has also been enumerated. In addition, an assessment of the achievements and limitations of the basic genetics and conventional breeding efforts has been presented.

It is a hard truth that in many instances we depend too much on a few advanced technologies, we are trained in, for creating and using novel or alien genes but forget the infinite wealth of desirable genes in the indigenous cultivars and wild allied species besides the available germplasms in national and international institutes or centers. Exploring as broad as possible natural genetic diversity not only provides information on availability of target donor genes but also on genetically divergent genotypes, botanical varieties, subspecies, species and even genera to be used as potential parents in crosses to realize optimum genetic polymorphism required for mapping and breeding. Genetic divergence has been evaluated using the available tools at a particular point of time. We included discussions on phenotype-based strategies employing morphological markers, genotype-based strategies employing molecular markers; the statistical procedures utilized; their utilities for evaluation of genetic divergence among genotypes, local landraces, species and genera; and also on the effects of breeding pedigrees and geographical locations on the degree of genetic diversity.

Association mapping using molecular markers is a recent strategy to utilize the natural genetic variability to detect marker-trait association and to validate the genomic locations of genes, particularly those controlling the quantitative traits. Association mapping has been employed effectively in genetic studies in human and other animal models and those have inspired the plant scientists to take advantage of this tool. We included examples of its use and implication in some of the volumes that devote to the plants for which this technique has been successfully employed for assessment of the degree of linkage disequilibrium related to a particular gene or genome, and for germplasm enhancement.

Genetic linkage mapping using molecular markers have been discussed in many books, reviews and book series. However, in this series, genetic mapping has been discussed at length with more elaborations and examples on diverse markers including the anonymous type 2 markers such as RFLPs, RAPDs, AFLPs, etc. and the gene-specific type 1 markers such as EST-SSRs, SNPs, etc.; various mapping populations including F_2, backcross, recombinant inbred, doubled haploid, near-isogenic and pseudotestcross; computer software including MapMaker, JoinMap, etc. used; and different types of genetic maps including preliminary, high-resolution, high-density, saturated, reference, consensus and integrated developed so far.

Mapping of simply inherited traits and quantitative traits controlled by oligogenes and polygenes, respectively has been deliberated in the earlier literature crop-wise or crop group-wise. However, more detailed information on mapping or tagging oligogenes by linkage mapping or bulked segregant analysis, mapping polygenes by QTL analysis, and different computer software employed such as MapMaker, JoinMap, QTL Cartographer, Map Manager, etc. for these purposes have been discussed at more depth in the present volumes.

The strategies and achievements of marker-assisted or molecular breeding have been discussed in a few books and reviews earlier. However, those mostly deliberated on the general aspects with examples drawn mainly from major plants. In this series, we included comprehensive descriptions on the use of molecular markers for germplasm characterization, detection and maintenance of distinctiveness, uniformity and stability of genotypes, introgression and pyramiding of genes. We have also included elucidations on the strategies and achievements of transgenic breeding for developing genotypes particularly with resistance to herbicide, biotic and abiotic stresses; for biofuel production, biopharming, phytoremediation; and also for producing resources for functional genomics.

A number of desirable genes and QTLs have been cloned in plants since 1992 and 2000, respectively using different strategies, mainly positional cloning and transposon tagging. We included enumeration of these and other strategies for isolation of genes and QTLs, testing of their expression and their effective utilization in the relevant volumes.

Physical maps and integrated physical-genetic maps are now available in most of the leading crop and model plants owing mainly to the BAC, YAC, EST and cDNA libraries. Similar libraries and other required genomic resources have also been developed for the remaining crops. We have devoted a section on the library development and sequencing of these resources; detection, validation and utilization of gene-based molecular markers; and impact of new generation sequencing technologies on structural genomics.

As mentioned earlier, whole genome sequencing has been completed in one model plant (Arabidopsis) and seven economic plants (rice, poplar, peach, papaya, grapes, soybean and sorghum) and is progressing in an array of model and economic plants. Advent of massively parallel DNA sequencing using 454-pyrosequencing, Solexa Genome Analyzer, SOLiD system, Heliscope and SMRT have facilitated whole genome sequencing in many other plants more rapidly, cheaply and precisely. We have included extensive coverage on the level (national or international) of collaboration and the strategies and status of whole genome sequencing in plants for which sequencing efforts have been completed or are progressing currently. We have also included critical assessment of the impact of these genome initiatives in the respective volumes.

Comparative genome mapping based on molecular markers and map positions of genes and QTLs practiced during the last two decades of the last century provided answers to many basic questions related to evolution, origin and phylogenetic relationship of close plant taxa. Enrichment of genomic resources has reinforced the study of genome homology and synteny of genes among plants not only in the same family but also of taxonomically distant families. Comparative genomics is not only delivering answers to the questions of academic interest but also providing many candidate genes for plant genetic improvement.

The 'central dogma' enunciated in 1958 provided a simple picture of gene function—gene to mRNA to transcripts to proteins (enzymes) to metabolites. The enormous amount of information generated on characterization of transcripts, proteins and metabolites now have led to the emergence of individual disciplines including functional genomics, transcriptomics, proteomics and metabolomics. Although all of them ultimately strengthen the analysis and improvement of a genome, they deserve individual deliberations for each plant species. For example, microarrays, SAGE, MPSS for transcriptome analysis; and 2D gel electrophoresis, MALDI, NMR, MS for proteomics and metabolomics studies require elaboration. Besides transcriptome, proteome or metabolome QTL mapping and application of transcriptomics, proteomics and metabolomics in genomics-assisted breeding are frontier fields now. We included discussions on them in the relevant volumes.

The databases for storage, search and utilization on the genomes, genes, gene products and their sequences are growing enormously in each second and they require robust bioinformatics tools plant-wise and purpose-wise. We included a section on databases on the gene and genomes, gene expression, comparative genomes, molecular marker and genetic maps, protein and metabolomes, and their integration.

Notwithstanding the progress made so far, each crop or model plant species requires more pragmatic retrospect. For the model plants we need to answer how much they have been utilized to answer the basic questions of genetics and genomics as compared to other wild and domesticated species. For the economic plants we need to answer as to whether they have been genetically tailored perfectly for expanded geographical regions and current requirements for green fuel, plant-based bioproducts and for improvements of ecology and environment. These futuristic explanations have been addressed finally in the volumes.

We are aware of exclusions of some plants for which we have comprehensive compilations on genetics, genomics and breeding in hard copy or digital format and also some other plants which will have enough achievements to claim for individual book volume only in distant future. However, we feel satisfied that we could present comprehensive deliberations on genetics, genomics and breeding of 30 model and economic plants, and their groups in a few cases, in this series. I personally feel also happy that I could work with many internationally celebrated scientists who edited the book volumes on the leading plants and plant groups and included chapters authored by many scientists reputed globally for their contributions on the concerned plant or plant group.

We paid serious attention to reviewing, revising and updating of the manuscripts of all the chapters of this book series, but some technical and formatting mistakes will remain for sure. As the series editor, I take complete responsibility for all these mistakes and will look forward to the readers for corrections of these mistakes and also for their suggestions for further improvement of the volumes and the series so that future editions can serve better the purposes of the students, scientists, industries, and the society of this and future generations.

Science publishers, Inc. has been serving the requirements of science and society for a long time with publications of books devoted to advanced concepts, strategies, tools, methodologies and achievements of various science disciplines. Myself as the editor and also on behalf of the volume editors, chapter authors and the ultimate beneficiaries of the volumes take this opportunity to acknowledge the publisher for presenting these books that could be useful for teaching, research and extension of genetics, genomics and breeding.

Chittaranjan Kole

Preface to the Volume

We are living in the "Golden Age of Biology", fueled by advances in biology in general and molecular biology in particular. The current pace of discovery and interpretation of biological data is unprecedented. The –omics revolution, in particular, has touched nearly all aspects of biological research. This is certainly true of agricultural research. Molecular biology, genomics, and bioinformatics have enabled rapid improvement of the plants that feed the world by providing researchers with an improved understanding of phylogenetic relationships between crop species and their wild relatives, strategies for marker-assisted breeding, transgenic approaches to crop improvement, and whole-genome sequence information. The common cultivated potato, *Solanum tuberosum* L., a staple of many households and cuisines, is a prime example of a crop species that has benefited from the –omics revolution.

Potato, eaten by humankind for at least 13,000 years, has been credited with fueling a population boom that coincided with the Industrial Revolution, and was so wholly adopted as a food in some parts of Europe that crop failure in the mid-1840s resulted in the starvation of a million people. Potato promises higher calorie per acre production potential than any grain and can be produced, stored, and consumed without major technological inputs. Potato is also highly nutritious. In recognition of these attributes, potato was the first crop to be grown in space. These experiments aimed to establish self-sufficiency on long space voyages and in eventual space colonies. On Earth, potato's attributes make it a nearly ideal food for both developed and developing countries. Recent production trends indicate that potato production in densely populated developing nations is on the rise. Today, potato ranks as the world's fourth most important human food crop. Recognizing the worldwide significance and potential of the potato, the United Nations declared 2008 the International Year of the Potato.

Shifting potato production and global climate change make the continued genetic improvement of the potato a top priority for plant scientists. In recent years, research efforts have provided an improved understanding of the potato transcriptome and the physical organization of the potato genome and have allowed unprecedented mapping and cloning of useful genes. These efforts have been translated into new breeding approaches, including the use of marker-aided selection and the transgenic development of disease resistant potato cultivars. In this volume, leading potato scientists from around the world detail recent advances in potato biology and assess their impact on potato improvement. Collectively, we are excited about the current pace of scientific discovery and the progress that has been made in recent years. We are equally excited about what the future holds for the study of potato. As we prepare this volume, the scientific community eagerly awaits the final release of the complete genome sequences of the potato and the tomato, a close relative. Both efforts are the result of international collaborations and both promise to significantly and positively impact potato research in the coming years. Potato has been an important food for humankind since before the dawn of modern agriculture. Recent production trends and scientific advancements guarantee that the potato will continue to feed the world.

A word of deep gratitude and appreciation goes to all who have contributed to this effort. This volume represents the cumulative work of 20 authors from 11 research institutes in eight different countries. Their insights, hard work, and dedication made this project possible and are gratefully acknowledged. Thanks are also due to Science Publishers for their ongoing support.

James M. Bradeen
Chittaranjan Kole

Contents

List of Contributors

James M. Bradeen
University of Minnesota, Department of Plant Pathology, 495 Borlaug, Hall/1991 Upper Buford Circle, St. Paul, MN 55108, USA.
Email: *jbradeen@umn.edu*

Glenn J. Bryan
Genetics Programme, Scottish Crop Research Institute, Invergowrie, Dundee DD2 5DA, UK.
Email: *glenn.bryan@scri.ac.uk*

Domenico Carputo
University of Naples, Federico II, Department of Soil, Plant, Environmental and Animal Production Sciences, Via Università 100, 80055 Portici, Italy.
Email: *carputo@unina.it*

Haixia Chen
Potato Research Centre, Agriculture and Agri-Food Canada, 850 Lincoln Road, Fredericton, NB, Canada, E3B 4Z7.
and
Hunan Provincial Key Laboratory of Crop Germplasm Innovation and Utilization, Hunan Agricultural University, Changsha, Hunan, P.R. China, 410128.
Email: *chemhx1996@qq.com*

Maria Luisa Chiusano
University of Naples, Federico II, Department of Soil, Plant, Environmental and Animal Production Sciences, Via Università 100, 80055 Portici, Italy.
Email: *chiusano@unina.it*

Nunzio D'Agostino
Department of Soil, Plant, Environmental and Animal Production Sciences, Via Università 100, 80055 Portici, Italy.
Email: *nunzio.dagostino@gmail.com*

David De Koeyer
Potato Research Centre, Agriculture and Agri-Food Canada, 850 Lincoln Road, Fredericton, NB, Canada, E3B 4Z7.
Email: *david.dekoeyer@agr.gc.ca*

Luigi Frusciante
University of Naples, Federico II, Department of Soil, Plant, Environmental and Animal Production Sciences, Via Università 100, 80055 Portici, Italy.
Email: *fruscian@unina.it*

Liangliang Gao
Department of Plant Pathology, University of Minnesota, 495 Borlaug, Hall/1991 Upper Buford Circle, St. Paul, MN 55108, USA.
Email: *gaoxx092@umn.edu*

Tatjana Gavrilenko
N.I. Vavilov Institute of Plant Industry, B. Morskaya Str. 42/44, 190000, St. Petersburg, Russia.
Email: *tatjana9972@yandex.ru*

Christiane Gebhardt
MPI for Plant Breeding Research, Carl von Linné Weg 10, 50829 Köln, Germany.
Email: *gebhardt@mpiz-koeln.mpg.de*

Vicki Gustafson
Solanum Genomics International Inc., 921 College Hill Road, Fredericton, NB, Canada, E3B 6Z9.
Email: *vgustafson@bioatlantech.nb.ca*

Kathleen G. Haynes
U.S. Department of Agriculture, Agricultural Research Service, Genetic Improvement of Fruits and Vegetables Laboratory, Beltsville, MD 20705, USA.
Email: *kathleen.haynes@ars.usda.gov*

Adrian D. Hegeman
Departments of Horticultural Science and Plant Biology, Microbial and Plant Genomics Institute, University of Minnesota–Twin Cites, 305 Alderman Hall, 1970 Folwell Avenue, Saint Paul, MN 55108, USA.
Email: *hegem007@umn.edu*

Massimo Iorizzo
University of Naples, Federico II, Department of Soil, Plant, Environmental and Animal Production Sciences, Via Università 100, 80055 Portici, Italy.
Email: *massimoiorizzo@libero.it*

Joseph C. Kuhl
Department of Plant, Soil, and Entomological Sciences, University of Idaho, P.O. Box 442339, Moscow, ID 83844, USA.
Email: *jkuhl@uidaho.edu*

Xiu-Qing Li
Potato Research Centre, Agriculture and Agri-Food Canada, 850 Lincoln Road, P.O. Box 20280, Fredericton, New Brunswick, E3B 4Z7, Canada, Xiu-Qing.Li@AGR.GC.CA or.
Email: *lixiuqing2008@gmail.com*

Harpartap Mann
University of Minnesota, Department of Plant Pathology, 495 Borlaug Hall/1991 Upper Buford Circle, St. Paul, MN 55108, USA.
Email: *mann0188@umn.edu*

Ewen Mullins
Plant Biotechnology Unit, Teagasc Crops Research Centre, Oak Park, Carlow, Ireland.
Email: *ewen.mullins@teagasc.ie*

Toni Wendt
Plant Biotechnology Unit, Teagasc Crops Research Centre, Oak Park, Carlow, Ireland.
Email: *Toni.Wendt@teagasc.ie*

Abbreviations

2DE	2-Dimensional gel electrophoresis
ABRF	Association of Biomolecular Resource Facilities
AFLP	Amplified fragment length polymorphism
ALD	Acetyleptinidine
ASA	Allele specific amplification
BAC	Bacterial artificial chromosome
BC	Backcross
BSA	Bulked segregant analysis
BSP	Bilateral sexual polyploidization
CAPS	Cleaved amplified polymorphic sequence
cDNA	Complementary DNA
CE-MS	Capillary electrophoresis mass spectrometry
CHI	Chalcone isomerase
CHS	Chalcone synthase
CID	Collisionally induced dissociation
cM	Centi-Morgan
CPB	Colorado potato beetle
CPC	Commonwealth Potato Collection
CPGP	Canadian Potato Genome Project
CSCDM	Chromosome-specific cytogenetic DNA marker
DArT	Diversity arrays technology
dfr	Dihydroflavonol 4-reductase
DIGE	Difference gel electrophoresis
DNA	Deoxyribonucleic acid
EBN	Endosperm balance number
EC	European Commission
EI	Electron impact ionization
ELISA	Enzyme linked immunosorbant serologic assay
EMBL	European Molecular Biology Laboratory
eQTL	Expression-QTL
ESI	Electrospray ionization
EST	Expressed sequence tag
ETD	Electron transfer dissociation
EU	European Union

F	Filial
FAA	Functional allele activity
FAD	Functional allele DNA
FDR	First division restitution
FIE-MS	Flow infusion electrospray ionization mass spectrometry
FISH	Fluorescence in situ hybridization
GC	Gas chromatography
GCA	General combining ability
GILB	Global Initiative on Late Blight
GISH	Genomic in situ hybridization
GM	Genetically modified
GMO	Genetically modified organism
GO	Gene ontogeny
HILIC	Hydrophilic interaction chromatography
HPLC	High performance liquid chromatography
HRM	High resolution DNA melting
HTC	High-throughput cDNA
ICA	Independent component analysis
IGS	Intergenic spacer
IMAC	Immobilized metal affinity chromatography
Indel	Insertion/deletion
IR	Infrared
ISSR	Inter-simple sequence repeat
ITR	Interstitial telomeric repeat
iTRAQ	Isobaric tag for relative and absolute quantitation
kb	Kilobase
LC-MS	Liquid chromatography mass spectrometry
LD	Linkage disequilibrium
LE	Linkage equilibrium
LR-PCR	Long range-PCR
LRR	Leucine rich repeat
LTR	Long terminal repeat
MALDI	Matrix assisted laser desorption ionization
MAS	Marker-assisted selection
MCR	Maturity corrected resistance
miRNA	MicroRNA
mRNA	Messenger RNA
MS	Mass spectrometry
MS/MS	Tandem mass spectrometry
MSTFA	N-Methyl-N-(trimethylsilyl)trifluoro acetamide
NBS	Nucleotide binding site
NMR	Nuclear magnetic resonance
nt	Nucleotide

PCA	Principal component analysis
PCN	Potato cyst nematode
PCR	Polymerase chain reaction
PGSC	Potato genome sequencing consortium
PLRV	Potato leaf roll virus
POCI	Potato oligo chip initiative
PTM	Post-translational modification
PVA	Potato virus A
PVM	Potato virus M
PVS	Potato virus S
PVX	Potato virus X
PVY	Potato virus Y
PVYNTN	Potato virus Y tuber necrotic strain
QRL	Quantitative resistance locus
qRT-PCR	Quantitative reverse transcription-PCR
QTL	Quantitative trait loci
R gene	Resistance gene
RACE	Rapid amplification of cDNA ends
RAPD	Random amplified polymorphic DNA
rDNA	ribosomal DNA
RFLP	Restriction fragment length polymorphism
RGA	Resistance gene analog
RGH	Resistance gene homologs
RGL	Resistance gene-like
RNA	Ribonucleic acid
RNAi	RNA-interference
RP	Reverse phase
SAG	Steroid alkaloid glycoside
SAGE	Serial analysis of gene expression
SCA	Specific combining ability
SCAR	Sequence characterized amplified regions
SDR	Second division restitution
SG	Specific gravity
SNP	Single nucleotide polymorphism
SNuPE	Single nucleotide primer extension
SOL	Solanaceae genome project
SSCP	Single strand conformation polymorphism
ssp	Subspecies
SSR	Simple sequence repeat
STS	Sequence tagged site
TCS	Transcript consensus sequence
TEV	Tobacco etch virus
TIGR	The Institute for Genomic Research

tiRNA	Transcription initiation RNA
TOF	Time of flight
TPS	True potato seed
UHD	Ultra high density
USP	Unilateral sexual polyploidization

Introduction to Potato

James M. Bradeen[1,*] and *Kathleen G. Haynes*[2]

ABSTRACT

Potato (*Solanum tuberosum* L.) has been consumed by humankind for thousands of years. Molecular and historical evidence suggests that potato probably originated in the Andean highlands of southern Peru. From there, it spread to Europe and other parts of the world beginning in the 16th century. By the mid-1800s, peasant populations in Ireland and other parts of Europe were so dependent on the potato as a primary food source that severe epidemics of potato late blight in 1845 and 1846 resulted in the starvation of one million people and emigration of a million more. Today, potato ranks among the world's most important human food crops. In 2007, China led the world in potato production. Potato yields more calories per acre than any grain. The potato tuber is composed primarily of carbohydrates, but is also a significant dietary source of potassium and other minerals, fiber, vitamins C and B6, and essential amino acids. Efforts are underway in numerous research programs to improve the nutritional value of potatoes by enhancing tuber concentrations of carotenoids, anthocyanins, and minerals. Cultivated potato is one of approximately 200 tuber-bearing *Solanum* species. Many of these wild potato species have been used by breeders for crop improvement. Researchers use phylogenetic relationships, endosperm balance number, and the genepool concept to predict crossability amongst potato species. The cultivated potato is an autotetraploid ($2n = 4x = 48$ chromosomes) and, although it produces true botanical seed, is usually propagated asexually. Potato is likely to continue to play a major role in feeding a growing world population and genetic improvement of the potato remains a top priority for plant researchers.

Keywords: History, nutrition, origins, potato, production statistics, taxonomy

[1]University of Minnesota, Department of Plant Pathology, 495 Borlaug Hall/1991 Upper Buford Circle, St. Paul, MN 55108, USA.
[2]U.S. Department of Agriculture, Agricultural Research Service, Genetic Improvement of Fruits and Vegetables Laboratory, Beltsville, MD 20705, USA.
*Corresponding author: *jbradeen@umn.edu*

1.1 A Brief History of the Potato

Potato (*Solanum tuberosum* L.) is a New World crop species. Remains of potato peels recovered from ancient fire pits in southern Chile reveal that potato species have been consumed in that portion of the world for at least 13,000 years (Ugent et al. 1987). Although which potato species is represented in this instance has never been determined, researchers agree that this archeological evidence points to consumption of potato tubers at a time that predates modern agriculture (Ugent et al. 1987; Hawkes 1990). Fossilized potato tubers suggest that potato has been used as a food source in Peru for at least 10,000 years (Engel 1970). Importantly, microscopic analysis of starch grains recovered from these samples allowed the classification of the tubers as *S. tuberosum,* the same species as the modern cultivated potato (Ugent et al. 1982). Whether the recovered tubers were collected in the wild or grown by early farmers is unclear. While irrigation agriculture was practiced in ancient times in coastal Peru (Pozorski and Pozorski 1979), some authors suggest, based on meager evidence of potato production along the coast, that potato was most likely produced or harvested in the Andean highlands and traded with coastal inhabitants (Sauer 1993). Evidence of potato trade includes the depiction of *chuño,* a freeze-dried potato product, on pottery produced by coastal Peruvian potters. Because *chuño* requires cold temperatures for production, it cannot be produced along the coasts. Presumably, *chuño*, produced in the Andean highlands, would have been available to coastal inhabitants only through trade (Sauer 1993).

Consistent with archeological evidence that suggests potato was first grown in South America, Vavilov (1951) identified the New World region of Mexico, Peru, and adjoining countries as the center of origin for potato. A more exact geographic origin for potato has been the topic of scientific study in recent decades. Leading hypotheses suggest that potato originated either in the Andean highlands of Peru in northern South America, or the area of Chile, Argentina, and Bolivia in southern and central South America. A related question posed by researchers is whether potato was domesticated once or multiple times. Some authors have advocated for a southern South American origin for cultivated potato (Bukasov 1978; Hawkes 1990; Ochoa 1990). Several others have argued for multiple domestication events, sometimes involving complex interspecific hybridization of various *Solanum* species (Ugent 1970; Grun 1990; Hosaka 1995; van den Berg et al. 1998; Huaman and Spooner 2002). Spooner et al. (2005) provided molecular evidence that refuted these claims. Applying amplified fragment length polymorphism (AFLP) marker technology to a collection of 98 potato landraces and 261 wild potato accessions, these authors demonstrated a clear distinction between "Northern" (Peru) and "Southern" (Bolivia and Argentina) wild potato species. Importantly, all cultivated types clustered together, suggesting a single domestication event. In this analysis, cultivated potatoes were more closely related to the "Northern" species and Spooner et al. (2005) concluded that potato was domesticated in what is now southern Peru.

The introduction of potato from South America into Europe has been steeped in folklore. Although some accounts suggested that Sir Francis Drake or Sir Walter Raleigh may have been responsible, most experts agree the involvement of these explorers in the introduction of potato into Europe was unlikely (McNeill 1949; Hawkes 1990; Sauer 1993). Salaman (1937), Laufer (1938), and Hawkes (1990) suggested that potato may have first been introduced into Spain in about 1570. From there, it moved to Italy and then to Belgium by 1587 (Laufer 1938). By about 1590, potato was introduced into England, perhaps directly from South America (Hawkes 1990). By whatever means potato was first introduced into Europe, it spread throughout Europe as a botanical novelty and as fodder for livestock. Initially treated with suspicion (potato is a relative of the poisonous nightshade, *S. nigrum*), potato was eventually adopted as a human food source and gradually gained popularity. Potato yields more calories per acre than any grain (Rubatzky and Yamaguchi 1997) and can be produced and stored even on a small scale. These attributes combined to fuel a population boom in much of Europe, coinciding with the Industrial Revolution. For example, in Ireland the population doubled between 1780 and 1841 as peasant farmers embraced potato as a food source. In the same country, due to a combination of social, political, and environmental factors, this growing Irish peasant population was dependent upon potato as a primary food source by the mid-1800s.

However, potato wasn't the only New World immigrant to find a happy home in Europe. *Phytophthora infestans* is an oomycete that causes potato (and tomato) late blight disease. Under cool, damp conditions, complete destruction of a potato field is possible in as little as 7–10 days. Even seemingly healthy tubers can develop late blight in storage, opening the door for secondary infections that can result in complete loss of a stored crop. By the mid-1800s, *P. infestans* had been introduced from the New World into Europe. The summers of 1845 and 1846 were cool and damp in Ireland and much of Europe—conditions that were conducive for late blight epidemics. Lacking knowledge of disease epidemiology, growers commonly discarded infected tubers and other plant tissues in heaps alongside production fields. These discarded heaps served as a source of inoculum for re-infestation of field plots and as an overwintering site for the pathogen. In what has become known as one of the most tragic and notorious events in the history of plant pathology and production agriculture, Ireland suffered nearly complete potato crop loss in 1845 and 1846. One million people starved to death. The human suffering of the time is immortalized in drawings published in newspapers of the day and in *The Potato Eaters*, a painting by Dutch artist Vincent van Gogh. In addition to the starvation deaths, it is estimated that a million more people emigrated from Ireland as a result of this potato disease. The 1841 Irish censuses reported 8.2 million people living in Ireland. By 1851, that number had decreased to 6.5 million. Today Ireland is home to only 4.2 million people.

As the potato's popularity and significance rose in Europe, the food source was introduced, from Europe, into other parts of the world including India by 1610 (Sauer 1993), Bermuda by 1613 and mainland North America by 1621 (Laufer 1938; Hawkes 1990), China by 1700 (Sauer 1993), and New Zealand in 1769 (Sauer 1993). Today potatoes are grown on six continents and rank as the world's fourth most important food crop behind only maize, rice, and wheat in terms of total production (FAOStat Data 2007; faostat.fao.org).

1.2 Potato Morphology

Potato is a short-lived perennial species. The economically significant part of the plant, the potato tuber, is a site for storage of carbohydrates. Tubers are produced underground as swellings along modified stems known as stolons. The tuber is both the organ of commercial production and the primary propagule, although potato can produce true botanical seed. Potato leaves are pinnately compound and alternate along aboveground stems that are generally less than 1m in length. Potato flowers are hypogynous and actinomorphic with fused, five-lobed corollas. Corolla color varies from white to deep purple to pinkish (Fig. 1-1). True potato seed, the product of fertilization, is small (~1–2 mm in maximum dimension), flat, oval- to kidney-shaped, and yellowish to tan in color. The potato fruit is a spherical berry and potato fruits and leaves may produce high levels of glycoalkaloid toxins. Figure 1-1 illustrates key potato morphology.

1.3 Potato Taxonomy, Related Species, and Crossability

The potato of worldwide economic significance is botanically classified as *Solanum tuberosum* subsp. *tuberosum.* The detailed taxonomy of cultivated potato is presented in Table 1-1.

Potato is just one member of the large plant family Solanaceae. The Solanaceae includes 95 genera, with *Solanum,* from which the family gets its name, being the largest and most economically significant. In addition to potato, the Solanaceae includes other significant crop plants including tomato (*Solanum lycopersicum*), eggplant (*Solanum melongena*), pepper (*Capsicum* spp.), and tobacco (*Nicotiana tabacum*). Other Solanaceous species include ornamentals (e.g., *Brugmansia* spp., *Petunia x hybrida, Nicotiana alata, Nolana* spp., *Cestrum nocturnum, Salpiglossis sinuata, Solanum crispum,* etc.), species of medicinal and cultural significance [e.g., *Datura* spp., mandrake

Figure 1-1 Potato morphology. (a) The flower of primitive wild potato species classified as superseries *Stellata* lack a fused corolla, giving the flower a star-shaped or stellate appearance. (b) In contrast, the five petals of the flower of the cultivated potato (*Solanum tuberosum* ssp. *tuberosum*) and other species classified as superseries *Rotata* are fused, giving the flower a wheel-shaped or rotate appearance. Potato flowers range in color from white to pinkish to purple. (c) The potato tuber is an underground stem modified for carbohydrate storage. Potato tubers come in a variety of shapes, sizes, and colors.

Color image of this figure appears in the color plate section at the end of the book.

Table 1-1 Botanical classification of the cultivated potato.

Family	Solanaceae
Subfamily	Solanoideae
Tribe	Solaneae
Genus	*Solanum* L.
Subgenus	*Potatoe* (G. Don) D'Arcy
Section	*Petota* Dumortier
Subsection	*Potatoe* G. Don
Superseries	*Rotata* Hawkes
Series	*Tuberosa* (Rydb.) Hawkes
Species	*Solanum tuberosum* L.
Subspecies	*tuberosum*

(*Mandragora* spp.), belladonna (*Atropa belladonna*), etc.], food crops of minor and regional significance [e.g., tomatillo (*Physalis philadelphica*), tamarillo (*Solanum betaceum*), pepino (*Solanum muricatum*), garden huckleberry (*Solanum melanocerasum*), etc.], and important weeds (e.g., *Solanum nigrum, Solanum rostratum, Solanum sisymbriifolium,* etc.).

The genus *Solanum* ranks among the five largest of plant genera, with an estimated 1,000–1,700 species (D'Arcy 1991; Mabberley 1997). Of these, approximately 200 tuber-bearing potato species are classified as section *Petota* of the subgenus *Potatoe*. All of these species are of New World origin and can be found naturally distributed throughout southwestern North America, Central America, and most of South America. The Andean Mountains are particularly rich in potato species. The potatoes have a base chromosome number of 12 and the majority of potato species are diploid, although tetraploid and hexaploid species are known. Because of the propensity for tuber-bearing potatoes to reproduce asexually, stable triploid and pentaploid populations are possible. The cultivated potato (*S. tuberosum* ssp. *tuberosum*) is an autotetraploid ($2n = 4x = 48$ chromosomes).

The approximately 200 wild potato species are potentially rich sources of genes useful for the improvement of cultivated potato. Accordingly, large potato genebank collections are maintained in several countries including Peru, Argentina, Germany, The Netherlands, Scotland, Russia, and the US. The Association of Potato Inter-genebank Collaborators provides researcher access to potato germplasm and passport information (*research.cip.cgiar.org/confluence/display/IPD/Home*). While all wild potato germplasm has potential use for the improvement of the cultivated potato, owing to biological differences and varying levels of species relatedness, some species have been of greater utility to potato researchers and breeders. To make efficient use of potato germplasm, three systems of germplasm organization have been employed by potato researchers. These systems, the Genepool Concept, the Endosperm Balance Number (EBN) model, and phylogenetic classification, assist in predicting successful strategies for gene transfer between species.

1.3.1 The Genepool Concept

Harlan and de Wet (1971) proposed a system of germplasm utilization that defines three "genepools" based on the crossability of germplasm to a cultivated species (Fig. 1-2). The first, the Primary Genepool, consists of the cultivated crop (*S. tuberosum* ssp. *tuberosum*) including all landraces and cultivars. Within the Primary Genepool, there are no barriers to gene flow; one potato cultivar can be freely crossed with another. Thus, all genetic materials within the Primary Genepool are accessible by potato breeders. The Secondary Genepool includes wild species that can be sexually crossed with the cultivated form, although manipulation of ploidy levels, 2*n* gametes, or modified breeding techniques must sometimes be employed. Thus, breeders can use wild species from the Secondary Genepool for potato improvement. The potato Secondary Genepool includes about 180 tuber-bearing species. Some Secondary Genepool species (e.g., the late blight resistant wild potato *S. demissum*) have been widely used for potato improvement, but most have not. Finally, the Tertiary Genepool includes wild species that are distantly related to the cultivated crop. Due to cross incompatibilities, wild species in the Tertiary Genepool are sexually isolated from the cultivated crop. Thus, genes found in Tertiary Genepool species are not normally accessible to plant breeders. In potato, the Tertiary Genepool consists of 18–20 diploid species. Each of these is a rich potential source of genes for potato improvement. For example, one Tertiary Genepool species, *S. bulbocastanum,* has long been recognized as a source of potentially durable late blight resistance genes (Reddick 1939; Niederhauser and Mills 1953; Black and Gallegly 1957; Graham et al. 1959). Consistent with its classification as a Tertiary Genepool species, *S. bulbocastanum* is sexually isolated from *S. tuberosum* (Jackson and Hanneman 1999). However, late blight resistance genes have been transferred to the cultivated potato using a series of complicated bridging species crosses (Hermsen and Ramanna 1973) and somatic hybridization (Helgeson et al. 1998). More recently, disease resistance gene cloning from *S. bulbocastanum* and transformation of potato cultivars has been possible (Song et al. 2003; Vossen et al. 2003; Park et al. 2005; Vossen et al. 2005; Bradeen et al. 2009; Lokossou et al. 2009; Oosumi et al. 2009).

1.3.2 The Endosperm Balance Number (EBN)

Later, Johnston et al. (1980) described a genetic model for predicting the crossability of potato species. This model suggests that each potato

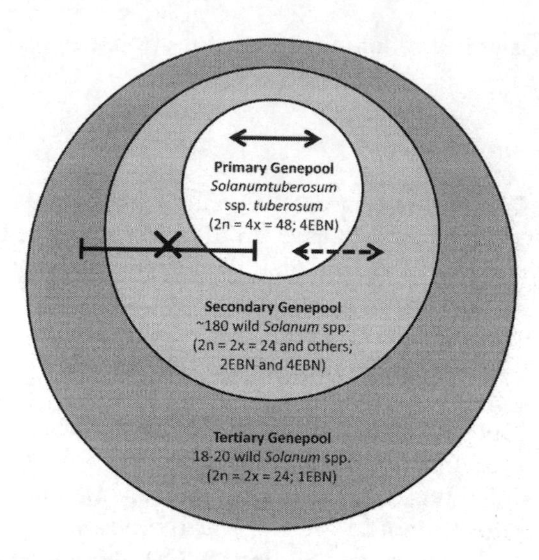

Figure 1-2 Systems of potato germplasm classification relevant to the improvement of cultivated potato (*Solanum tuberosum* ssp. *tuberosum*). Harlan and de Wet (1971) defined three genepools based upon crossability with crop plants. The Primary Genepool of potato (white circle) encompasses *S. tuberosum* ssp. *tuberosum*, including all landraces and cultivars. Cultivated potato is a tetraploid ($2n = 4x = 48$) and Johnston et al. (1980) assigned the Endosperm Balance Number (EBN) of 4. Within the Primary Genepool there are no biological barriers to cross pollination, allowing free gene flow (depicted as a solid double-ended arrow). The potato Secondary Genepool (light gray circle) includes approximately 180 wild potato species. Most of these are diploid ($2n = 2x = 24$), although this group also includes triploids, tetraploids, pentaploids, and hexaploids. Secondary Genepool species have an EBN of 2 or, less commonly, 4. Species in the Secondary Genepool may be crossed with cultivated potato, although manipulation of ploidy or use of $2n$ gametes is sometimes necessary to achieve matched EBNs. These species can be used for potato improvement since geneflow between Secondary Genepool species and cultivated potato (depicted as a dashed double-ended arrow) is possible. The potato Tertiary Genepool (dark gray circle) includes 18-20 wild potato species. These are diploids ($2n = 2x = 24$) with an assigned EBN of 1. Sexual incompatibilities between Tertiary Genepool species and cultivated potato have prevented gene flow (depicted as an X'd out solid black line bounded by bars), severely limiting their use for potato improvement. Genes can be transferred from Tertiary Genepool species using bridge crosses (Hermsen and Ramanna 1973; Jansky and Hamernik 2009), somatic hybridization (Helgeson et al. 1998), and gene cloning and transformation (Song et al. 2003; Vossen et al. 2003; Park et al. 2005; Vossen et al. 2005; Bradeen et al. 2009; Lokossou et al. 2009; Oosumi et al. 2009). The Genepool Concept and the EBN model approximate predicted species phylogenies and taxonomy. The Primary and Secondary Genepool of potato encompass all species classified as superseries *Rotata* plus species housed in series *Yungasensa* of superseries *Stellata* (Hawkes 1990). The Tertiary Genepool of potato encompasses all species of superseries *Stellata* with the exception of those in series *Yungasensa* (Hawkes 1990).

species has associated with it a characteristic Endosperm Balance Number (EBN) ranging from 1 to 4. The EBN defines crossability between species such that successful cross-species hybridizations require a 2:1 ratio of maternal:paternal genetic contribution to the developing endosperm.

This ratio is achieved only when the two parents have matched EBNs. Manipulating ploidy levels effectively manipulates EBN as well, providing a mechanism for achieving successful crosses in some instances when EBNs are not properly matched. For example, cultivated potato is 4EBN. Most diploid wild potatoes are 2EBN. Directly crossing cultivated potato with a 2EBN wild species is predicted, on the basis of mismatched EBNs, to be unsuccessful due to collapse of the developing endosperm. However, the creation of *S. tuberosum* dihaploids ($2n = 2x = 24$) results in cultivated potato material with an effective EBN of 2. A cross between *S. tuberosum* dihaploids and a 2EBN wild species would be expected to yield F_1 progeny. Similarly, doubling the chromosome number of a 2EBN wild species or the use of $2n$ gametes derived from this species results in an effective EBN of 4, allowing these materials to be crossed with tetraploid cultivated potato. Although the genetic basis of the EBN system has not been fully elucidated, this model has been widely used by potato breeders and researchers.

The Genepool Concept is largely consistent with assigned EBN values (Fig. 1-2). The Primary Genepool for potato consists of the cultivated potato (4EBN). The Secondary Genepool consists of wild potato species, most of which are 2EBN. However, some species in the Secondary Genepool are 4EBN. The EBN model predicts that these species are directly crossable to cultivated potato without ploidy manipulation or the use of $2n$ gametes. The Tertiary Genepool consists of wild potato species that are 1EBN.

1.3.3 Predicted Phylogenetic Relationships

The EBN model and the Genepool Concept also approximate predicted phylogenetic relationships. In the most recent comprehensive taxonomic treatment of *Solanum* section *Petota*, Hawkes (1990) recognized the informal taxon "superseries" below the level of section but above the level of series (Table 1-1). Based on morphological similarities, particularly of floral structure, Hawkes (1990) classified cultivated potato and the vast majority of wild potato species as superseries *Rotata*. These materials share a fused corolla, giving the flower a wheel-shaped or rotate appearance (Fig. 1-1). Morphologically primitive potatoes, those lacking a fused corolla and thus having star-shaped or stellate flowers (Fig. 1-1), are classified as superseries *Stellata*. Molecular data have provided general support for this distinction between superseries *Rotata* and superseries *Stellata*, although it is clear that further refinement of potato taxonomy is needed (e.g., Rodriguez and Spooner 1997; Spooner and Castillo 1997; Lara-Cabrera and Spooner 2004; Spooner et al. 2004). With few exceptions, superseries *Rotata*, as defined by Hawkes (1990) encompasses the 4EBN and 2EBN potatoes and the potato Primary and Secondary Genepools. The majority of species classified as superseries *Stellata*, in contrast, are 1EBN and comprise the Tertiary

Genepool of potato. The notable exception to this rule is the inclusion of the Secondary Genepool species of series *Yungasensa* ($2n = 2x = 24$; 2EBN) in superseries *Stellata* (Hawkes 1990).

Researchers are advised that while taxonomic relationships, the EBN model, and the Genepool Concept offer guidance in the utilization of wild germplasm for potato improvement, none is foolproof and refinements are ongoing. Studies of species crossability such as that of Jackson and Hanneman (1999) provide more definitive evidence.

1.4 Potato Production Statistics

In 2007, the year for which the most recent production statistics are available, the world produced 309,344,247 metric tons of potato on 18,531,194 hectares (FAOStat data 2007). In that same year, China was the world leader in potato production (56,196,000 metric tons) followed by the Russian Federation (36,784,200 metric tons), India (22,090,000 metric tons), and the United States (20,373,267 metric tons) (FAOSTAT data 2007). Table 1-2 summarizes average potato productions statistics for the 46-year period 1962–2007 for top-rated countries. A noteworthy trend in recent decades has been the rapid increase in potato production in parts of Asia (Fig. 1-3a). This is especially true for China (407% increase in potato production between 1962 and 2007) and India (903% increase), which together accounted for 67.2% of potato production in Asia and 25.3% of worldwide potato production in 2007. These increases appear to have resulted predominantly from an increase in production area (Fig. 1-3b). During the same period, potato production area has decreased in some regions of the world, most notably Europe (Fig. 1-3b). While potato yield/hectare has remained largely constant between 1962 and 2007 in much of the world, yield gains are evident in North and Central America (Fig. 1-3c) and Oceania (Australia and New Zealand; not shown).

Of significant note are increases in potato production among developing countries. During the 46-year period 1962–2007, while potato production in Europe fell 35.9%, potato production in developing nations rose 632% (FAOSTAT data 2007). China and India, ranked among the top potato producers in the world (Table 1-2), are both developing nations with large and growing population bases. In 2007, seven of the top 15 potato-producing developing nations are Asian countries [China (#1), India (#2), Bangladesh (#3), Pakistan (#6), Nepal (#7), South Korea (#8), and Indonesia (#11)]. While Africa has traditionally been only a modest producer of potatoes, eight of the top 15 potato-producing developing nations are African countries [Malawi (#4), Egypt (#5), Morocco (#9), Rwanda (#10), Kenya (#12), Nigeria (#13), Tanzania (#14), and Uganda (#15)]. Each of these countries shows an increase in potato production of between 412% (Kenya) and 4,611%

Table 1-2 Average potato production (Mt) between 1962 and 2007 for top potato-producing countries[a].

Region	Rank[b]	% World production	1962–1971	1972–1981	1982–1991	1992–2001	2002–2007
World	--	100.0	281 646 991	278 415 907	273 631 914	299 350 125	315 139 821
China	1	20.7	17 249 998	25 512 204	28 555 809	53 559 994	65 299 528
Russian Federation	2	11.5	—	—	—	35 711 526	36 361 375
India	3	7.4	3 728 510	7 110 470	12 663 270	19 793 160	23 312 566
United States	4	6.4	13 313 034	15 091 084	17 033 113	21 021 183	20 270 482
Ukraine	5	6.0	—	—	—	17 253 761	18 976 466
Poland	6	3.9	45 253 944	45 188 216	35 014 435	24 508 651	12 399 392
Germany	7	3.6	31 909 857	21 671 925	17 047 042	12 040 619	11 291 589
Belarus	8	2.7	—	—	—	8 774 550	8 538 464
Netherlands	9	2.2	4 525 779	5 836 850	6 737 029	7 469 834	6 922 633
France	10	2.1	10 892 766	6 870 360	6 434 110	6 199 000	6 773 771
United Kingdom	11	1.9	7 374 300	6 444 300	6 612 100	6 902 590	6 113 083
Canada	12	1.6	2 244 453	2 390 367	2 834 993	4 007 416	4 953 081
Turkey	13	1.5	1 773 500	2 643 500	3 896 000	5 002 000	4 672 252
Iran	14	1.4	371 800	783 721	2 036 264	3 262 055	4 451 511
Bangladesh	15	1.3	575 793	849 950	1 137 283	1 913 338	4 078 401

[a]Data from FAOStat 2007.
[b]Rank and % world production are based on average production between 2002 and 2007.

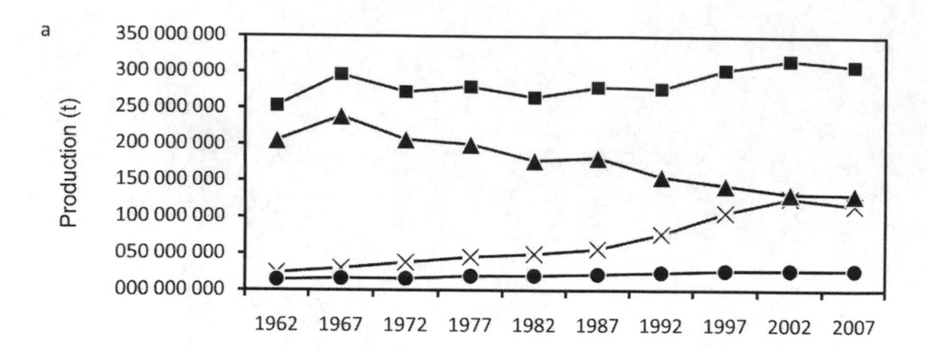

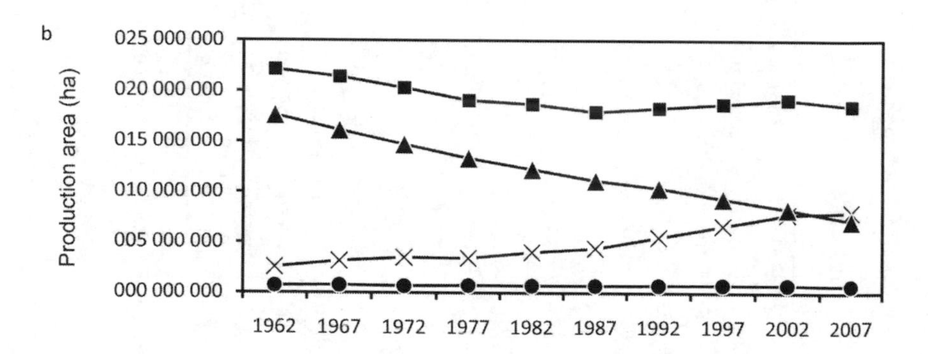

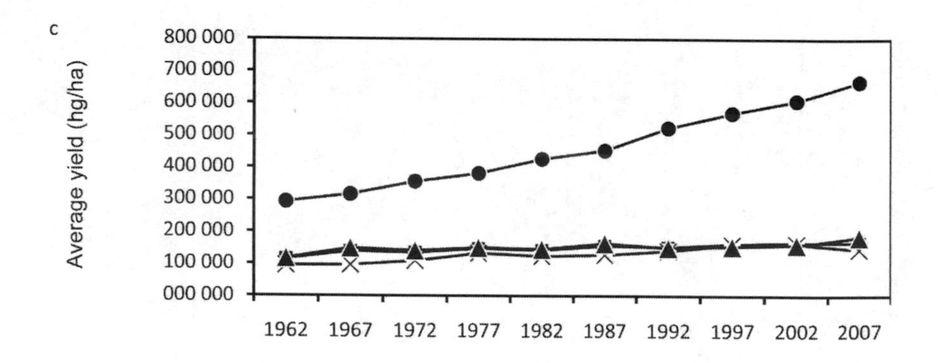

Figure 1-3 Trends in world potato production (1962–2007). Potato production statistics for the world (■), Asia (✕), Europe (▲), and North and Central America (●). Combined, Asia, Europe, and Central America produced 89.1% of the world's potatoes in 2007. (a) Potato production in metric tons (t). While world potato production has remained large constant between 1962 and 2007, notable increases in production in Asia are evident. During the same period, Europe has experienced a decrease in potato production. (b) Potato production area

Figure 1-3 contd....

(Malawi) between 1962 and 2007. On average, potato production in Africa has risen 837% between 1962 and 2007. However, all African countries combined produced only 5.8% of the world's potatoes in 2007. Nevertheless, production trends underscore the potential for potato to feed the hungry population of the world's developing countries in Asia as well as Africa.

1.5 Potato Nutritional Content

Potatoes are about 75% water, 21% carbohydrates (of which about 82% is starch), 2.5% protein, and less than 1% fat (*www.nal.usda.gov/fnic/foodcomp/cgi-bin/list_nut_edit.pl*). Often looked upon as primarily a starchy vegetable, potatoes are actually highly nutritious. Potatoes are a good source of vitamins C and B6. A large (299 g) baked potato provides daily values (DV) of about 48% for vitamin C and 46% for vitamin B6. They are also good sources of fiber (26% DV), minerals, and proteins. A large baked potato provides DVs of 46% for potassium, 33% for manganese, 21% for magnesium, and 21% for phosphorus. Potatoes are a well balanced and complete source of the essential amino acids necessary in the adult human diet (*www.nutritiondata.com/facts/vegetables-and-vegetable-products/2770/2*).

Potatoes come in a wide variety of flesh colors besides white. Two main classes of pigments impart flesh color: the carotenoids, responsible for yellow and orange colors, and anthocyanins, responsible for red and purple colors. White-fleshed potatoes are low (< 100 μg per 100 g fresh weight) in carotenoids, whereas, the carotenoid content of yellow-fleshed varieties may be as high as 560 μg per 100 g FW (Iwanzik et al. 1983; Tevini et al. 1984; Tevini and Schonecker 1986; Lu et al. 2001; Hale 2003; Nesterenko and Sink 2003). Intense yellow- to near orange-flesh color, associated with carotenoid concentrations > 2,000 μg per 100 g FW, have been reported in diploid *Solanum* germplasm (Brown et al. 1993; Lu et al. 2001). The primary tuber carotenoids are lutein, zeaxanthin, and violanxanthin, although some studies have also reported finding neoxanthin, and antheraxanthin (Brown et al. 1993; Lu et al. 2001). All carotenoids are antioxidants (Palozza

---

Figure 1-3 contd....

in hectares (ha). World potato production area has decreased slightly from 1962 to 2007, reflecting dramatic decreases in production area in Europe. While production area has not changed much in North and Central America, Asia has experienced a large increase in potato production area in the same time period. (c) Average potato yield in hectograms/hectare (hg/ha) has remained nearly constant in Europe and Asia for the period 1962–2007, while North and Central America have experienced steady increases. Owing to the significant contribution of Europe and Asia to world potato production, note that average worldwide potato yield closely mirrors the average yield of these regions. Taken together, production increases in Asia and corresponding decreases in Europe between 1962 and 2007 appear to reflect changes in production area rather than increases or decreases in average potato yield. (Adapted from FAOStat data 2007).

and Krinsky 1992). Lutein and zeaxanthin are found in the macula of the human eye and may play a role in reducing the risk of age-related macular degeneration (Mares-Perlman et al. 1995; Seddon et al. 1994). In addition, increased intake of zeaxanthin may improve mental acuity in the elderly (Akbaraly et al. 2007). Tuber anthocyanins are found in concentrations on the order of a hundred fold greater than carotenoids. The primary anthocyanins in red-fleshed tubers are pelogonidin-3-(*p*-coumaroyl-tutinoside)-5-glucoside (200–2,000 μg per gram FW) and peonidin-3-(*p*-coumaroyl-rutinoside)-5-glucoside (20–400 μg per gram FW), whereas, the primary anthocyanins in dark purple-fleshed potatoes are petunidin-3-(*p*-coumaroyl-rutinoside)-5-glucoside (1,000–2,000 μg per gram FW) and malvidin-3-(*p*-coumaroyl-rutinoside)-5-glucoside (2,000–5,000 μg per gram FW) (Lewis et al. 1998). There is increasing interest in anthocyanins in potato tubers because of their perceived higher antioxidant content (Tsuda et al. 2000; Ross and Kasum 2002; Brown et al. 2005; Brown et al. 2007), and ability to combat both prostate cancer (Reddivari et al. 2007) and breast cancer (Thompson et al. 2009).

The high glycemic index of potatoes is of concern for diabetics. Foods with a high glycemic index are quickly broken down in the intestines and cause blood sugar levels to rise rapidly. Potato starch is about 20–25% amylose and 75–80% amylopectin. After cooking, a portion of amylose starch recrystallizes to form resistant starch. Resistant starch acts as a dietary fiber (Karlsson et al. 2007). It passes through the small intestines and once in the large intestines microbial fermentation results in the production of small chain fatty acids, which enhance colon health and lower the risk of colorectal cancer and diverticulosis (van Amelsvoort and Weststrate 1992; Brouns et al. 2002). In preliminary studies examining cultivated diploid potato species adapted to long-day growing conditions, the concentration of amylose in starch ranged from 25–36% (Shelley Jansky, pers. comm.). Jansen et al. (2001) found even greater variation for amylose content in wild potato species, which ranged from 22–43%. These studies suggest that genetic variation exists in *Solanum* germplasm and raise the possibility of developing varieties with higher levels of resistant starch in the future.

There is considerable interest in the potential of potato to lessen micronutrient malnutrition, particularly for iron and zinc. About 40% of the world's population is iron deficient (Welch and Graham 1999), however, no reliable estimate is available for the number of people with zinc deficiency. The iron content of potato ranges from 15–20 μg per gram DW (O'Neill 2005; USDA 2006). However, iron concentrations ranging from 9–158 μg per gram DW in Andean potato cultivars (Andre et al. 2007; Burgos et al. 2007) and 18–65 μg per gram DW in American potato cultivars and advanced selections (Brown 2008) have been reported. Variation in zinc concentrations has also been reported: ranging from 8–29 μg per gram DW in Andean potato cultivars (Andre et al. 2007; Burgos et al. 2007) and 12.5–20 μg per gram DW in American potato cultivars and advanced selections (Brown 2008). Genetic variation for iron and zinc content in potatoes is estimated to represent 0–73% (Brown et al. 2010) and 19–61% (Charles Brown, pers. comm.), respectively, of the total variation. The reported variation in iron content in cultivated potatoes and the proportion of that variation under genetic control again suggests that breeding efforts to improve iron content should be undertaken. Less variation exists for zinc concentrations in cultivated potatoes, but the high estimate of the proportion of variation under genetic control is reason to undertake similar breeding efforts to improve zinc concentrations.

1.6 Future Prospects for Potato Research and Production

The potato is a nutritious food that can be grown in a wide variety of climates on scales both large and small, yielding more calories per acre than any grain. The ability of the potato tuber to be stored for months with minimal technological inputs makes potato an ideal food for both developed and developing nations. Accordingly, recent decades have seen a marked increase in potato production in populous developing countries. This trend is likely to continue in coming years. With shifting production and a changing global climate, there is a clear need for continued genetic improvement of the potato. Vast genetic resources for potato improvement, including genebank collections of approximately 200 related wild species will undoubtedly be important for continued potato improvement. Researchers continue to study the genetic control of significant potato traits and to develop molecular and non-molecular strategies for the efficient utilization of genes derived from cultivated potato and wild potato relatives. The impending release of complete genome sequence for potato and tomato will undoubtedly speed these efforts in coming years. Potato has a long history of cultivation and utilization by humankind; ongoing research and recent potato production trends ensure potato will continue to play an important role in feeding the world.

References

Akbaraly N, Faure H, Gourlet V, Favier A, Berr C (2007) Plasma carotenoid levels and cognitive performance in an elderly population: results on an EVA study. J Gerontol: Med Sci 62A: 308–316.

Andre C, Ghislain M, Bertin P, Oufir M, del Rosario M, Hoffmann L, Hausman J, Larondelle Y, Evers D (2007) Andean potato cultivars (*Solanum tuberosum* L.) as a source of antioxidant and mineral micronutrients. J Agri Food Chem 55: 366–378.

Black W, Gallegly ME (1957) Screening of *Solanum* species for resistance to physiologic races of *Phytophthora infestans*. Am Potato J 34: 273–281.

Bradeen JM, Iorizzo M, Mollov DS, Raasch J, Colton Kramer L, Millett BP, Austin-Phillips S, Jiang J, Carputo D (2009) Higher copy numbers of the potato *RB* transgene correspond to enhanced transcript and late blight resistance levels. Mol Plant-Microbe Interact 22: 437–446.

Brouns F, Kettlitz B, Arrigoni E (2002) Resistant starch and "the butyrate revolution". Trends Food Sci Technol 13: 251–261.

Brown C (2008) Breeding for phytonutrient enhancement of potato. Am J Pot Res 85: 298–307.

Brown C, Edwards C, Yang C-P, Dean B (1993) Orange flesh trait in potato: Inheritance and carotenoid content. J Am Soc Hort Sci 118: 145–150.

Brown C, Culley D, Yang C, Durst R, Wrolstad R (2005) Variation of anthocyanin and carotenoid contents and associated antioxidant values in potato breeding lines. J Am Soc Hort Sci 130: 174–180.

Brown C, Culley D, Bonierbale M, Amoros W (2007) Anthocyanin, carotenoid content, and antioxidant values in native South American potato cultivars. HortScience 42: 1733–1736.

Brown C, Haynes K, Moore M, Pavek M, Hane D, Love S, Novy R, Miller JJ (2010) Stability and broad-sense heritability of mineral content in potato: iron. Am J Pot Res 87: 390–396.

Bukasov SM (1978) Systematics of the potato. Systematics, Breeding, and Seed Production of Potatoes. Bulletin of Applied Botany, Genetics and Breeding, Leningrad, USSR, pp 1–42.

Burgos G, Amoros W, Morote M, Stangoulis J, Bonierbale M (2007) Iron and zinc concentration of native Andean potato cultivars from a human nutrition perspective. J Sci Food Agri 87: 668–675.

D'Arcy WG (1991) The Solanaceae since 1976, with a review of its biogeography. In: JG Hawkes, RN Lester , M Nee , N Estrada (eds) Solanaceae III: Taxonomy, Chemistry, Evolution. Royal Botanical Gardens, Kew, UK, pp 75–137.

Engel F (1970) Exploration of the Chilca Canyon, Peru. Curr Anthropol 11: 55–58.

Graham KM, Niederhauser JS, Servin L (1959) Studies on fertility and late blight resistance in *Solanum bulbocastanum* Dun. in Mexico. Can J Bot 37: 41–49.

Grun P (1990) The evolution of cultivated potatoes. Econ Bot 44 (3 Suppl): 39–55.

Hale A (2003) Screening potato genotypes for antioxidant activity, identification of the responsible compounds, and differentiating Russet Norkotah strains using AFLP and microsatellite marker analysis. Texas A&M Univ, College Station, TX, USA.

Harlan JR, de Wet JMJ (1971) Toward a rational classification of cultivated plants. Taxon 20: 509–517.

Hawkes JG (1990) The Potato: Evolution, Biodiversity and Genetic Resources. Smithsonian Institution Press, Washington DC, USA.

Helgeson JP, Pohlman JD, Austin S, Haberlach GT, Wielgus SM, Ronis D, Zambolim L, Tooley P, McGrath JM, James RV (1998) Somatic hybrids between *Solanum bulbocastanum* and potato: a new source of resistance to late blight. Theor Appl Genet 96: 738–742.

Hermsen JGT, Ramanna MS (1973) Double-bridge hybrids of Solanum bulbocastanum and cultivars of Solanum tuberosum. Euphytica 22: 457–466.

Hosaka K (1995) Successive domestication and evolution of the Andean potatoes as revealed by chloroplast DNA restriction endonuclease analysis. Theor Appl Genet 90: 356–363.

Huaman Z, Spooner DM (2002) Reclassification of landrace populations of cultivated potatoes (*Solanum* sect. *Petota*). Am J Bot 89: 947–965.

Iwanzik W, Tevini M, Stute R, Hilbert R (1983) Carotinoidgehalt undzusammensetzung verschiedener deutscher Kartoffelsorten und deren Bedeutung für die Fleischfarbe der Knolle. Potato Res 26: 149–162.

Jackson SA, Hanneman RE, Jr. (1999) Crossability between cultivated and wild tuber- and non-tuber-bearing Solanums. Euphytica 109: 51-67.

Jansen G, Flamme W, Schuler K, Vandrey M (2001) Tuber and starch quality of wild and cultivated potato species and cultivars. Potato Res 44: 137–146.

Jansky S, Hamernik A (2009) The introgression of 2X 1EBN *Solanum* species into the cultivated potato using *Solanum verrucosum* as a bridge. Genet Resour Crop Evol 56: 1107–1115.

Johnston SA, den Nijs TM, Peloquin SJ, Hanneman RE, Jr (1980) The significance of genic balance to endosperm development in interspecific crosses. Theor Appl Genet 57: 5–9.

Karlsson M, Leeman A, Bjorck I, Eliasson A (2007) Some physical and nutritional characteristics of genetically modified potatoes varying in amylose/amylopectin ratios. Food Chem 100: 136–146.

Lara-Cabrera SI, Spooner DM (2004) Taxonomy of North and Central American diploid wild potato (*Solanum* sect. *Petota*) species: AFLP data. Plant Syst Evol 248: 129–142.

Laufer B (1938) The American plant migration. Part I: the potato. Field Museum of Natural History Anthropological Series 28: 1–132.

Lewis C, Walker J, Lancaster J, Sutton K (1998) Determination of anthocyanins, flavonoids and phenolic acids in potatoes. I. Coloured cultivars of *Solanum tuberosum* L. J Sci Food Agri 77: 45–57.

Lokossou AA, Park T, van Arkel G, Arens M, Ruyter-Spira C, Morales J, Whisson SC, Birch PRJ, Visser RGF, Jacobsen E, van der Vossen EAG (2009) Exploiting knowledge of R/Avr genes to rapidly clone a new LZ-NBS-LRR family of late blight resistance genes from potato linkage group IV. Mol Plant-Microbe Interact 22: 630–641.

Lu W, Haynes K, Wiley E, Clevidence B (2001) Carotenoid content and color in diploid potatoes. J Am Soc Hort Sci 126: 722–726.

Mabberley DJ (1997) The Plant Book. 2nd edn. Cambridge Univ Press, Cambridge, UK.

Mares-Perlman J, Brady W, Klein R, Klein B, Bowen P, Stacewiczsapuntsakis M, Palta M (1995) Serum antioxidants and age-related macular degeneration in a population-based case-control study. Arch Ophthalmol 113: 1518–1523.

McNeill WH (1949) The introduction of the potato into Ireland. J Mod Histor 21: 218–222.

Nesterenko S, Sink K (2003) Carotenoid profiles of potato breeding lines and selected cultivars. HortScience 38: 1173–1177.

Niederhauser JS, Mills WR (1953) Resistance of *Solanum* species to *Phytophthora infestans* in Mexico. Phytopathology 43: 456–457.

O'Neill K (2005) Statistical Derivation of Nutrition Label for Raw Potatoes from 2001–2002 United States Department of Agriculture (USDA) Agricultural Research Service (ARS) Nutrient Data Laboratory (NDL) Data for Raw Russet, White, and Red Potatoes by Weighting by Market Share. In: Center for Food Safety and Applied Nutrition, FDA.

Ochoa CM (1990) The Potatoes of South America: Bolivia. Cambridge Univ Press, Cambridge, UK.

Oosumi T, Rockhold DR, Maccree MM, Deahl KL, McCue KF, Belknap WR (2009) Gene *Rpi-bt1* from *Solanum bulbocastanum* confers resistance to late blight in transgenic potatoes. Am J Potato Res 86: 456–465.

Palozza P, Krinsky N (1992) Antioxidant effects of carotenoids in vivo and in vitro: an overview. Meth Enzymol 213: 403–420.

Park TH, Gros J, Sikkema A, Vleeshouwers VGAA, Muskens M, Allefs S, Jacobsen E, Visser RGF, Vossen EAGvd (2005) The late blight resistance locus *Rpi-blb3* from *Solanum bulbocastanum* belongs to a major late blight *R* gene cluster on chromosome 4 of potato. Mol Plant-Microbe interact 18: 722–729.

Pozorski S, Pozorski T (1979) An early subsistence exchange system in the Moche Valley, Peru. J Field Archaeol 6: 413–432.

Reddick D (1939) Whence came Phytophthora infestans? Chron Bot 4: 410–412.

Reddivari L, Vanamala J, Chintharlapalli S, Safe S, Miller J (2007) Anthocyanin fraction from potato extracts is cytotoxic to prostate cancer cells through activation of caspase-dependent and caspase-independent pathways. Carcinogenesis 28: 2227–2235.

Rodriguez A, Spooner DM (1997) Chloroplast DNA analysis of *Solanum bulbocastanum* and *S. cardiophyllum*, and evidence for the distinctiveness of *S. cardiophyllum* subsp. *ehrenbergii* (sect. *Petota*). Syst Bot 22: 31–43.

Ross J, Kasum C (2002) Dietary flavonoids: bioavailability, metabolic effects, and safety. Annu Rev Nutr 22: 19–34.

Rubatzky VE, Yamaguchi M (1997) World Vegetables: Principles, Production, and Nutritive Values. 2nd edn. Chapman and Hall, NY, USA.

Salaman RN (1937) The potato in its early home and its introduction into Europe. J Roy Hort Soc 62: 61–77, 112–123, 153–162, 253–266.

Sauer JD (1993) Historical Geography of Crop Plants: A Select Roster. CRC Press, Boca Raton, FL, USA.

Seddon J, Ajani U, Speruto R, Hiller R, Blair N, Burton T, Farber M, Gragoudas E, Haller J, Miller D, Yannuzzi L, Willett W (1994) Dietary carotenoids, vitamin A, vitamin C and vitamin E and advanced age-related macular degeneration. J Am Med Assoc 272: 1413–1420.

Song J, Bradeen JM, Naess SK, Raasch JA, Wielgus SM, Haberlach GT, Liu J, Kuang H, Austin-Phillips S, Buell CR, Helgeson JP, Jiang J (2003) Gene *RB* from *Solanum bulbocastanum* confers broad spectrum resistance against potato late blight pathogen *Phytophthora infestans*. Proc Natl Acad Sci USA 100: 9128–9133.

Spooner DM, Castillo R (1997) Reexamination of series relationships of South American wild potatoes (Solanaceae: *Solanum* sect. *Petota)*: evidence from chloroplast DNA restriction site variation. Am J Bot 84: 671–685.

Spooner DM, van den Berg RG, Rodriguez A, Bamberg J, Hijmans RJ, Lara-Cabrera SI (2004) Wild Potato (*Solanum* Section *Petota;* Solanaceae) of North and Central America. American Society of Plant Taxonomists, Ann Arbor, MI, USA.

Spooner DM, McLean K, Ramsay G, Waugh R, Bryan GJ (2005) A single domestication for potato based on multilocus amplified fragment length polymorphism genotyping. Proc Natl Acad Sci USA 102: 14694–14699.

Tevini M, Schonecker G (1986) Occurrence, properties, and characterization of potato carotenoids. Potato Res 29: 265.

Tevini M, Iwanzik W, Schonecker G (1984) Analyse vorkommen und nerhalten von carotinoiden in kartoffeln und kartoffelprodukten. Jahrbuch Forschungskreis Ernahrungsindustrie V 5: 36–53.

Thompson M, Thompson H, McGinley J, Neil E, Rush D, Holm D, Stushnoff C (2009) Functional food characteristics of potato cultivars (*Solanum tuberosum* L.): phytochemical composition and inhibition of 1-methyl-1-nitrosourea induced breast cancer in rats. J Food Compos Anal 22: 571–576.

Tsuda T, Horio F, Osawa T (2000) The role of anthocyanins as an antioxidant under oxidative stress in rats. Biofactors 13: 133–139.

US Department of Agriculture ARS (2006) USDA National Nutrient Database for Standard Reference, Release 19. Nutrient Data Laboratory Home Page.

Ugent D (1970) The potato. Science 170: 1161–1166.

Ugent D, Pozorski S, Pozorski T (1982) Archaeological potato tuber remains from the Casma Valley of Peru. Econ Bot 36: 182–192.

Ugent D, Dillehay T, Ramirez C (1987) Potato remains from a late Pleistocene settlement in southcentral Chile. Econ Bot 41: 16–27.

van Amelsvoort J, Weststrate J (1992) Amylose-amylopectin ratio in a meal affects postprandial variables in male volunteers. Am J Clin Nutr 55: 712–718.

van den Berg RG, Miller JT, Ugarte ML, Kardolus JP, Villand J, Nienhuis J, Spooner DM (1998) Collapse of morphological species in the wild potato *Solanum brevicaule* complex (Solanaceae: sect. *Petota*). Am J Bot 85: 92–109.

Vavilov NI (1951) The origin, variation, immunity and breeding of cultivated plants. Chron Bot 13: 1–366 [Translated from Russian by S Chester].

Vossen EAGvd, Gros J, Sikkema A, Muskens M, Wouters D, Wolters P, Pereira A, Allefs S (2005) The *Rpi-blb2* gene from *Solanum bulbocastanum* is an *Mi-1* gene homolog conferring broad-spectrum late blight resistance in potato. Plant J 44: 208–222.

Vossen Evd, Sikkema A, Hekkert BTL, Gros J, Stevens P, Muskens M, Wouters D, Pereira A, Stiekema W, Allefs S (2003) An ancient R gene from the wild potato species *Solanum bulbocastanum* confers broad-spectrum resistance to *Phytophthora infestans* in cultivated potato and tomato. Plant J 37: 867–882.

Welch R, Graham R (1999) A new paradigm for world agriculture: meeting human needs. Productive, sustainable, nutritious. Field Crops Res 60: 1–10.

Classical Genetics and Traditional Breeding

*Domenico Carputo** and *Luigi Frusciante*

ABSTRACT

Conventional breeding of the potato is not an easy task due to its polyploid ($2n = 4x = 48$) nature. Traits display tetrasomic inheritance patterns and chromatid segregation is possible. In addition, for several traits the genetic variance is almost entirely non-additive, so that breeders must maximize allelic diversity at any given locus to optimize heterozigosity and, consequently, heterosis. Breeding efforts are very much dependent on some reproductive features that are typical of the cultivated potato as well as its tuber and non-tuber related species. Among these characteristics, the production of $2n$ gametes and barriers to interspecific hybridization are probably the most important that breeders should consider when designing specific crossing schemes. After presenting the genetic and reproductive characteristics of the potato, this chapter describes the main breeding objectives, especially in terms of quality and resistance traits. It also focuses on methods and breeding schemes based on intra- and interspecific sexual hybridization, providing essential highlights for each of them.

Keywords: allelic diversity, polyploidy, resistance traits, sexual polyploidization, $2n$ gametes, $4x$ x $2x$ crosses, $4x$ x $4x$ crosses

University of Naples, Federico II, Department of Soil, Plant, Environmental and Animal Production Sciences, Via Università 100, 80055 Portici, Italy.
*Corresponding author: *carputo@unina.it*

2.1 Introduction

2.1.1 Genetic Characteristics

The cultivated commercial potato *Solanum tuberosum* is a polysomic tetraploid ($2n = 4x = 48$) crop with basic chromosome number $x = 12$, and a haploid genome size of approximately 840 Mbp, which makes it a medium-sized plant genome. Each of the four copies of homologous chromosomes segregates from multivalents or random bivalents. Due to the small size (about 1–3.5 µm at mitotic metaphase) of individual chromosomes, it has been generally regarded as a cytologically difficult species to study. Indeed, potato mitotic chromosomes cannot be identified based on their morphology. Mohanty et al. (2004) found high variability in cytological characteristics (e.g., total chromosome length and volume, number of secondary restricted chromosomes) between various cultivars and breeding lines. On the basis of size and position of constrictions, they classified potato mitotic chromosomes into four groups. Advanced approaches are now the most suitable alternatives to the strategies traditionally used for potato chromosome analysis. Reliable identification of the potato somatic metaphase chromosomes has been achieved using a set of restriction fragment length polymorphism (RFLP) marker-anchored bacterial artificial chromosome (BAC) clones as florescence in situ hybridization (FISH) probes (Dong et al. 2000). This work enabled researchers to assign the genetic linkage groups of potato to specific chromosomes, and the potato chromosomes were consequently numbered according to their correspondent linkage group (this numbering system is generally accepted and was also adopted by the International Potato Genome Sequence Consortium).

Potato karyotypes based on pachytene chromosomes have been described (e.g., Mok et al. 1974; Pijnacker and Ferwerda 1984) and, as done in other species and for conventional karyotyping, chromosomes were numbered based on their descending length. A recent treatment by Peloquin et al. (2008) described the pachytene-based karyotype of the *S. andigena* haploid, which consisted of three chromosomes with median primary constrictions, three with submedian constrictions, five with subterminal constrictions and one with a terminal constriction. The amount of euchromatin of each chromosome varied from 21 to 75%. Although the distribution patterns of euchromatin and heterochromatin and the differential staining of the centromeric regions enabled researchers to distinguish each chromosome (Yeh and Peloquin 1965), pachytene analysis and chromosome identification (based on a structural basis Fig. 2-1) remains very difficult in potato.

The genetics of the tetraploid potato is very unique and more complicated than that of either diploid crops or polyploids with disomic inheritance. At

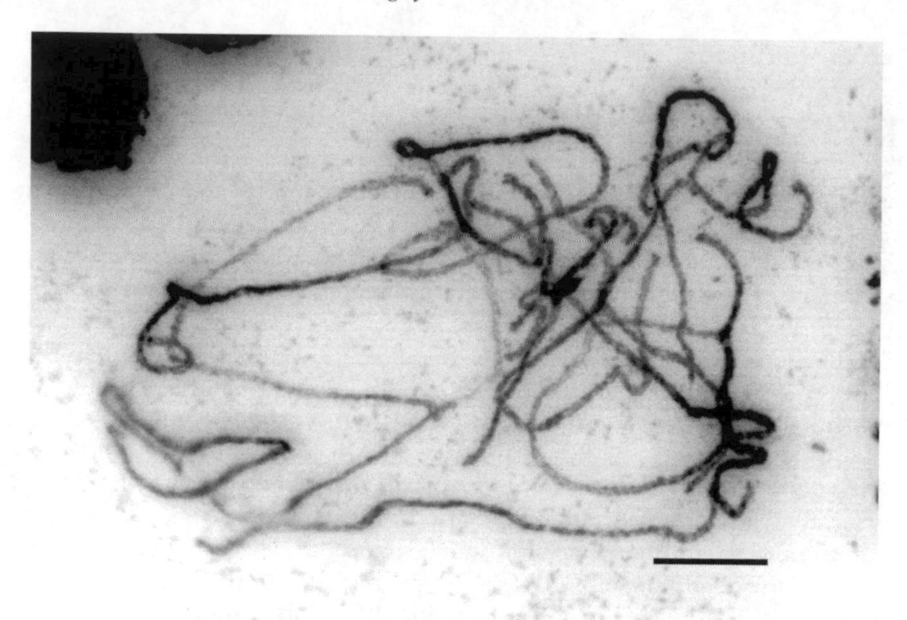

Figure 2-1 Pachytene chromosomes of a *S. tuberosum* haploid ($2n = 2x = 24$). Courtesy of Dr. M. Iovene and Dr. J. Jiang, University of Wisconsin-Madison (scale bar = 10 micron).

a given locus, five genotypes (aaaa = nulliplex, Aaaa = simplex, AAaa = duplex, AAAa = triplex and AAAA = quadruplex) are possible. In reality, there could be as many as four alleles per locus, providing the opportunity for rich interaction patterns. In polysomic polyploids like the potato, the more diverse the alleles are within a locus, the higher the heterozygosity and the greater the number of increased interlocus and epistatic interactions. While diploids hold only one non-additive genetic effect at a locus, in the tetraploid potato there may be three non-additive genetic effects: first order (between two alleles), second order (between three alleles) and third order (between four alleles) (Mendiburu and Peloquin 1977). In a tetrallelic genotype this corresponds to 11 heteroallelic interactions. A unique advantage for breeders is that genotypes displaying tetrallelic genic interactions can be immediately fixed through clonal propagation. For traits whose genetic variance is almost entirely non-additive, this is a very important issue, for the breeder must strive to obtain maximum heterozygosity to optimize heterotic combinations. Indeed, in the tetraploid potato a drastic decline in yield and fertility occurs upon self-fertilization and in progenies from closely related parents. This can be attributed not only to increased homozygosity but also to the loss of favorable allelic intralocus interactions and of favorable epistatic combinations required to maximize heterosis. In designing breeding schemes aimed at maximizing heterozygosity, breeders need to be very careful in the appropriate use of parents. High levels of heterozygosity can be reached when new genetic materials are synthetized through controlled crosses involving genetically diverse parents or gametes with unreduced chromosome numbers, also referred as 2n gametes (see next section). Multiple allelism in potato has been demonstrated for genes affecting either metabolic pathways or morphological traits. van Eck et al. (1994) reported the presence of three alleles at the *Ro* locus controlling tuber shape. van de Wal et al. (2001) found at least eight different alleles coding for granule-bound starch synthase I, responsible for amylose biosynthesis. Recently, through SNPs genotyping, Achenbach et al. (2009) demonstrated the existence of multiple alleles either at a single locus or at various linked loci controlling resistance to *Globodera pallida*.

Complexity to inhertance is added by the fact that, depending on gene-centromere distance, alternative segregation patterns in the gametes are possible: chromosome segregation, random chromatid segregation and maximum equational segregation (see Burnham 1962). Chromosome segregation occurs when chromatids derived from a particular multivalent belong to different chromosomes of that multivalent. This type of segregation is assumed for genes completely linked to the centromere. In this scenario, the duplex genotype AAaa, forms AA, Aa and aa gametes in a 1:4:1 ratio. If either random chromatid segregation or maximum equational segregation occurs, aa gametes derived from sister chromatids are also expected. This phenomenon has been termed double reduction, characterized by a coefficient of double reduction ($\propto$) defined as the probability of two sister chromatids ending up in the same gamete. The cytological bases for double reduction include multivalent pairing with one cross over event between the locus and the centromere and the two pairs of chromatids from that cross over event passing to the same pole in anaphase I, and random separation of chromatids in anaphase II. The theoretical extreme values for $\propto$ generally reported in the literature are: 0 for chromosome segregation, 1/6 for maximum equational segregation and 1/7 for random chromatid segregation. If the same duplex genotype AAaa mentioned before is considered, AA, Aa and aa gametes are formed in a 2:5:2 ratio with maximum equational segregation, in a 3:8:3 ratio with random chromatid segregation if $\propto = 0$, and in a $1+2\propto/6$: $4-4\propto/6$: $1+2\propto/6$ ratio with both types of chromatid segregation if $\propto$ is different from 0 (Burnham 1962).

Some other genetic features that are important in this context should be underlined. As with every clonally propagating crop, the cultivated potato is highly heterozygous. Consequently, segregation of traits is expected in the F_1 generation following hybridization. Homozygous genotypes are difficult to obtain upon selfing due to inbreeding depression. The cultivated potato also has a narrow genetic base compared with that at its center of origin and with wild species. This is mainly due to the limited number of genotypes that were used to develop the *S. tuberosum* form of the cultivated potato (Bradshaw and Ramsay 2005).

2.1.2 Reproductive Characteristics Relevant for Breeding

2.1.2.1 Production of 2n Gametes

A noteworthy reproductive characteristic with important potato breeding applications (see Section 2.2.2) is that several genotypes of cultivated and wild potatoes produce 2*n* gametes. They are the result of meiotic anomalies affecting various nuclear and cytoplasmic aspects of sporogenesis (chromosome pairing, centromere division, spindle formation or cytokinesis). For a review on meiotic variants leading to 2*n* gamete formation in potato see Peloquin et al. (1999) and Carputo et al. (2000). Three distinct meiotic mutations leading to 2*n* pollen production have been described: parallel spindles (ps), including also fused and tripolar spindles, premature cytokinesis 1 (pc1), and premature cytokinesis 2 (pc2), all inherited as simple Mendelian recessive traits. When these mutations are present, at the end of meiosis dyads with two 2*n* microspores are formed. Watanabe and Peloquin (1991) reported that ps is the most common mutation leading to 2*n* pollen production in the potato. Meiotic mutations affecting 2*n* egg production have also been found and described. Werner and Peloquin (1987) reported that omission of second meiotic division (os) was the most common mutation of 2*n* egg formation. When os is present, two 2*n* megaspores are produced, one degenerating and the other yielding the functional spore. Another meiotic variant resulting in 2*n* eggs is the failure of cytokinesis after the second division, which results in the fusion of nuclear products of this division. Other mutations leading to 2*n* egg formation may affect megasporogenesis. These cause either an irregular anaphase II and delayed meiotic divisions or an irregular spindle axis formation at metaphase I. Each meiotic mutation leading to 2*n* gametes is genetically equivalent to either first division restitution (FDR) or second division restitution (SDR) mechanisms (Peloquin et al. 1999; Fig. 2-2). It should be pointed out that genetic mode of 2*n* gamete formation and cytological event are two separated entities. Thus, for example, ps occurs at second meiotic division but it is genetically equivalent to a FDR mechanism, whereas os (occurring during first meiotic division) is genetically equivalent to SDR. Due to the genetic consequences associated with the mode of 2*n* gamete formation, various levels of variability, fitness, and heterozygosity are expected in newly formed polyploids (see below). The genetic control of 2*n* gamete production has been generally attributed to the action of single recessive genes (Bretagnolle and Thompson 1995; Peloquin et al. 1999). These genes exhibit incomplete penetrance and variable expressivity, and their phenotypic expression may be significantly modified by genetic,

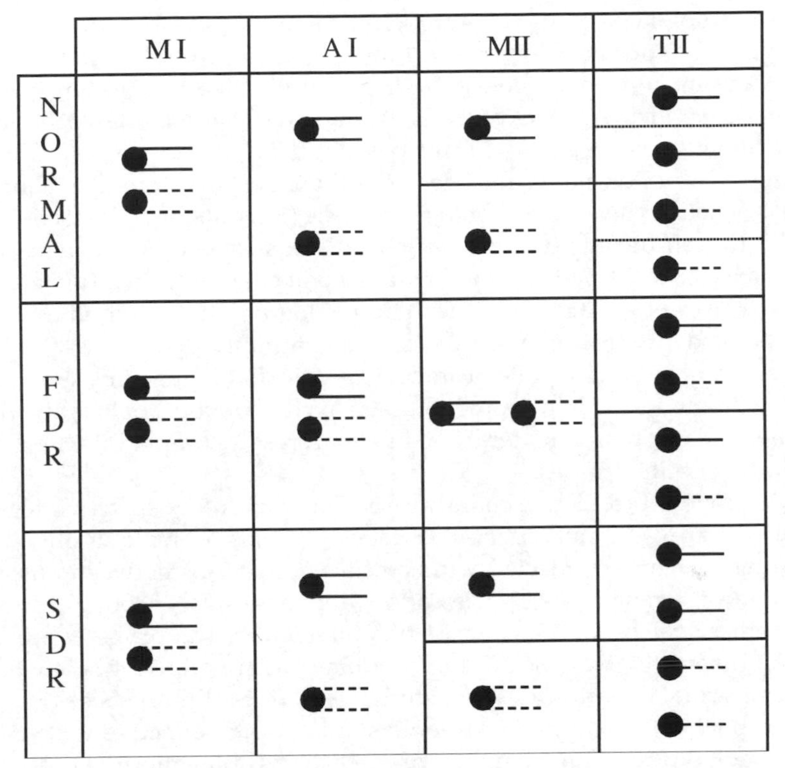

Figure 2-2 Representation of metaphase I (MI), anaphase I (AI), metaphase II (MII), and telophase II (TII) of a meiocyte with two homologous chromosomes undergoing: normal meiosis, first division restitution (FDR), and second division restitution (SDR) mechanisms.

environmental, and developmental factors. Recently, the first gene (*AtPS1*) implicated in the parallel orientation of spindles and 2*n* pollen production in *Arabidopsis thaliana* has been isolated and characterized at the molecular level (d'Erfurth et al. 2008). This gene encodes a protein of about 1,500 amino acids conserved throughout the plant kingdom, and it is a good candidate for the *ps* locus in potato.

2.1.2.2 Barriers to Interspecific Hybridization

There are three additional reproductive features of the potato that must be underlined: pollen-pistil incompatibility, nuclear-cytoplasmic male sterility, and endosperm development. They have important implications because there is strong and circumstantial evidence that, rather than chromosome differentiation, they provide effective barriers to interspecific hybridization at the pre-zygotic and/or post-zygotic level (Camadro et al. 2004).

Most diploid potato species are self-incompatible because they possess a single polymorphic *S* locus determining gametophytic incompatibility. This locus includes two genes, *S-RNAse* and *SLF/SFB*, controlling female and male specificity, respectively (Takayama and Isogai 2005). Glycoproteins with ribonuclease activity control pistil specificity. By contrast, *S* gene products in pollen have not been identified so far (Stone and Goring 2001). Genetic analysis suggested that a gene (*Sli*) outside the *S* locus may inhibit the self-incompatibility mechanism (Hosaka and Hanneman 1998; Phumichai et al. 2006). Unilateral and bilateral cross incongruity may also occur (Hayes et al. 2005). Camadro and Peloquin (1981) hypothesized a genetic model with dominant cross incompatibility genes in styles that prevent fertilization by pollen carrying specific dominant complementary genes. The model, which assumes segregation for both type of loci, accounts for the results of both unilateral and bilateral cross incompatibility in inter- and intra-specific crosses.

A second reproductive characteristic that guarantees species integrity in tuber-bearing *Solanum* species is nuclear-cytoplasmic male sterility. This isolating barrier is connected with specific interactions between nuclear-encoded proteins of one species and cytoplasmically-encoded proteins of another species. As a result of this interaction, the progeny is male sterile. [Interestingly, male sterility combines with pollinator behavior to achieve sexual isolation of male sterile genotypes. Bumblebees, known potato pollinators, do not visit male sterile plants, effectively isolating these genotypes.] The existence of nuclear-cytoplasmic male sterility has been reported as a result of either interactions between *S. tuberosum* cytoplasm and nuclear genetic factors of wild species (e.g., *S. sanctae-rosae, S. infundibuliforme, S. raphanifolium, S. commersonii*) or species cytoplasm and nuclear genes of *S. tuberosum*. The importance of this type of sterility in Solanums was emphasized by Grun (1990), who described a number of genes affecting anther indehiscence, sporad formation, anther style fusion, etc. Information about the genetic bases of nuclear-cytoplasmic male sterility is limited. Iwanaga et al. (1991) suggested the action of male fertility restorer genes. Ortiz (1998) reported that one or a few nuclear genes can interact in incompatible cytoplasms to cause male sterility. Molecular analysis in tuber-bearing *Solanum* species suggested that the nuclear-cytoplasmic male sterility is controlled by mitochondrial DNA (Breiman and Galun 1990; Cardi et al. 1999).

In potato, as well as in several other angiosperms, endosperm breakdown is considered the primary reason for seed failure after double fertilization. Genetically, the nuclear constitution of the endosperm differs from the embryo only in having one extra set of maternal chromosomes. Thus, endosperm development is dependent on the same genes as the embryo but in different doses. To explain failures in interspecific crosses in *Solanum*, Johnston et al. (1980) developed a model that extended the 2:1 maternal to paternal genome dosage in the endosperm described in corn. This model relies on a balance of qualitative genetic factors (Endosperm Balance Number, EBN) and not genomes for normal endosperm development. In this model, each species has an effective number (the EBN), which is not necessarily a direct reflection of its ploidy. It is the EBN which must be in a 2:1 maternal to paternal ratio in the hybrid endosperm for normal development of this tissue and, consequently, of the normal hybrid embryo. Thus, successful interspecific hybridization occurs only when parents produce gametes with the same EBN (Fig. 2-3). The EBN is an arbitrary value experimentally assigned to each *Solanum* species based on its behavior in crosses with EBN standards, and assuming the 2:1 ratio as a prerequisite for normal endosperm development. Cultivated *S. tuberosum* has been assigned 4EBN. Diploid species have been assigned either 1EBN (e.g., those belonging to *Etuberosa* and *Commersoniana* series) or 2EBN (e.g., all diploid cultivated species and most wild species, like *S. multidissectum*, *S. chacoense*, *S. verrucosum*). Tetraploid species have been assigned either 2EBN or 4EBN. All the hexaploid species, such as *S. demissum* and *S. oplocense*, have been assigned 4EBN. Several genetic studies provided evidence that an oligogenic control of EBN exists in potato species (Ehlenfeldt and Hanneman 1988; Camadro and Masuelli 1995; Johnston and Hanneman 1996). The essential role of EBN as an internal reproductive barrier to interspecific hybridization is that it creates effective barriers between sympatric species, leaving intact their genotypic integrity. For this reason, for example, 2x(1EBN) *S. commersonii* cannot be crossed with *S. chacoense*, which is also diploid but with 2EBN. On the other hand, *S. commersonii* hybridizes with other diploid (1EBN) species. The EBN represents a powerful screen for 2*n* gametes during sexual polyploidization events that lead to the polyploid evolution of *Solanum* species (Carputo et al. 2003). Inter-EBN crosses can occur, but they are only sporadic events

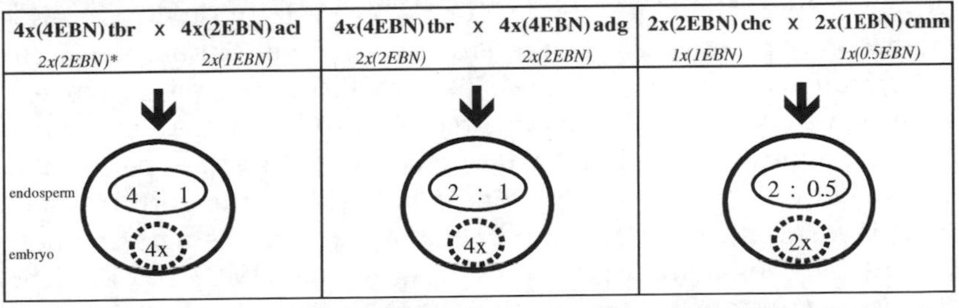

Figure 2-3 Prediction of female:male EBN ratio in the hybrid endosperm (solid oval) and ploidy level of the hybrid embryo (hatched circle) resulting from interspecific crosses involving *Solanum* species with different EBN. Tbr = *S. tuberosum*, acl = *S. acaule*, adg = *S. andigena*, chc = *S. chacoense*, cmm = *S. commersonii*. Only cross combinations resulting in a 2:1 female to male EBN ratio [e.g., 4*x*(4EBN)tbr x 4*x*(4EBN)adg] in the hybrid endosperm are compatible.
* Gamete ploidy (EBN).

that probably result from non-heritable random events, such as multiple fertilizations of the central cell, mitotic abnormalities in the gametophyte, endomitosis of the polar nuclei in the endosperm, or increase in the number of polar nuclei (Hanneman 1999).

2.2 Conventional Breeding Methods

Potato enjoys an array of crossing procedures that makes it a model crop for breeders. It is amenable to various in vitro culture techniques such as somatic embryo development, protoplast fusion, cell, anther and pollen culture. It is also an ideal species for *Agrobacterium*-mediated transformation protocols that allow the development of trans- and cis-genic genotypes. However, in this section only conventional breeding methods based on sexual hybridization will be considered. These can be classified into two levels based upon ploidy and EBN: (a) the tetraploid 4EBN level, and (b) the diploid 2EBN level.

2.2.1 *The Tetraploid 4EBN Level*

Tetraploid breeding is essentially based on phenotypic recurrent selection, involving crosses between 4*x*(4EBN) genotypes (often varieties) and then field evaluation and selection of descendants for several years. Most currently used varieties derive from this breeding approach. Its genetic and reproductive bases have been presented in previous sections of this chapter. The procedure starts with the selection of diverse, desirable 4*x*(4EBN) parents, a crucial aspect to prevent homozygosity and guarantee high allelic diversity of offspring. Parents are often selected based on their phenotype rather than genotype, and crosses are made between pairs with complementary characteristics. However, due to the non-additive gene action for several traits of interest, phenotypic parental selection is often ineffective. Therefore, the use of test-crosses to identify desirable combination of parents has been reported by various authors (see Tarn et al. 1992; Bradshaw and Mackay 1994; Mackay 2005, and references therein). Test-crosses evaluate both general combining ability (GCA) and specific combining ability (SCA). The former refers to the average performance of a clone in crosses with other clones, the latter to the performance of a clone in a specific combination in comparison with its performance in other specific combinations. Reviews by Tarn et al. (1992) and Bradshaw and Mackay (1994) provide a detailed list of reports on combining ability analysis using *S. tuberosum* genotypes as parents. There are essentially two traditional approaches to calculate combining abilities, one based on crosses in all directions between two groups of complementary clones, and the other method based on diallel crosses (Bradshaw 1994). Recently, to reduce the number of matings normally used in test-crosses to estimate GCA, Gopal et al. (2008) proposed the use of top-crosses (matings with selected single testers). Tarn et al. (1992) suggested that when GCA is predominant, valuable cross combinations can be determined simply based on the GCA information of parents. When SCA is larger, breeders should carry out progeny tests on samples of progenies produced within the breeding program. It should be pointed out that superior parents may also derive from pre-breeding activities aimed at producing improved tetraploid parents.

Due to the high levels of heterozygosity of tetraploid potatoes, segregation of traits is expected in the F_1 generation derived from $4x$ x $4x$ crosses. In a typical program, the year after crossing large populations (tens of thousands) of F_1 seedlings are grown for visual selection. A single tuber is usually taken from each selected F_1 plant for planting in the next season in a single-hill plot. This represents the first clonal generation. During the next two to three seasons, clones are cultivated and screened in larger unreplicated plots with plants spaced for further evaluations and selection and to increase the number of seed tubers. After a number of years of such selection, superior clones are tested in replicated trials in different locations to estimate the genotype x environment interaction. In the breeding flow described, a high percentage of clones is discarded in the first clonal generations based on field evaluations (early generation selection). Resistance and quality traits are usually screened during the final stages of selection, when the number of clones under evaluation has been drastically reduced (late generation selection). At later stages, breeders generally select simultaneously for a very high number of traits. Tarn et al. (1992) outlined that while at later stages of selection emphasis is given to characteristics of individual clones, at early stages of selection parents and crosses play a significant role in breeders' decisions.

For traits with low heritability, stringent visual selection at early generations is usually ineffective and increases the risk of losing valuable clones. To raise the efficiency of early generation selection, various authors have developed modified selection procedures. Gopal et al. (1992) suggested elimination of seedlings with poor vigor before transplanting; the rejection of clones with undesired color, shape, eye depth and tuber cracking at the seedling stage; negative selection for tuber yield and weight at the first clonal generation and for low tuber number at the second clonal generation. Negative selection involves the elimination of clones with poor traits and the retention of those with no obvious serious defects. Bradshaw et al. (2003) proposed the combined use of mid-parent values to select the best parents and progeny tests to discard whole progenies before starting the within progeny selection. They also suggested increasing the number of clones on which to practice selection by sowing additional seeds from best progenies (resowing). Among several other reports outlining strategies of early generation selection for important traits, are those by Thill and Peloquin (1995) for chipping ability, Bisognin and Douches (2002) for tuber appearance, specific gravity and chip color, and Bae et al. (2008) for resistance to Verticillium wilt. In sum, classical breeding at the tetraploid level is not very different in principle from that performed at the beginning of the 20th century. Its rationale is essentially based on the evaluation of a decreasing number of selected clones in increasingly elaborate field/laboratory trials in a wide range of environments (Fig. 2-4). The last part of the breeding program includes official regional variety testing, healthy maintenance and propagation of seed tubers and marketing strategies.

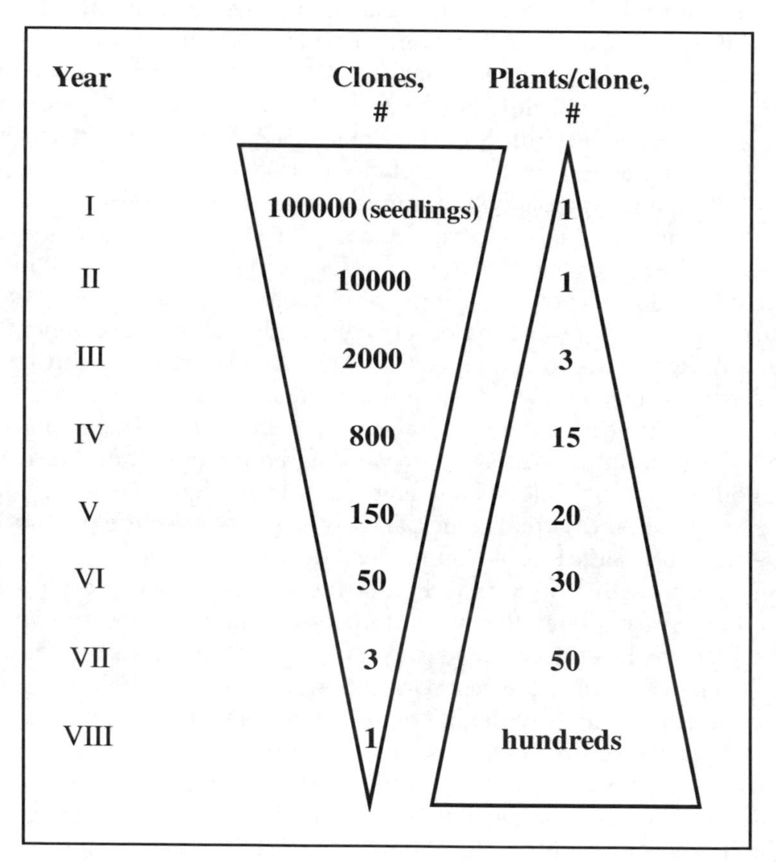

Figure 2-4 Schematic representation of a traditional potato breeding scheme. A hypothetical number of clones are evaluated each year and a hypothetical number of plants are grown for each clone. Following crosses between selected parental lines, each year a decreasing number of selected clones is evaluated in increasingly sophisticated field/laboratory trials.

2.2.2 The Diploid 2EBN Level

In modern potato breeding, several programs operate at least in part at the diploid (2EBN) level. This not only takes advantage of disomic inheritance, which is less complicated than the typical tetrasomic patterns of the 4*x*(4EBN) level, but also requires a smaller population size at the start. In addition, this approach allows the most efficient exploitation of diploid potato germplasm. The first essential ingredient for potato breeding at the diploid 2EBN level is haploids (sporophytes with the gametic chromosome number). *S. tuberosum* maternal haploids can be easily obtained following crosses with diploid clones of *S. phureja*. Starting from the pioneering work of Hougas and Peloquin (1960), *S. tuberosum* haploids have been routinely produced for breeding purposes, and also for genetic, cytogenetic and evolutionary studies (Rokka 2009). It is believed that haploids originate via parthenogenesis, and that both sperm nuclei fuse with the polar nuclei of the central cell of the female gametophyte. Clulow et al. (1991) hypothesized that *S. tuberosum* haploids may also originate from a process of chromosome elimination from zygotes deriving from *S. tuberosum* x *S. phureja* crosses. Wilkinson et al. (1995) and Ercolano et al. (2004) demonstrated stable incorporation of *S. phureja* DNA into haploids obtained through crosses with *S. phureja*. Potato haploids may also be obtained through in vitro microspore/anther culture, but this method is more complicated and strongly genotype-dependent compared to the *S. tuberosum* x *S. phureja* approach. From the breeding standpoint, *S. tuberosum* haploids are widely used in sexual interspecific hybridization programs. They often represent essential ingredients also for the production of somatic hybrids through protoplast fusion (Orczyc 2003). Since potato haploids have a 24-chromosome complement and an EBN = 2, they are essential tools to efficiently capture the genetic diversity of 2*x*(2EBN) *Solanum* species through haploid x wild species crosses. In addition, they also allow breeders to indirectly evaluate tuber characteristics of wild species (several potato species do not tuberize under long-day conditions). It should be pointed out that haploid-wild species hybrids have 50% wild genome. Therefore, after population improvement at the 2*x* level for traits of interest, further breeding is needed to re-establish the chromosome number of the tetraploid cultivated potato and gradually reduce the wild genome content. This is accomplished through 2*n* gametes, representing the second essential feature of breeding at the diploid 2EBN level.

Haploid-wild species hybrids producing 2*n* gametes and with the desired traits are selected for the production of tetraploid hybrids through bilateral and unilateral sexual polyploidization (BSP and USP, respectively; Fig. 2-5). BSP involves crosses in which the female parent forms 2*n* eggs and the male one 2*n* pollen. By contrast, USP is based on crosses where only one

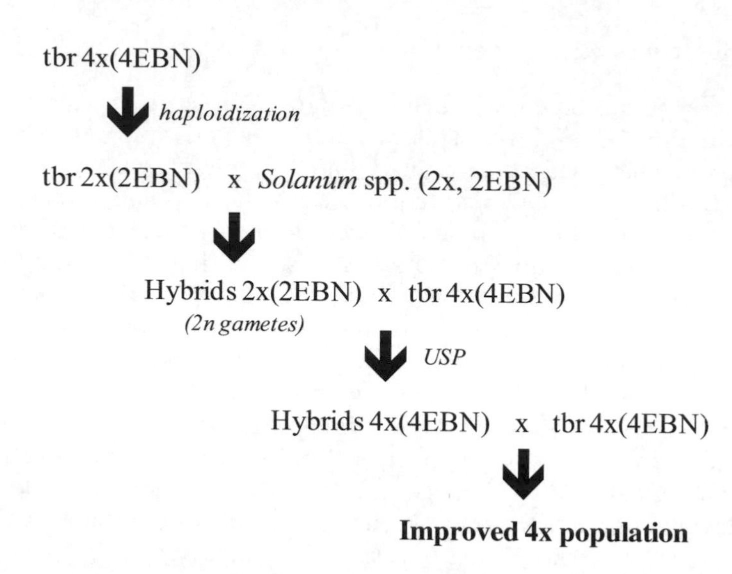

Figure 2-5 Breeding at the diploid (2EBN) level involving the production of *Solanum tuberosum* (tbr) haploids and the selection of superior 2*x*(2EBN) interspecific hybrids producing 2*n* gametes. In this outline, the unilateral sexual polyploidization crossing scheme (USP) is employed.

parent produces 2*n* gametes (either 2*n* pollen or 2*n* eggs). The exploitation of 2*n* gametes and the asexual reproduction system of the potato provide the breeders the unique possibility of accumulating non additive genetic effects at the tetraploid level of the offspring and the immediate fixation of heterosis, respectively. The extent of allelic interaction transmission from haploid-species hybrids to the tetraploid progeny depends mainly on the mode of 2*n* gamete production (FDR vs. SDR), whereas the creation of new interactions depends mainly on the allelic diversity of parents involved. It has been estimated that with FDR all loci from the centromere to the first crossover that are heterozygous in the parent will be heterozygous in the gamete, and one half of heterozygous parental loci beyond the first chiasma will be heterozygous in the gametes. By contrast, with SDR mechanisms, all the loci from the centromere to the first chiasma will be homozygous in the gametes, and all loci past the first crossover, which are heterozygous in the parent, will be heterozygous in the gametes. Based on the positions of centromeres and chiasmata and on the fact that in the potato generally only one chiasma per chromosome arm occurs, it has been estimated that the percent of heterozygosity transmitted by gametes is about 80% with FDR and 40% with SDR. FDR gametes are expected to strongly resemble each other and the parental clone from which they originate. They also maintain and transmit a large amount of inter-locus epistatic interactions, which are important in maximizing heterosis. By contrast, SDR is expected to produce a heterogeneous population of highly homozygous gametes (Fig. 2-2). The choice of USP vs. BSP depends on the mode of 2*n* gamete formation in the parent and, in the case of USP, on the fertility of the 4x parents. Due to the high level of allelic interactions transmitted by FDR gametes and on the wide occurrence of 2*n* pollen, the use of the 4*x* x 2*x*-FDR approach has been documented thoroughly (Tai 1994; Ortiz 1998; Buso et al. 1999; Buso et al. 2003; Alberino et al. 2004). Ortiz et al. (1991) reported that besides the allelic diversity transmitted, the 4*x* x 2*x* breeding scheme is more effective than the traditional 4*x* x 4*x* scheme also because fewer replications and locations are necessary for progeny evaluation. Another advantage of the 4*x* x 2*x* breeding scheme is that it can increase the efficiency of cultivar development (Buso et al. 1999). The authors found that the percentage of clones combining high yield and good tuber quality traits selected in relatively small 4*x* x 2*x* populations was much higher than that reported using 4*x* x 4*x* crosses. BSP can be employed as an alternative to USP when superior 2*n* egg and 2*n* pollen producing 2*x* parents are available. This method potentially allows the combination of multiple traits from two diploid interspecific hybrids into tetraploid genotypes through 2*x* x 2*x* crosses. It also allows transmission of high levels of heterozygosity into 4*x* progeny, especially when 2*x* hybrids used have unrelated genetic backgrounds. Werner and Peloquin (1991) reported a high increase in tuber yield of 4*x* families from 2*x* x 2*x* crosses in comparison to control cultivars and advanced selections. In 2*x* (SDR) x 2*x* (FDR) crosses, SDR and FDR complement each other in terms of transmission of heterozygosity. Species exploited in potato breeding programs at the diploid 2EBN level include, among others, *S. berthaultii*, *S. tarijense*, *S. bukasovii*, *S. sparsipilum*, *S. phureja*, *S. multidissectum*, and *S. chacoense* (reviewed in Jansky 2006). In Poland, Scotland, Canada, China, the USA and other countries with potato breeding programs, haploid–species hybrids with multiple resistances and tuber quality are developed and used in 4*x* x 2*x* crossing schemes to produce parental lines and cultivars.

It should be pointed out that crossing barriers sometimes do not allow the exploitation of breeding at the diploid 2EBN level. For example, 2*x*(1EBN) species cannot be hybridized with 2*x*(2EBN) *S. tuberosum* haploids. Similarly, crosses between 4*x*(2EBN) species and 2*x*(2EBN) potato haploids produce triploid offsprings. In these cases, breeding schemes based on ploidy manipulations and bridge crosses can be designed. These result in intermediate hybrids that allow the successful exploitation of potato species (Carputo and Barone 2005). Figure 2-6 illustrates the breeding schemes used by Carputo et al. (1997) to transfer useful genes and genetic diversity from 2*x*(1EBN) *S. commersonii* to *S. tuberosum*. Additional strategies to overcome hybridization barriers in potato include mentor pollinations and embryo rescue, hormone treatment, reciprocal crosses, careful selection of parents and, of course, somatic hybridization (Jansky 2006).

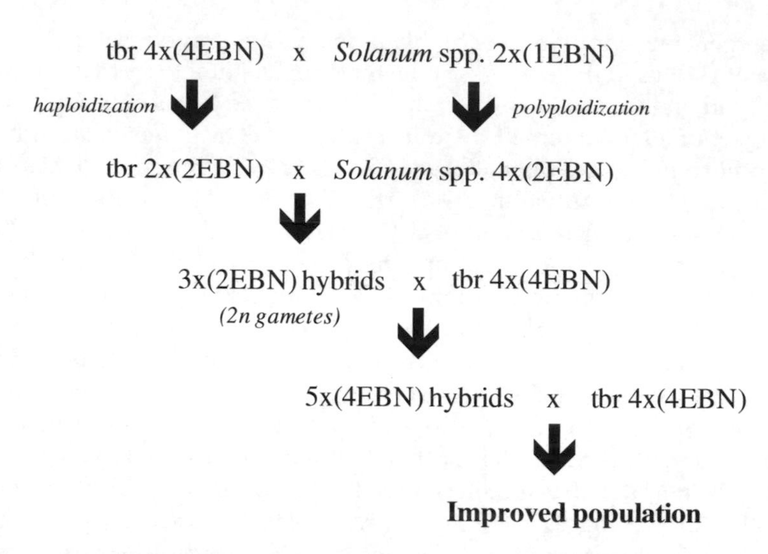

Figure 2-6 Breeding scheme aimed at the exploitation of incongruent 2*x*(1EBN) *Solanum* species. Ploidy manipulations are necessary to make parental EBNs compatible. Triploid and pentaploid bridge ploidies are produced. Polyploidization of the 2*x*(1EBN) species can be performed somatically through either antimitotic metabolites (e.g., colchicine) or in vitro spontaneous chromosome doubling. Alternatively, in a 2*n* gamete producer, the 2*x*(1EBN) species can be directly crossed with tbr 2*x*(2EBN) haploids.

2.3 Breeding Objectives

Recent world issues concerning environment protection, agriculture sustainability, food safety and quality have largely impacted potato breeding objectives. Furthermore, as recently outlined by Bonnel (2008), two additional aspects have affected breeders' activities: the increase in trade exchanges between countries traditionally involved in potato production and the rapid emergence of new countries that have become leaders in potato production (e.g., India and China). In light of this situation, breeders are focusing more and more on the development of regional or international varieties rather than local or national ones. At the same time, however, markets with specific requirements are also emerging. This is the case of organic potato production. Traditionally, breeders consider more than 50 traits during field and laboratory evaluation and selection. These can be grouped into three categories: 1) yield; 2) quality characteristics of tubers; and 3) resistance to biotic and abiotic environmental constraints.

Quality has become a challenging target for potato breeders worldwide. Quality parameters can be organized into two major categories. The first category includes external characteristics, comprising skin color, tuber size and shape, and eye depth. These traits are deemed very important for fresh consumption, where they are most likely to influence consumer's choice.

The second category comprises internal characteristics, including nutritional properties, cooking/after cooking properties, and processing quality. Internal quality encompasses traits such as dry matter content, flavor, sugar and protein content, starch quality, type and amount of glycoalkaloids. There are several factors affecting tuber quality: genotype, earliness, field practices, environmental conditions, storage temperatures, and the impact of pests and diseases. Of these, genotype is probably the most important. Traits that are genetically controlled include biological (proteins, carbohydrates, vitamins, minerals, reduced amounts of toxic glycoalkaloids); sensorial (flavor, texture, color); and industrial (tuber shape and size, dry matter content, cold sweetening, oil absorption, starch quality) traits. Breeding objectives related to quality for processed potatoes are normally different from those for fresh use. The former are well defined and are mainly related to dry matter content, level of reducing sugars during cold storage, starch type and, to some extent, tuber shape and size. The latter fall more into the category of external traits, even though eating quality traits such as after-cooking blackening are also important. Renatus (2005) pointed out that breeders need to consider two main types of potatoes for the fresh market. The first includes the standard merchandise at a discount price level, where breeders are required to provide cost efficient, high yielding, multi-use varieties. The second category comprises nice-looking, high price, top quality potatoes. This high-end segment needs the release of varieties with superior taste, increased nutritional value, and superior appearance. Among new goals of potato breeding for quality, is the development of varieties that do not accumulate asparagine. This aminoacid is the precursor of acrylamide, formed during the Maillard reaction as part of the cooking process. It may cause damage to the nervous system and has carcinogenic effects. Equally important is the need to control tuber dormancy, in order to avoid the use of expensive chemical suppressants that leave chemical residues, and the identification of varieties with low glycoalkaloid accumulation in the tubers. It is widely known that potato glycoalkaloids are extremely toxic to humans and animals, having possible teratogenic, embryotoxic, genotoxic, and carcinogenic effects (Korpan et al. 2004).

In terms of biotic stress resistance, of primary importance is *Phytophthora infestans,* the causal agent of late blight. Since its first outbreak in Europe during the 19th century, late blight is considered the most devastating fungal disease of potato, causing decay of both foliage and tubers. Race-specific resistance has been introgressed into the cultivated gene pool from wild *S. demissum*. However, new *P. infestans* races have overcome it. In recent years, breeding activities on late blight resistance have been brought together by the Global Initiative on Late Blight (GILB), a network of researchers and technology developers from all over the world. In recent years, resistance to various pathogens and pests (e.g., nematodes, *Colletotrichum coccodes,*

Leptinotarsa decemlineata, Streptomyces scabies, Rizochtonia solani, Verticillium spp., *Clavibacter michiganensis* and *Erwinia chrysantemii*) has become a major objective for potato breeding. Despite the very high number of other biotic constraints that may hamper potato cultivation, not all pathogens and pests are present in all areas where potato is grown (Mackay 2005). Resistance to nematodes, for example, is an important breeding target in Europe but not in the USA. In Europe, the potato is a major host for potato cyst nematodes *Globodera pallida* and *G. rostochiensis* and for root-knot nematodes (*Meloidogyne* spp.). Here, the availability of varieties resistant to *G. rostochiensis* has accentuated the presence of *G. pallida*, and therefore breeding efforts are being directed towards the development of genotypes resistant to this cyst nematode.

Among abiotic stresses, resistance to high temperatures and water stress is deemed very important due to the trend of increased potato cultivation in warm climates (Levy and Veilleux 2007). It should be pointed out that in these areas cultivation is carried out using varieties that were developed for cool climates. Also noteworthy is resistance to low temperatures in environments where the potato can be grown all year around. In the Mediterranean region, for example, the potato is planted starting in November, in a cycle that is much earlier than the typical spring-summer cycle (Frusciante et al. 1999). Due to the lack of cold resistant varieties, this off-season production can be seriously damaged by frost events.

Specific objectives are required in potato programs aimed at producing true potato seed (TPS). TPS technology represents a possible alternative to the traditional use of seed tubers. It has several advantages over the traditional cultivation, including a reduction in production/management costs and in the incidence of diseases (Chujoy and Cabello 2007). TPS varieties represent a collection of genotypes deriving from a cross between two heterozygous parents that produce a phenotipically uniform progeny. Therefore, in a typical TPS breeding program, the selection of parental lines is an important prerequisite. Both GCA and SCA may be important in parental source identification (Ortiz and Golmirzaie 2004). TPS parents must guarantee satisfactory botanical seed production when crossed. At the same time, they must result in progenies with outstanding tuber yield and traits. Therefore, along with breeding objectives common with other production systems (e.g., tuber yield and quality, resistances), specific objectives related to flower duration and intensity, number of flowers per inflorescence, vigorous early growth of seedlings, emergence uniformity, etc. are also important (Golmirzaie and Ortiz 2004). In other words, antagonism should not occur between tuber growth from one side and viable botanical seed production from the other.

Acknowledgments

The authors dedicate this chapter in fondest memory of Stan Peloquin, whose research achievements represent one of the most impressive and successful applications of genetics, cytogenetics and reproductive biology to breeding. This is DISSPAPA-book contribution no. 5 from the Department of Soil, Plant, Environmental and Animal Production Sciences.

References

Achenbach U, Paulo J, Ilarionova E, Lübeck J, Strahwald J, Tacke E, Hofferbert H-R, Gebhardt C (2009) Using SNP markers to dissect linkage disequilibrium at a major quantitative trait locus for resistance to the potato cyst nematode *Globodera pallida* on potato chromosome V. Theor Appl Genet 118: 619–629.

Alberino S, Carputo D, Caruso G, Ercolano MR, Frusciante L (2004) Field performance of families and clones obtained through unilateral sexual polyploidization in potato (*Solanum tuberosum*). Adv Hort Sci 18: 47–52.

Bae JJ, Janski SH, Rouse DI (2008) The potential of early generation selection to identify potato clones with resistance to Verticillium wilt. Euphytica 164: 385–393.

Bisognin DA, Douches DS (2002) Early generation selection for potato tuber quality in progenies of late blight resistant parents. Euphytica 127: 1–9.

Bonnel E (2008) Potato Breeding: a Challenge, as ever! Potato Res 51: 327–332.

Bradshaw JE (1994) Quantitative genetics theory for tetrasomic inheritance. In: JE Bradshaw, GR Mackay (eds) Potato Genetics, CABI Publ, Wallingford, Oxon, UK, pp 71–100.

Bradshaw JE, Mackay GR (1994) Breeding strategies for clonally propagated potatoes. In: JE Bradshaw, GR Mackay (eds) Potato Genetics. CABI Publ, Wallingford, Oxon, UK, pp 467–497.

Bradshaw JE, Ramsay G (2005) Utilisation of the Commonwealth potato collection in potato breeding. Euphytica 146: 9–19.

Bradshaw JE, Dale MFB, Mackay GR (2003) Use of mid-parent values and progeny tests to increase the efficiency of potato breeding for combined processing quality and disease and pest resistance. Theor Appl Genet 107: 36–42.

Breiman A, Galun E (1990) Nuclear-mitochondrial interrelation in angiosperms. Plant Sci 71: 3–19.

Bretagnolle F, Thompson JD (1995) Tansley review No. 78. Gametes with somatic chromosome number: mechanisms of their formation and role in the evolution of autopolyploid plants. New Phytol 129: 1–22.

Burnham CR (1962) Discussions in Cytogenetics. Burgess Publ, Minneapolis, MN, USA.

Buso JA, Boiteux LS, Peloquin SJ (1999) Multitrait selection system using populations with a small number of interploid (4x x 2x) hybrid seedlings in potato: degree of high-parent heterosis for yield and frequency of clones combining quantitative agronomic traits. Theor Appl Genet 99: 81–91.

Buso JA, Boiteux LS, Peloquin SJ (2003) Tuber yield and quality of 4x-2x (FDR) potato progenies derived from the wild diploid species *Solanum berthaultii* and *Solanum tarijense*. Plant Breed 122: 229–232.

Camadro EL, Peloquin SJ (1981) Cross-incompatibility between two sympatric polyploid *Solanum* species. Theor Appl Genet 60: 65–70.

Camadro EL, Masuelli RW (1995) A genetic model for the endosperm balance number (EBN) in the wild potato *Solanum acaule* Bitt. and two related diploid species. Sex Plant Reprod 8: 283–288.

Camadro EL, Carputo D, Peloquin SJ (2004) Substitutes for genome differentiation in tuber-bearing *Solanum*: interspecific pollen-pistil incompatibility, nuclear-cytoplasmic male sterility, and endosperm. Theor Appl Genet 109: 1369–1376.

Cardi T, Bastia T, Monti L, Earle ED (1999) Organelle DNA and male fertility variation in *Solanum* spp. and interspecific somatic hybrids. Theor Appl Genet 99: 819–828.

Carputo D, Barone A (2005) Ploidy manipulation in potato through sexual hybridization. Ann Appl Biol 146: 71–79.

Carputo D, Barone A, Cardi T, Sebastiano A, Frusciante L, Peloquin SJ (1997) Endosperm Balance Number manipulation for direct *in vivo* germplasm introgression to potato from a sexually isolated relative (*Solanum commersonii* Dun). Proc Natl Acad Sci USA 94: 12013–12017.

Carputo D, Barone A, Frusciante L (2000) 2n gametes in the potato: essential ingredients for breeding and germplasm transfer. Theor Appl Genet 101: 805–813.

Carputo D, Frusciante L, Peloquin SJ (2003) The role of 2n gametes and endosperm balance number in the origin and evolution of polyploids in the tuber-bearing Solanums. Genetics 163: 287–294.

Chujoy E, Cabello R (2007) The Canon of Potato Science: 29 True Potato Seed (TPS). Potato Res 50: 323–325.

Clulow SA, Wilkinson MJ, Waugh R, Baird E, De Maine MJ, Powell W (1991) Cytological and molecular observations on *Solanum phureja*-induced dihaploid potatoes. Theor Appl Genet 82: 545–551.

d'Erfurth I, Jolivet S, Froger N, Catrice O, Novatchkova M, Simon M, Jenczewski E, Mercier R (2008) Mutations in *AtPS1 (Arabidopsis thaliana Parallel Spindle 1)* lead to the production of diploid pollen grains. PLoSos Genet 4 (11): 1–9.

Dong F, Song J, Naess SK, Helgeson JP, Gebhardt C, Jiang J (2000) Development and applications of a set of chromosome-specific cytogenetic DNA markers in potato. Theor Appl Genet 101: 1001–1007.

Ehlenfeldt MK, Hanneman RE Jr (1988) Genetic control of Endosperm Balance Number (EBN): three additive loci in a threshold-like system. Theor Appl Genet 75: 825–832.

Ercolano MR, Carputo D, Li J, Monti L, Barone A, Frusciante L (2004) Assessment of genetic variability of haploids extracted from tetraploid (2n = 4x = 48) *Solanum tuberosum*. Genome 47: 633–638.

Frusciante L, Barone A, Carputo D, Ranalli P (1999) Breeding and physiological aspects of potato cultivation in the Mediterranean region. Potato Res 42: 265–277.

Golmirzaie AM, Ortiz R (2004) Diversity in reproductive characteristics of potato landraces and cultivars for producing true seed. Genet Resour Crop Evol 51: 759–763.

Gopal J, Gaur PC, Rana MS (1992) Early generation selection for agronomic characters in a potato breeding programme. Theor Appl Genet 84: 709–713.

Gopal J, Kumar V, Luthra SK (2008) Top-cross vs. poly-cross as alternative to test-cross for estimating the general combining ability in potato. Plant Breed 127: 441–445.

Grun P (1990) The evolution of cultivated potatoes. Econ Bot 44: 39–55.

Hanneman Jr RE (1999) The reproductive biology of the potato and its implications for breeding. Potato Res 42: 283–312.

Hayes RJ, Dinu II, Thill CA (2005) Unilateral and bilateral hybridization barriers in inter-series crosses of 4x 2EBN *Solanum stoloniferum*, *S. pinnatisectum*, *S. cardiophyllum*, and 2x 2EBN *S. tuberosum* haploids and haploid-species hybrids. Sex Plant Reprod 17: 303–311.

Hosaka K, Hanneman Jr RE (1998) Genetics of self-compatibility in a self-incompatible wild diploid potato species *Solanum chacoense*. 2. Localization of an *S* locus inhibitor (*Sli*) gene on the potato genome using DNA markers. Euphytica 103: 265–271.

Hougas RW, Peloquin SJ (1960) Crossability of *Solanum tuberosum* haploids with diploid *Solanum* species. Eur Potato J 3: 325–329.

Iwanaga M, Ortiz R, Cipar MS, Peloquin SJ (1991) A restorer gen for genetic-cytoplasmic male sterility in cultivated potatoes. Am Potato J 68: 19–28.

Jansky SH (2006) Overcoming hybridization barriers in potato. Plant Breed 125: 1–12.

Johnston SA, Hanneman RE (1996) Genetic control of Endosperm Balance Number (EBN) in the Solanaceae based on trisomic and mutation analysis. Genome 39: 314–321.

Johnston SA, den Nijs TPM, Peloquin SJ, Hannemann RE (1980) The significance of genic balance to endosperm development in interspecific crosses. Theor Appl Genet 57: 5–9.

Korpan YI, Nazarenko EA, Skryshevskaya IV, Martelet C, Jaffrezic-Renault N, El'skaya AV (2004) Potato glycoalkaloids: true safety or false sense of security? Trends Biotechnol 22: 147–151.

Levy D, Veilleux RE (2007) Adaptation of Potato to High Temperatures and Salinity-A Review. Am J Potato Res 84: 487–506.

Mackay GR (2005) Propagation by Traditional Breeding Methods. In: MK Razdan, AK Mattoo (eds) Genetic Improvement of Solanaceous Crops, vol I Potato. Science Publ, Enfield, NH, USA, pp 65–81.

Mendiburu AO, Peloquin SJ (1977) The significance of 2n gametes in potato breeding. Theor Appl Genet 49: 53–61.

Mok DWS, Heiyoung KL, Peloquin SJ (1974) Identification of potato chromosomes with Giemsa. Am Potato J 51: 337–341.

Mohanty IC, Mahapatra D, Mohanty S, Das AB (2004) Karyotype analyses and studies on the nuclear DNA content in 30 genotypes of potato (*Solanum tuberosum* L.) Cell Biol Intern 28: 625–633.

Orczyc W, Przetakiewicz J, Nadolska-Orczyc A (2003) Somatic hybrids of *Solanum tuberosum*—application to genetics and breeding. Plant Cell Tiss Org Cult 74: 1–13.

Ortiz R (1998) Potato breeding via ploidy manipulations. In: J Janick (ed) Plant Breeding Reviews, vol 16. John Wiley, New York, USA, pp 15–86.

Ortiz R, Golmirzaie AM (2004) Combining ability analysis and correlation between breeding values in true potato seed. Plant Breed 123: 564–567.

Ortiz R, Peloquin SJ, Freyre R, Iwanaga M (1991) Efficiency of 4x x 2x breeding scheme in potato for multitrait selection and progeny testing. Theor Appl Genet 82: 602–608.

Ortiz R, Franco J, Iwanaga M (1997) Transfer of resistance to potato cyst nematode (*Globodera pallida*) into cultivated potato *Solanum tuberosum* through first division restitution 2n pollen. Euphytica 96: 339–344.

Peloquin SJ, Boiteux LS, Carputo D (1999) Meiotic mutants in potato: Valuable variants. Genetics 153: 1493–1499.

Peloquin SJ, Boiteux LS, Simon PW, Jansky SH (2008) A chromosome-specific estimate of transmission of heterozygosity by 2*n* gametes in potato. J Hered 99: 177–181.

Pijnacker LP, Ferwerda MA (1984) Giemsa C-banding of potato chromosomes. Can J Genet 26: 415–419.

Phumichai C, Ikeguchi-samitsu Y, Fujimatsu M, Kitanishi S, Kobayashi A, Mori M, Hosaka K (2006) Expression of *S*-locus inhibitor gene (*Sli*) in various diploid potatoes. Euphytica 148: 227–234.

Renatus J (2005) General trends in the European potato trade. In: AJ Haverkort ,PC Struik (eds) Potato in progress. Wageningen Academic Publ, Wageningen, The Netherlands, pp 341–347.

Rokka VM (2009) Potato haploids and breeding. In: A Touraev, BP Forster, SM Jain (eds) Advances in Haploid Production in Higher Plants. Springer, New York, USA, pp 199–208.

Stone SL, Goring DR (2001) The molecular biology of self-incompatibility systems in flowering plants. Plant Cell Tiss Org Cult 67: 93–114.

Takayama S, Isogai A (2005) Self-incompatibility in plants. Annu Rev Plant Biol 56: 467–489.

Tai GCC (1994) Use of 2n gametes. In: JE Bradshaw, GR Mackay (eds) Potato Genetics. CABI Publ, Wallingford, Oxon, UK, pp 109–132.

Tarn TR, Tai GCC, De Jong H, Murphy A.M., Seabrook JEA (1992) Breeding potatoes for long day temperate climates. Plant Breed Rev 9: 217–332.

Thill CA, Peloquin SJ (1995) A breeding method for accelerated development of cold chipping clones in potato. Euphytica 84: 73–80.

van de Wal MHBJ, Jacobsen E, Visser RGF (2001) Multiple allelism as a control mechanism in metabolic pathways: GBSSI allelic composition affects the activity of granule-bound starch synthase I and starch composition in potato. Mol Genet Genom 265: 1011–1021.

van Eck HJ, Jacobs JME, Stam P, Ton J, Stiekema WJ, Jacobsen E (1994) Multiple alleles for tuber shape in diploid potato detected by qualitative and quantitative genetic analysis using RFLPs. Genetics 137: 303–309.

Watanabe K, Peloquin SJ (1991) Occurrence and frequency of 2n pollen in 2x, 4x and 6x wild-tuber-bearing *Solanum* species from Mexico and Central and South Americas. Theor Appl Genet 82: 621–626.

Werner JE, Peloquin SJ (1987) Frequency and mechanisms of 2n egg formation in haploid *tuberosum*-wild species F_1 hybrids. Am Potato J 64: 641–654.

Werner JE, Peloquin SJ (1991) Yield and tuber characteristics of 4x progeny from 2x x 2x crosses. Potato Res 34: 261–267.

Wilkinson MJ, Bennett ST, Clulow SA, Allainguillaume J, Harding K, Bennett MD (1995) Evidence for somatic translocation during potato dihaploid induction. Heredity 74: 146–151.

Yeh BP, Peloquin SJ (1965) Pachytene chromosomes of the potato (*Solanum tuberosum* Group Andigena). Am J Bot 52: 1014–1020.

3

Molecular Breeding for Potato Improvement

David De Koeyer,[1,*] *Haixia Chen*[1,2] and *Vicki Gustafson*[3]

ABSTRACT

The immense amount of diversity within cultivated potato and its relatives provides both opportunities and challenges. The demands of industry and the public for rapid introduction of new and better varieties contrasts with the long-term efforts that are needed to incorporate beneficial diversity not present in the adapted germplasm. In this chapter, the application of genomics technology within potato germplasm evaluation and breeding programs is reviewed and new opportunities identified.

DNA marker and sequencing information has been critical in taxonomic studies within wild and cultivated potato species. Markers have helped elucidate the phylogenetic relationships within the primary and secondary gene pools of cultivated potato and determine the most-likely progenitors of cultivated *S. tuberosum*. The number of markers reported to be linked to important traits in potato greatly exceeds the number being used for marker-assisted selection. Various strategies for incorporating markers into potato breeding programs are discussed. The greatest impact of molecular breeding in potato will likely be for pre-breeding and parental development activities.

The typical model of genetic modification of existing cultivars may not be sufficient as a means to make continual breeding progress to address the varied requirements of industry and consumers. Cisgenic

[1]Potato Research Centre, Agriculture and Agri-Food Canada, 850 Lincoln Road, Fredericton, NB, Canada, E3B 4Z7.
[2]Hunan Provincial Key Laboratory of Crop Germplasm Innovation and Utilization, Hunan Agricultural University, Changsha, Hunan, P.R. China, 410128.
[3]Solanum Genomics International Inc., 921 College Hill Road, Fredericton, NB, Canada, E3B 6Z9.
*Corresponding author: *david.dekoeyer@agr.gc.ca*

technology has potential to revolutionize how breeders access genes for pre-breeding and parent development. Enhanced knowledge of the genetic factors controlling traits important to industry as well as those traits important for breeding success will make the coming years exciting for potato improvement. Molecular breeding provides the means to transfer this knowledge to the development of new cultivars for a variety of uses.

Keywords: cisgenic, DNA marker, high-resolution DNA melting, marker-assisted selection, parent development, plant transformation, tetraploid potato

3.1 Introduction: Molecular Breeding—The Beginning of a New Era

We have entered a rapidly changing period where new technology and new analysis techniques provide opportunities for crop improvement. Molecular markers have long held the promise of improving the efficiency and effectiveness of plant breeding. Due to the autotetraploid state of potato (*Solanum tuberosum* L.), researchers have primarily utilized diploid genetic materials for genetic map construction and the identification of markers linked to important traits (reviewed by Gebhardt 2005). These markers have proven useful for enhancing our understanding of cultivated potato and wild *Solanum* species used as sources of beneficial diversity by potato breeders. Application of association genetics to tetraploid breeding populations (see Chapter 7) will further contribute to the identification of valuable markers for potato improvement. Gene cloning and transformation technology offer a route to overcome the weaknesses inherent to widely grown potato cultivars; however, lack of consumer and industry acceptance of genetically modified organisms has limited the impact of this technology to date. Advances in potato, and the closely related tomato (*Solanum lycopersicon*), genome sequencing will greatly improve our knowledge of the gene content and genome organization of potato. Next-generation sequencing, high-throughput genotyping, and bioinformatics will further contribute to the evaluation of allelic diversity existing in *S. tuberosum* and related species. The question that has yet to be answered is whether this wealth of genomics information will have an impact on potato breeding.

The goal of this chapter is to develop realistic expectations for molecular breeding and to place molecular breeding in the proper context within a larger potato improvement program. Fundamental to determining the role of markers for plant improvement is an understanding of the basic principles of selection (reviewed by Sharma 1994 and Walker 1969). The three main principles are: 1) selection operates on existing variation; 2) selection acts only through heritable differences; and 3) selection promotes reproduction of some individuals at the expense of others. The use of markers will only be useful in crop improvement when the target traits have these selection criteria. It should also be obvious then that there are breeding targets for which markers will have little or no impact—either due to a lack of understanding of the trait, a lack of heritable variation for the trait, or a negative relationship between the target trait and other critical traits. In some cases, the role of transgenics-assisted breeding should be considered as a more viable option for meeting breeding goals.

Molecular breeding also needs to be viewed in a crop-dependent manner. Potato has unique cultural requirements, reproduction methods, and breeding strategies. Therefore, this chapter will take a potato-orientated approach and review issues that may be more important for potato than for other crops. We will review the application of genomics technology within potato germplasm evaluation and breeding programs and identify new opportunities. This chapter is written from the perspective of a scientist within a tetraploid potato improvement program in a "developed" country with a long-day growing season.

3.2 Germplasm and Variety Characterization

DNA marker and sequencing information is playing a significant role in taxonomic studies. While taxonomic investigations do not have an obvious relationship to breeding, enhanced understanding of gene pools available to geneticists and breeders will lead to better utilization of these resources. Random amplified polymorphic DNA (RAPD), amplified fragment length polymorphism (AFLP) and simple sequence repeat (SSR) markers have been used to generate phylogenetic data that have contributed to the revision of the relationships within cultivated and wild potato species germplasm. Two topics of particular interest to potato breeders are the phylogenetic relationships within the primary and secondary gene pools of cultivated potato and the elucidation of the progenitors of cultivated *S. tuberosum*.

3.2.1 Characterization of Solanum *Gene Pools*

Molecular data, generated by sequencing or DNA markers, has generally led to fewer unique species or taxonomic groups compared to the initial morphology-based studies summarized by Hawkes (1990). Currently, there are 187 recognized tuber-bearing species (Spooner and Salas 2006). Within *Solanum* section *Petota*, the grouping of species into 21 different series was proposed (Hawkes 1990). Molecular evidence does not support this number, instead supporting between four (Spooner and Salas 2006) and 10 (Jacobs et al. 2008) clades. Various types of markers have contributed to the classification changes within section *Petota*, including AFLPs (Lara-Cabrera and Spooner 2004; Jacobs et al. 2008; Jiménez et al. 2008), SSRs (Lara-Cabrera and Spooner 2005), or a combination of AFLPs, RAPDs, and SSRs (van den Berg et al. 2002).

More recently, sequencing data from single-copy genes has been used to elucidate the relationships within *Solanum*. Genes that have been studied include *GBSSI* (Spooner et al. 2008) and nitrate reductase (Rodriguez and Spooner 2009). This new molecular evidence has been particularly informative in studying polyploids within the genus. Polyploidy is thought to have contributed to the expansion of the range of species in the genus. Of the 189 potato species, 64% are diploid ($2x = 24$), and 36% are partially or completely polyploid (Hijmans et al. 2007). The importance of polyploidy within this taxonomic group was illustrated in this study by the amount of geographic area containing only polyploid species (28%) and that the two most widespread species are tetraploids (*S. stoloniferum* and *S. acaule*). Two important aspects of polyploidy within *Solanum* section *Petota* revealed by the study of single gene sequence data include molecular evidence for allopolyploidy and the lower level of diversity present in diploid versus polyploidy species (Spooner et al. 2008; Rodriguez and Spooner 2009). These authors also examined the relationship between genome type and endosperm balance number (EBN). Potato species have been typically classified by the type of genome they possess based on meiotic pairing studies of interspecific hybrids. The classifications include A-E and P, with *S. tuberosum* being an A genome species (Matsubayashi 1991) (see also Chapter 9). While there isn't a perfect relationship between EBN and genome type, it is apparent that genome-specific markers would allow breeders to more effectively access diversity within non-A genome species. In the Rodriguez and Spooner (2009) and Spooner et al. (2008) studies, the level of diversity within a species was significantly higher in polyploids compared to diploids. These polymorphisms were in the form of single nucleotide differences and insertion/deletions. The issue of ploidy level is also very important in the study of cultivated potatoes and requires a critical assessment of the choice of marker platform for these studies.

3.2.2 *Genealogy and Diversity of Modern Varieties*

Molecular markers have been utilized to help unravel the mystery of the origin of the modern cultivated potato. The hypothesis of rapid adaptation of short-day Andigena types of *S. tuberosum* to the long-day environment of Europe has been challenged recently by proponents of the introduction of *S. tuberosum* ssp. types from Chile. The molecular evidence strongly indicates that modern cultivars arose from a Chilean ancestor rather than from Andigena (Spooner et al. 2005a, b, 2007; Rios et al. 2007; Ghislain et al. 2009b). The implications of this paradigm shift include a re-evaluation of past and future uses of Andigena materials in long-day adapted breeding programs. The questionable origin of "Neo-tuberosum" materials generally suggests that potato pedigree information is often unreliable and that the negative selection pressure against Andigena type potatoes (short-day adaptation, deep eyes, etc.) needs to be factored into breeding schemes designed to enhance diversity. The study by Ghislain et al. (2009b) indicates that modern varieties and breeding lines possess considerable diversity (93.8% SSR alleles in common with landraces).

In a series of publications, cultivated accessions and modern varieties have been characterized using RAPDs (Demeke et al. 1996; Paz and Veilleux 1997; Ghislain et al. 1999; Pattanayak et al. 2002; Sun et al. 2003; Yasmin et al. 2006), RAPDs and SSRs (Ghislain et al. 2006), SSRs (Ashkenazi et al. 2001; Bisognin and Douches 2002; Raker and Spooner 2002; Ghislain et al. 2004; Sukhotu et al. 2005; Ispizúa et al. 2007; Fu et al. 2009; Ghislain et al. 2009a, b), and AFLPs (Veteläinen et al. 2005; Aversano et al. 2007; Solano Solis et al. 2007; D'Hoop et al. 2008). Comparison of AFLP, RAPD, and SSR markers for diversity analysis of tetraploid potato has not revealed a distinct advantage of one marker type over another (Milbourne et al. 1997; McGregor et al. 2000). Standardized diversity analysis within cultivated potato species has been aided by the development of a microsatellite genetic identity kit by researchers at the International Potato Center (Ghislain et al. 2004, 2009a). The various studies examining molecular diversity in cultivated potato have not demonstrated that diversity is a limiting factor for breeding progress. However, the high levels of polymorphism at the sequence level are generally not captured by the commonly used marker systems. These markers are also not sensitive enough to quantify allele dosage on a routine basis. Rather than marker assessments of genetic relationships and diversity, it may be more useful to measure the diversity of genomic regions that have undergone selection within breeding programs.

3.2.3 Distinctiveness, Uniformity, and Stability

Genome-wide marker evaluation can also be used to identify redundancy within a germplasm collection or conversely be used as a measure of distinctiveness of cultivars prior to commercialization. Published studies have documented duplication within potato genebanks (McGregor et al. 2002; van Treuren et al. 2004; Veteläinen et al. 2005; Ispizúa et al. 2007; Fu et al. 2009; Ghislain et al. 2009a) and it is expected that markers will become more important in prioritization of accessions. Ghislain et al. (2006) illustrated the importance of utilizing more than one marker system for the establishment of core collections. Several researchers have documented the utility of molecular markers for discriminating potato cultivars using RAPD (Ford and Taylor 1997; Isenegger et al. 2001) or SSR (Norero et al.

2002; Coombs et al. 2004; Li et al. 2008b) markers. The point at which genetic distinctiveness can be ascertained to declare a variety essentially derived is unclear. An assessment of genetic variability will need to be combined with morphological assessments, since the epigenetic variation plays a role in controlling phenotypic variation is now a topic of great interest in model plant species. AFLP and SSR markers failed to discriminate between intraclonal variants of "Russet Norkotah" (Hale et al. 2005) and RAPD markers failed to distinguish intraclonal variants of four varieties (Isenegger et al. 2001), suggesting that epigenetic variation occurs in *S. tuberosum*.

3.3 Marker-assisted Gene Introgression

There is a growing collection of markers linked to important traits in potato. The genome location of pathogen resistance and tuber traits has been used to create potato function maps for pathogen resistance and tuber traits (Gebhardt 2005). An update of the location of important genes is provided in other chapters of this book (see Chapters 5 and 6). In crops it is generally viewed that markers have a role in both positive and negative selection within a germplasm enhancement program. These approaches in potato were reviewed by Barone (2004).

3.3.1 Genome-wide Markers for Negative Selection

Selection against the donor genome (negative selection) is generally considered for pre-breeding applications in combination with phenotypic selection for a novel trait derived from a species distantly related to *S. tuberosum*. The strategy of introgression depends on the extent the donor species is reproductively isolated from the cultivated potato gene pool. Strategies for overcoming crossing barriers were reviewed by Jansky (2006) and include somatic hybridization and the use of bridge species. Once an initial hybrid plant is produced, backcrossing is typically employed as the breeding method. However, it is not uncommon to observe severe sterility or incompatibility in the early generation hybrids. When fertile hybrids are produced, genome-wide marker genotyping may be employed to preferentially select clones with a higher level of *tuberosum*-type markers, as long as the selected clones maintain the desirable trait. Most published examples of this marker-assisted selection (MAS) scheme in potato involve the diploid wild donor species *S. commersonii* introgression into a tetraploid *S. tuberosum* background (Barone et al. 2001; Barone 2004; Carputo et al. 2002; Iovene et al. 2004).

3.3.2 *The Toolbox of Available Markers*

Positive selection with DNA markers has been more widely published, involving markers linked to several different traits. PCR markers closely linked to genes or quantitative trait loci (QTL) of interest have been available to potato breeders for quite some time. These include markers linked to nematode resistance genes, including *Gro1*, *Gro1-4*, *H1*, and *Grp1* genes for resistance to golden cyst nematode [*Globodera rostochiensis* (Wollenweber) Behrens] (Niewöhner et al. 1995; Paal et al. 2004; Finkers-Tomczak et al. 2009); *Grp1* and other QTLs for resistance to the cyst nematode *G. pallida* (Bradshaw et al. 1998; Bryan et al. 2002; Sattarzadeh et al. 2006; Achenbach et al. 2009; Finkers-Tomczak et al. 2009); and the *Mc1* gene for resistance to the root-knot nematode (*Meloidogyne chitwoodi*) (van der Voort et al. 1999; Zhang et al. 2007).

Several PCR markers linked to virus resistance genes have been reported and include the Ry_{adg} and Na_{adg} genes on chromosome XI for extreme resistance to Potato virus Y (PVY) resistance and hypersensitive resistance to Potato virus A (Hämäläinen et al. 1997; Sorri et al. 1999; Kasai et al. 2000); the Ry_{sto} gene for extreme resistance to PVY (Brigneti et al. 1997; Flis et al. 2005; Černák et al. 2008b; Song and Schwarzfischer 2008; Witek et al. 2006); the *Rx1* gene for Potato virus X resistance (Bendahmane et al. 1997); and the *Ns* gene for Potato virus S resistance (Marczewski et al. 2001; Witek et al. 2006). PCR markers linked to oomycete and fungal resistance include the late blight resistance genes *RB* (Colton et al. 2006) and *R1* (Gebhardt et al. 2004), and QTLs (Oberhagemann et al. 1999; Bormann et al. 2004); the *Verticillium* wilt resistance genes *StVe* and *Vc* (Simko et al. 2004a, b; Bae et al. 2008); and the potato wart resistance gene *Sen1* (Gebhardt et al. 2006). Several PCR-based candidate gene markers are associated with the tuber quality traits chip color and starch content (Li et al. 2005, 2008a) and potato tuber skin and flesh color (De Jong et al. 2003a, 2003b; Brown et al. 2006; Zhang et al. 2009).

3.3.3 *Marker-assisted Selection in Potato*

There have been relatively few studies documenting the use of MAS within potato breeding programs. This section will summarize the published work on this topic and mention some recent findings that will contribute to future MAS studies. The markers linked to Ry_{adg} and Ry_{sto} have been evaluated by several groups with varying degrees of success. The Ry_{adg} marker genotypes were in good agreement with phenotypic assessment of extreme resistance to PVY when the source of resistance is known (Hämäläinen et al. 1997; Sorri et al. 1999; Kasai et al. 2000; Dalla Rizza et al. 2006). In a recent study, Ottoman et al. (2009) validated the previously published markers for Ry_{adg}

by showing a 96.4% concordance with ELISA results for PVY in a full-sib tetraploid breeding population. The authors also proposed a breeding scheme to incorporate the markers for selection after the first year of field evaluation. The pedigree of the material used to identify the initial Ry_{sto} markers on chromosome XI (Brigneti et al. 1997) has been questioned (Gebhardt and Valkonen 2001). It is now generally recognized that Ry_{sto} is located on chromosome XII (Flis et al. 2005; Song et al. 2005; Cernák et al. 2008a; Valkonen et al. 2008). However, the previously published markers were of limited use in a tetraploid breeding population, necessitating the screening of RAPD markers (Cernák et al. 2008b) and development of sequence-characterized amplified region (SCAR) markers linked to the gene (Cernák et al. 2008a).

MAS for the *RB* gene in 110 breeding lines from different field generations was highly effective in selecting late blight resistance and less error-prone than detached leaf assays (Colton et al. 2006). Bormann et al. (2004) discussed a preliminary MAS study for late blight resistance based on three unlinked markers which resulted in a group of plants more resistant than a control group with unfavorable alleles at the same loci. They also reported a much higher frequency of susceptible alleles, which would necessitate much larger populations to identify sufficient progeny with the target alleles to allow for selection on other traits. For autotetraploid mapping and MAS experiments, Bradshaw and colleagues (1998) recommended the use of a minimum of 250 progeny per family. These observations could also be viewed as evidence for the use of MAS for parental selection to increase the frequency of desirable alleles segregating in a population. The genome region containing the *R1* locus showed a significant association with late blight resistance and plant maturity across a diverse set of germplasm and the markers could be traced back to the donor *S. demissum* (Gebhardt et al. 2004). This important finding suggests that markers in this region may also contribute to developing lines with the appropriate phenotype. An association genetic study with 184 clones from two breeding programs identified nine single nucleotide polymorphisms (SNPs) significantly associated with field resistance to late blight (Pajerowska-Mukhtar et al. 2009). The allene oxide synthase 2 gene on chromosome XI appears to be important in controlling this trait and markers targeting the diagnostic SNPs may be useful for future MAS applications.

A diagnostic marker for resistance to *G. pallida* pathotypes 2 and 3 has been developed and holds promise for MAS (Sattarzadeh et al. 2006). Five sequence tagged site (STS) markers tightly linked to the $R_{mc(blb)}$ gene perfectly co-segregated with root-knot nematode resistance and also would be very suitable for use in a MAS program (Zhang et al. 2007). The predictive value of a marker associated with *Verticillium* wilt susceptibility was relatively high (0.72), but not high enough for MAS alone (Simko et al. 2004a). A microsatellite allele (STM1051-193) was tested on 139 breeding clones and shown to be significantly associated with resistance to *Verticillium* wilt (Simko et al. 2004b). This assay would be suitable for MAS, provided that maturity data are included in the selection decision.

Multi-trait MAS was applied to select breeding lines with combined resistance alleles for Ry_{adg}, *Rx1*, and *Gro1* or Ry_{adg} and *Sen1* (Gebhardt et al. 2006). The results from this study were better than expected, with selection of progeny containing resistance to PVY, PVX, and golden nematode, and unexpected resistance to *S. endobioticum* pathotypes 1, 2, and 6. In addition, twice as many clones with three target alleles were selected. These unusual findings were explained by the use of parental clones which were tetraploid, not diploid as assumed, so the crosses were in fact 2*x* by 4*x* not 2*x* by 2*x*. The contribution of wart resistance from the parents with extreme resistance to PVY explained the unexpected wart resistance. This study also supports the importance of parent development in potato molecular breeding, since parents producing 2*n* gametes transmitted the *Gro1* allele to 66% of the progeny. The use of tetraploid parents with more than one copy of a favorable allele would also significantly increase the proportion of progeny with the target traits. To increase the efficiency of MAS, Witek et al. (2006) developed a multiplex PCR assay that successfully identified the Ry_{sto} and *Ns* alleles for resistance to PVY and PVS, respectively.

For tuber quality characteristics, there are fewer applications of markers within breeding programs. Several marker-trait associations were detected with chip color and starch content in 243 clones using 36 microsatellite and candidate gene loci (Li et al. 2008a). This study also validated the previous associations of the invertase alleles *InvGE-6f* and *InvGF-4b* with chip color in independent breeding lines (Li et al. 2005). Tuber skin color is a trait controlled by three genes (De Jong 1991) and can be considered a model trait to examine the role of candidate gene markers within a potato breeding program. A cleaved amplified polymorphic sequence (CAPS) marker derived from the anthocyanin pathway gene dihydroflavonol 4-reductase (*dfr*) was present in all red-skinned clones (De Jong et al. 2003b). Polymorphisms present in the *dfr* allele associated with red color were used to develop a dosage sensitive Taqman assay for genotyping tetraploid potato lines (De Jong et al. 2003a). Recently, a high-resolution DNA melting (HRM) probe assay was developed for the *dfr* gene and this assay was able to quantify the number of copies of the haplotype associated with red pigment and identify novel haplotypes (De Koeyer et al. 2010). A unique allele was present in all chipping clones derived from "Lenape". It is also interesting to note that the *dfr* gene is located in close proximity to the sucrose synthase III gene on chromosome II, a gene that was found to be associated with light chip color in Canadian breeding lines (Kawchuk et al. 2008).

3.3.4 Advances in Marker Technology for Autotetraploids

DNA marker systems developed for diploid organisms are not well-suited for applications with autopolyploid organisms. The typical marker considerations for a MAS project include ease of use, robustness, cost, linkage to genes controlling the trait of interest, and the amount of phenotypic variation explained by the marker. In autotetraploid potato, a single genomic region may have up to four alleles or several possible combinations involving two or three alleles. Therefore, dosage-sensitivity and the ability to distinguish various haplotype combinations and their dosage should be considered when selecting a marker system for potato molecular breeding.

To address the limitations of electrophoresis-based marker systems, new SNP genotyping methods have been developed and applied to tetraploid potato. SNPs have been genotyped in tetraploids using pyrosequencing, dideoxy sequencing, and single nucleotide primer extension (Rickert et al. 2002, 2003), fluorogenic 5′ nuclease (Taqman) assays (De Jong et al. 2003a), ion-pair reversed-phase high-performance liquid chromatography-electrospray ionization mass spectrometry (Oberacher et al. 2004), and allele-specific oligonucleotide microarrays (Rickert et al. 2005). While these genotyping systems look promising, it is not apparent how they could be incorporated into a potato breeding program without considerable investment. The short analysis time, lack of post-PCR sample processing, and ability to quantify allele dosage indicate that HRM, reviewed by Erali (2008), would be appropriate for MAS. Recently, we demonstrated that high-resolution DNA melting can be applied to genotyping and variant scanning of both diploid and tetraploid potatoes (De Koeyer et al. 2010). One of the assays described in this study targets a haplotype of a marker (TG689) linked to the *H1* gene. This HRM probe assay is now routinely being used to select breeding lines with pathotype *Ro1* golden nematode resistance within the Agriculture Canada potato breeding program in New Brunswick.

3.4 Breeding Strategies Incorporating DNA Markers

The conventional view of marker-assisted selection needs to be evaluated within the context of the steps in a potato breeding program. It is difficult to generalize, since each potato breeding program is unique; however, there are common elements to all programs and these will be emphasized. Based on the review presented in previous sections of this chapter, application of molecular breeding with DNA markers will be discussed pertaining to pre-breeding, parent development, early generation selection, and advanced generation selection.

3.4.1 Pre-breeding with DNA Markers

Marker-assisted germplasm characterization could occur on the genome-wide or candidate gene level. The concept of taxonomic predictivity proposed by Jansky and co-workers to target species for trait evaluation (Jansky et al. 2006; Jansky et al. 2008, 2009) would be aided by molecular characterization within selected taxa. High density genotyping systems, which target known chromosomal locations, such as the Illumina Infinium SNP genotyping platform being developed by the SolCAP initiative (*solcap.msu.edu*) or markers that are more cost-effective for whole-genome profiling, such as the potato DArT array (*www.diversityarrays.com/index.html*), will contribute to germplasm characterization and selection of new genetic resources for pre-breeding activities. These genome-wide markers will also be useful for selection against donor genome content during a back-crossing program. Screening for variation in candidate genes will be greatly facilitated by the completion of the potato genome sequence (Visser et al. 2009), next-generation sequencing technologies, and novel approaches to variant scanning, such as the use of HRM (De Koeyer et al. 2010; Hofinger et al. 2009).

The introgression of beneficial traits from wild potato species has several biological and practical hurdles. Wild species offer a new source of variation that can be selected within a breeding program; however, this variation is often introgressed with reproductive or phenotypic characteristics that inhibit selection of individuals as progenitors of subsequent breeding cycles. A better understanding of the genetic mechanisms underpinning reproductive barriers and adaptation traits, such as photoperiod response, plant architecture, and tuber morphology may also provide a source of markers that can be used to screen early backcross generations. Conventional, marker, and transgenic approaches to enhance meiotic recombination should also be developed to enhance the probability of identifying individuals with smaller fragments of the donor genome. There are several opportunities to apply cisgenic technology to access beneficial diversity in wild species and these will be discussed in Section 3.5.

3.4.2 Parent Development with DNA Markers

Parental line breeding in hybrid potato cultivar development generally represents a smaller component of potato breeding programs relative to the selection of potential cultivars. For qualitatively inherited traits, such as the phenotype imparted by a dominant resistance allele, the benefits of developing parents with two or more copies of a favorable allele are evident by the predicted increase in proportion of progeny with the desired allele. Despite these advantages, the conventional development of parents with duplex genotypes is very difficult since it requires progeny testing before the parental genotype is known. Once a parent is identified with two or more copies of a resistance allele, there is a tendency among potato breeders to re-use these parents, even when these parents produce progeny which are discarded due to undesirable characteristics. Dosage-sensitive markers allow for selection of parents with multiple copies of a desirable allele without the need for progeny testing. An example of the use of a HRM unlabeled probe assay to detect allele dosage at the *dfr* locus is given in Fig. 3-1. Such markers should promote the notion of forward selection of parents within potato breeding programs. The use of markers will also allow crosses to be designed that pyramid resistance genes for a single pathogen, thereby improving the chances of selecting cultivars with more durable resistance.

Quantitative genetics theory predicts that cultivars with the best phenotypic performance will not necessarily be the best parents for subsequent generations in a heterozygous autotetraploid crop. Genomics technology may help clarify the distinction between parental line development and cultivar development by determining the molecular basis of heterosis and the relative roles of additive, dominant, and epistatic genetic controls. As gene action in autotetraploid potato is better understood, markers will likely play an even greater role in parental development in the future. The greatest impact of markers will be for traits predominantly under the control of additive genetic variation, since they will contribute the most to heritable differences upon which selection may act.

One potential application of MAS to parent development is the replacement or complementation of pedigree-based relationship assessments with one derived from genome-wide marker information. This information has been used in association genetics studies to generate mixed-model equations to identify markers linked to quantitative traits (Yu et al. 2006; Malosetti et al. 2007). Best linear unbiased predictors of breeding value that rely on pedigree-based calculation of a relationship matrix are routinely utilized in animal breeding to select superior parents. This methodology has also been shown to have great potential for identifying superior parents within a potato breeding program (Tai et al. 2009). As evidenced by the unexpected results from analysis of the origin of long-day adapted Neo-tuberosum germplasm (Ghislain et al. 2009b), early pedigrees from progenitor clones may be very inaccurate. It is also known that the covariance among relatives in autotetraploids is much more complicated than in diploids (Gallais 2003), so it is conceivable that marker-based relationship matrices may be superior for breeding value calculations than those derived from pedigrees.

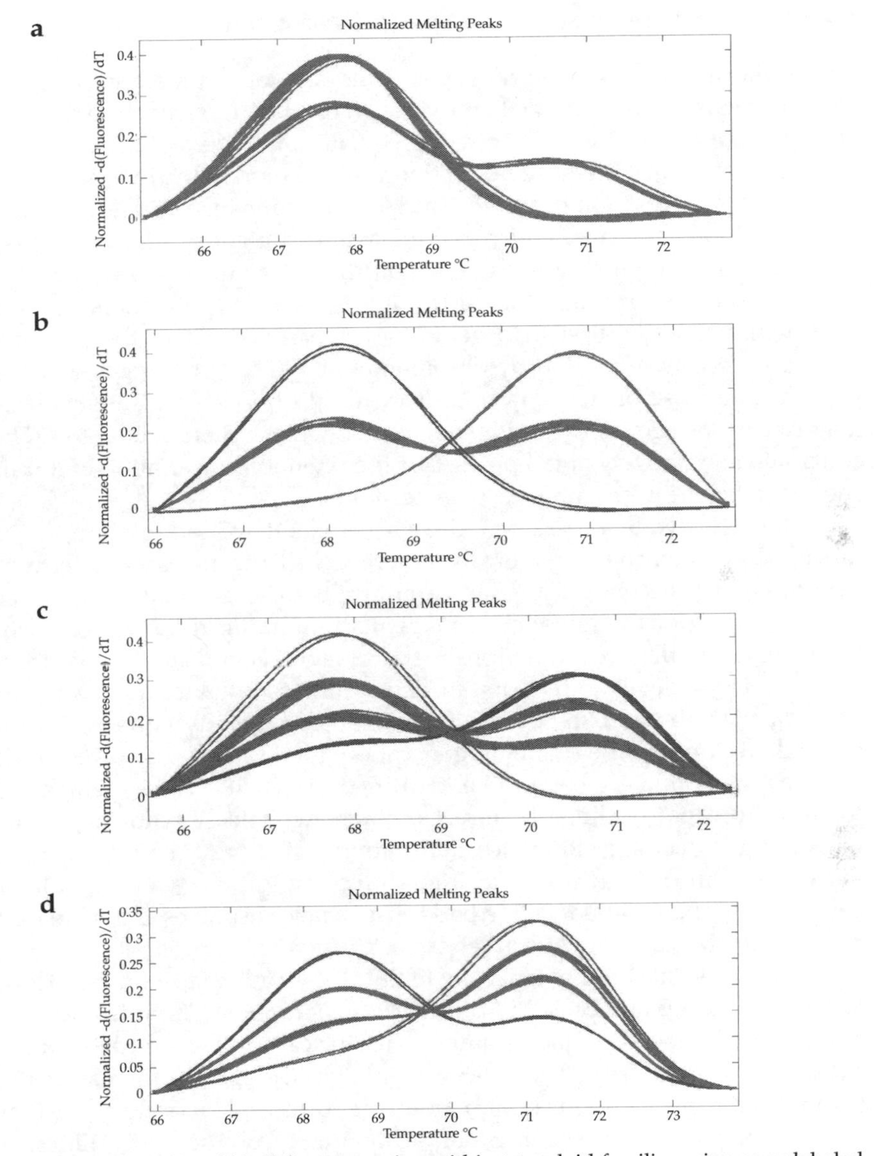

Figure 3-1 Examples of bi-allelic segregation within tetraploid families using an unlabeled probe HRM assay for the *dfr* locus, described in De Koeyer et al. (2010). The peak at 71°C represents the perfect match to the dominant R allele producing red pigment and the peak at 67.5°C represents a mismatch or non-red (r) pigment allele. The relative heights of the two peaks indicate the genotype at this locus. (A) rrrr (*grey* samples) x Rrrr (*red* samples) produce rrrr and Rrrr progeny in a 1:1 ratio; (B) RRRR (*red*) x rrrr (*blue*) produce all RRrr (*grey*) progeny; (C) RRrr (*red*) x Rrrr (*grey*) produce rrrr (*green*), Rrrr, RRrr, and RRRr (*blue*) progeny; (D) RRRR (*green*) x Rrrr (*blue*) produce RRrr (*grey*), and RRRr (*red*) progeny.

Color image of this figure appears in the color plate section at the end of the book.

3.4.3 Early-generation Selection with DNA Markers

The application of MAS in early generations of a potato breeding program requires a careful cost—benefit analysis. On one hand, selection in the first field generation is quite ineffective, especially for traits with mid to low heritability and marker-based selection may be more effective (heritable) than visual selection. On the other hand, the selection intensity is very high in the early generations, which would result in a considerable expense for marker genotyping clones that would be subsequently discarded for a critical trait. Therefore, it would be more practical to initiate marker genotyping after the first field generation when 90–95% of the progeny are discarded, as suggested by Ottoman et al. (2009). At this stage of the program, the use of markers linked to highly heritable traits, such as major resistance genes, would be recommended. Markers linked to QTLs controlling a relatively small portion of the phenotypic variation of a trait should not be utilized in the early generations.

Traits selected by breeders in the early generations (adaptation, tuber morphology, etc.) have generally not received the research attention devoted to more economically important traits (disease, quality). A better understanding of the genetic mechanisms regulating traits which form the basis of a "discard" decision in the first or second field generation may lead to the development of a sufficient number of markers to warrant implementation of MAS sooner. Data are not typically collected from individual clones in the first field generation; therefore, these marker-trait associations will likely have to be identified in genetic studies which are separate from the breeding program. Combined with the selection of parents with higher breeding values for adaptation traits, MAS could facilitate the increase in proportion of desirable progeny to be advanced to later generations. This should result in better use of field resources and increased population sizes to more effectively select difficult traits.

Advances in DNA extraction technology will also contribute to increasing the feasibility of MAS in early generations. The utilization of automated tissue grinding equipment with the capacity to simultaneously extract DNA from samples in 96-well plates has greatly increased the throughput capacity of our potato molecular breeding laboratory. Further capacity may be obtained using a liquid-handling robotics workstation.

3.4.4 Advanced-generation Selection with DNA Markers

At the advanced stages of a breeding program, substantially more phenotypic data are available for consideration compared to early generations. It is logical then that marker information will be combined with trait data and perhaps pedigree or marker derived relationship data to identify selections to be promoted and/or released to industry. Markers may also play a role in identifying clones which will undergo a particular evaluation. These decisions are often based on a limited amount of data from the parents (passport data, for example) since these trials can be very expensive or difficult to undertake. The use of the *H1*—linked HRM assay has identified parents in the Agriculture and Agri-Food Canada breeding program that putatively contain the *H1* allele for resistance to golden nematode pathotype *Ro1*. Progeny containing this *H1*—linked marker are entered into resistance evaluation trials, regardless of whether the phenotype of the parent is known.

Each breeding program will have its own way of incorporating marker data into selection decisions. Marker data would fit nicely into selection indices that incorporate information from multiple sources (Lande and Thompson 1990; Cerón-Rojas et al. 2008). The molecular eigen selection index method proposed by Cerón-Rojas et al. (2008) has some practical advantages, since it does not require economic weights of the traits under selection. For any marker-based selection approach, it will be important to take into account the proportion of phenotypic variation explained by a marker, the number of markers evaluated per trait, and the existing phenotypic performance data for the clone. Enhanced informatics capacity would allow breeders to have access to databases with marker, phenotype, and pedigree information. Such an information/data analysis system would facilitate "real-time" assessment of individual markers in the context of genetic relationships, environments and other data to develop a value that could be used to rank clones each year prior to planting.

At this time, the major factor limiting applications of MAS within breeding programs, especially at the advanced generation stage, is that there are relatively few validated, user-friendly markers linked to QTLs. Association genetics studies involving advanced breeding lines have contributed greatly to the tool-box of available markers for quality (D'Hoop et al. 2008; Li et al. 2008a) and disease resistance (Gebhardt et al. 2004; Malosetti et al. 2007; Pajerowska-Mukhtar et al. 2009). The Potato Genome Sequencing Consortium will enhance our knowledge of the potato genome and initiatives like SolCAP will further bridge the gap between breeders and biotechnologists.

3.5 Transgenic Breeding

Classical breeding is most often thought of as the crossing of two individuals of the same species or closely related species, with the resulting progeny having a mixture of genes from both parents and tens of thousands of genes involved. Classical breeding is often inefficient and imprecise as the breeder has very little control over which genes will be transferred to the next generation or how to track them; although the use of markers will improve precision. Multiple years of evaluation are often needed to select a few candidate lines for further analysis from the pool of progeny generated. Clonal propagation and a tetraploid genome add another layer of complexity to the process.

With transgenic breeding, a specific gene or genes from a related or widely divergent species can be moved into potato, often into the cultivar of choice without backcrossing. As the potato industry has been slow to adopt new germplasm, the ability to add the desired trait to a selected cultivar is an added benefit. However, the newly created transgenic line must undergo years of testing and regulatory hurdles to show that the introduced genes carry no deleterious effects to either the plant itself or the consumers and users of the plant. In this section, we will review the use of transgenic breeding in plants, and more specifically, in potatoes.

In potato, *Agrobacterium*-mediated transformation is the most common method for transferring genes in transgenic breeding. *Agrobacterium tumefaciens*, a common soil bacterium, transforms many dicot plants, including potato, with relative ease. Select bacterial genes are removed and the genes of interest are inserted in the bacterial plasmid (Ti-plasmid). The bacterium orchestrates the random insertion and integration of the selected DNA (T-DNA) into the genome of the plant. Expression of the gene or genes is dependent on many factors including location in the genome and type of promoter used to drive the gene. A screenable or selectable marker gene is often linked to the gene of interest to be able to determine if the plant is carrying the intended DNA. The most common selectable markers are usually antibiotic or herbicide resistance and bacterial in origin. For potato, the *nptII* gene is frequently used. *nptII* confers antibiotic resistance to the plant and allows the expressing plant to survive on kanamycin. Other genes and gene components are often bacterial (*Bt*) or viral (coat protein) in origin as are the promoters (35S) and gene terminators (mas). The use of widely divergent genes has allowed many unique traits to be moved into plants to improve quality, provide resistance, and provide stress tolerance.

There are a wide variety of traits in potato that are currently being targeted using transgenic breeding. A partial list of traits includes: starch (Trethewey et al. 1998; Edwards et al. 1999; Andersson et al. 2006); late blight resistance (Song et al. 2003; Halterman et al. 2008; Bradeen et al. 2009; Kramer et al. 2009); virus resistance (Xu et al. 1995; Lawson et al. 2001; Melander 2006); carbohydrate metabolism (Bachem et al. 2000); bacterial resistance (Osusky et al. 2005); insect tolerance [Colorado potato beetle (CPB), Cooper et al. (2006); (Kalushkov and Batchvarova 2005); potato tuber moth, Meiyalaghan et al. (2005); Davidson et al. (2004)]; molecular farming (Lauterslager et al. 2001; Kim and Langridge 2003; Joung et al. 2004; Pribylova et al. 2006); bioplastics (Romano et al. 2005); drought tolerance (Yeo et al. 2000); salt tolerance (Tang et al. 2008); improved nutrition (Chakraborty et al. 2000; Sevenier et al. 2002); and functional foods (Hellwege et al. 2000; Ducreux et al. 2005; Pribylova et al. 2006).

One of the most notable uses of transgenic breeding in potato was Monsanto's NewLeaf by NatureMark™ potatoes (Kaniewski and Thomas 2004). Monsanto released several varieties of the NewLeaf potato with CPB resistance alone (NewLeaf) or in combination with either Potato Virus Y (NewLeaf Y) or Potato Leaf Roll Virus (NewLeaf Plus) resistance. CPB resistance was conferred by a bacterial gene (*Bt*) and the first line was released in 1995, followed by lines with virus resistance (conferred by viral coat protein genes) over the next several years (Lawson et al. 2001). In 2000, McCain Foods announced that it would no longer accept NewLeaf potatoes due to concerns over consumer acceptance from the quick food industry. NatureMark was dissolved in 2001 and the potato project discontinued due an 80% decrease in sales of the NewLeaf varieties in 2000. Monsanto continues to license both the insect and viral resistance genes to other organizations outside of the US and Canada (Kalushkov and Batchvarova 2005).

BASF Plant Science Company currently has two transgenic potato varieties in development. "Amflora", an industrial modified starch variety, is under development for the past 12 years and awaiting approval since 2007. "Fortuna", a late blight resistant variety, is several years away from approval at the time of publication. The "Amflora" application has been reviewed and the variety has been deemed safe by the European Food Safety Authority. However, continual delays in approval from the European Commission (EC) have led the company to file a lawsuit against the EC that have resulted in the suspension of further research by the company to develop additional late blight resistant varieties until issues with the approval process have been resolved (*www.gmo-compass.org search word: BASF*).

Consumer acceptance of transgenic crops varies from country to country and from crop to crop. Many North American consumers are comfortable with the use of transgenic plants, also known as genetically modified organisms (GMOs), as a food ingredient. However, there is uncertainty regarding consumer acceptance of GMOs when consumed as a whole food such as potato, banana, and papaya; especially when the consumer sees no immediate personal benefit. Rather than risk losing any market share, the quick food industry has chosen to avoid using GMOs and transgenic potatoes have disappeared from the market place. Additionally, consumer response to GMOs in Europe and other parts of the world has been less favorable and has led to the exploration and development of "greener" methods of creating transgenics (transformation). Greener methods are expected to have greater consumer acceptance by using genes from more closely related species of plants or from the same species and, in some jurisdictions, reduce the regulatory hurdles to approval (Jacobsen 2007).

The word "transgenic" indicates a transfer of genes across species, such as a bacterial gene introduced into potato. New terminology, "cisgenic", has arisen recently to indicate that the movement of genes is occurring within the same species or plant family. There are variations on this theme, but the basic elements are that the genes being transferred into the plant come only from plant sources. There are now several reports in which *Agrobacterium*-mediated transformation has been used to create cisgenic plants (Conner et al. 2007; Rommens et al. 2005; Rommens et al. 2004).

Other ways to make the genetic transformation process more acceptable to the consumer is to avoid the use of a selectable marker, such as antibiotic resistance (de Vetten et al. 2003); to remove the selectable marker before the plant goes to the consumer (Lu et al. 2001; Schaart et al. 2004; Gidoni et al. 2008; Nanto and Ebinuma 2008); or to use a selectable marker that is plant based (Rommens et al. 2004).

Transgenic or cisgenic breeding methods have the potential to be another tool for use in potato improvement. However, there are limitations and benefits to both classical and transgenic breeding. Research on transgenic potatoes continues with a wide variety of biotic and abiotic stress resistance traits being developed even though the current market is limited. The ultimate goals of the breeder will determine which method will best help to achieve these goals.

3.6 Conclusions

Molecular breeding of potato is now a reality after several years of unfulfilled promise. Biotechnologists and breeders are working together to elucidate the genome sequence, to identify genetic factors controlling important economic traits within breeding populations, and to develop transformation technology that is more acceptable to the public. Molecular studies have contributed to our understanding of the taxonomic relationships of wild and cultivated potato. The immense amount of diversity within cultivated potato and its relatives provides both opportunities and challenges. The demands of industry and the public for the rapid introduction of new and better varieties contrasts with the long-term efforts that are needed to incorporate beneficial diversity not present in the adapted germplasm. Marker technology has advanced to the point where whole genome genotyping is now within the reach of many laboratories and dosage-sensitive markers are available for application within breeding programs. Markers will have an impact on early and advanced generation selection in potato breeding programs in a manner similar to that seen for other crops; however, the greatest impact of molecular breeding in potato will be for pre-breeding and parental development activities. Cisgenic technology has potential to revolutionize how breeders access genes for pre-breeding and parent development. The typical model of genetic modification of existing cultivars may not be sufficient as a means to make continual breeding progress to address the varied requirements of industry and consumers. Enhanced knowledge of the genetic factors controlling traits important to industry as well as those traits important for breeding success will make the coming years exciting for potato improvement. Molecular breeding provides the means to transfer this knowledge to the development of new cultivars for a variety of uses.

References

Achenbach U, Paulo J, Ilarionova E, Lubeck J, Strahwald J, Tacke E, Hofferbert H-R, Gebhardt C (2009) Using SNP markers to dissect linkage disequilibrium at a major quantitative trait locus for resistance to the potato cyst nematode *Globodera pallida* on potato chromosome V. Theor Appl Genet 118: 619–629.

Andersson M, Melander M, Pojmark P, Larsson H, Bulow L, Hofvander P (2006) Targeted gene suppression by RNA interference: an efficient method for production of high-amylose potato lines. J Biotechnol 123: 137–148.

Ashkenazi V, Chani E, Lavi U, Levy D, Hillel J, Veilleux RE (2001) Development of microsatellite markers in potato and their use in phylogenetic and fingerprinting analyses. Genome 44: 50–62.

Aversano R, Ercolano MR, Frusciante L, Monti L, Bradeen JM, Cristinzio G, Zoina A, Greco N, Vitale S, Carputo D (2007) Resistance traits and AFLP characterization of diploid primitive tuber-bearing potatoes. Genet Resour Crop Evol 54: 1797–1806.

Bachem CW, Oomen RJF, Kuyt S, Horvath BM, Claassens MM, Vreugdenhil D, Visser RG (2000) Antisense suppression of a potato alpha-SNAP homologue leads to alterations in cellular development and assimilate distribution. Plant Mol Biol 43: 473–482.

Bae JJ, Halterman D, Jansky S (2008) Development of a molecular marker associated with Verticillium wilt resistance in diploid interspecific potato hybrids. Mol Breed 22: 61–69.

Barone A (2004) Molecular marker-assisted selection for potato breeding. Am J Potato Res 81: 111–117.

Barone A, Sebastiano A, Carputo D, Della Rocca F, Frusciante L (2001) Molecular marker-assisted introgression of the wild *Solanum commersonii* genome into the cultivated *S. tuberosum* gene pool. Theor Appl Genet 102: 900–907.

Bendahmane A, Kanyuka K, Baulcombe DC (1997) High-resolution genetical and physical mapping of the *Rx* gene for extreme resistance to potato virus X in tetraploid potato. Theor Appl Genet 95: 153–162.

Bisognin DA, Douches DS (2002) Genetic diversity in diploid and tetraploid late blight resistant potato germplasm. HortScience 37: 178–183.

Bormann CA, Rickert AM, Ruiz RAC, Paal J, Lubeck J, Strahwald J, Buhr K, Gebhardt C (2004) Tagging quantitative trait loci for maturity-corrected late blight resistance in tetraploid potato with PCR-based candidate gene markers. Mol Plant-Microbe Interact 17: 1126–1138.

Bradeen JM, Iorizzo M, Mollov DS, Raasch J, Kramer LC, Millett BP, Austin-Phillips S, Jiang J, Carputo D (2009) Higher copy numbers of the potato *RB* transgene correspond to enhanced transcript and late blight resistance levels. Mol Plant-Microbe Interact 22: 437–446.

Bradshaw JE, Hackett CA, Meyer RC, Milbourne D, McNicol JW, Phillips MS, Waugh R (1998) Identification of AFLP and SSR markers associated with quantitative resistance to

Globodera pallida (Stone) in tetraploid potato (*Solanum tuberosum* subsp. *tuberosum*) with a view to marker-assisted selection. Theor Appl Genet 97: 202–210.

Brigneti G, Garcia-Mas J, Baulcombe DC (1997) Molecular mapping of the potato virus Y resistance gene Ry-sto in potato. Theor Appl Genet 94: 198–203.

Brown C, Kim T, Ganga Z, Haynes K, De Jong D, Jahn M, Paran I, De Jong W (2006) Segregation of total carotenoid in high level potato germplasm and its relationship to beta-carotene hydroxylase polymorphism. Am J Potato Res 83: 365–372.

Bryan GJ, McLean K, Bradshaw JE, De Jong WS, Phillips M, Castelli L, Waugh R (2002) Mapping QTLs for resistance to the cyst nematode *Globodera pallida* derived from the wild potato species *Solanum vernei*. Theor Appl Genet 105: 68–77.

Carputo D, Frusciante L, Monti L, Parisi M, Barone A (2002) Tuber quality and soft rot resistance of hybrids between *Solanum tuberosum* and the incongruent wild relative *S. commersonii*. Am J Potato Res 79: 345–352.

Cernák I, Decsi K, Nagy S, Wolf I, Polgár Z, Gulyás G, Hirata Y, Taller J (2008a) Development of a locus-specific marker and localization of the Ry_{sto} gene based on linkage to a catalase gene on chromosome XII in the tetraploid potato genome. Breed Sci 58: 309–314.

Cernák I, Taller J, Wolf I, Feher E, Babinszky G, Alfoldi Z, Csanadi G, Polgar Z (2008b) Analysis of the applicability of molecular markers linked to the PVY extreme resistance gene Ry_{sto}, and the identification of new markers. Acta Biol Hung 59: 195–203.

Cerón-Rojas JJ, Castillo-González F, Sahagún-Castellanos J, Santacruz-Varela A, Benítez-Riquelme I, Crossa J (2008) A molecular selection index method based on eigenanalysis. Genetics 180: 547–557.

Chakraborty S, Chakraborty N, Datta A (2000) Increased nutritive value of transgenic potato by expressing a nonallergenic seed albumin gene from *Amaranthus hypochondriacus*. Proc Natl Acad Sci USA 97: 3724–3729.

Colton LM, Groza HI, Wielgus SM, Jiang JM (2006) Marker-assisted selection for the broad-spectrum potato late blight resistance conferred by gene *RB* derived from a wild potato species. Crop Sci 46: 589–594.

Conner A, Barrell P, Baldwin S, Lokerse A, Cooper P, Erasmuson A, Nap J-P, Jacobs J (2007) Intragenic vectors for gene transfer without foreign DNA. Euphytica 154: 341–353.

Coombs JJ, Frank LM, Douches DS (2004) An applied fingerprinting system for cultivated potato using simple sequence repeats. Am J Potato Res 81: 243–250.

Cooper SG, Douches DS, Grafius EJ (2006) Insecticidal activity of avidin combined with genetically engineered and traditional host plant resistance against Colorado potato beetle (Coleoptera: Chrysomelidae) larvae. J Econ Entomol 99: 527–536.

D'Hoop BB, Paulo MJ, Mank RA, van Eck HJ, van Eeuwijk FA (2008) Association mapping of quality traits in potato (*Solanum tuberosum* L.). Euphytica 161: 47–60.

Dalla Rizza M, Vilaro FL, Torres DG, Maeso D (2006) Detection of PVY extreme resistance genes in potato germplasm from the Uruguayan breeding program. Am J Potato Res 83: 297–304.

Davidson MM, Takla MFG, Jacobs JME, Butler RC, Wratten SD, Conner AJ (2004) Transformation of potato (*Solanum tuberosum*) cultivars with a *cry1Ac9* gene confers resistance to potato tuber moth (*Phthorimaea operculella*). NZ J Crop Hort Sci 32: 39–50.

De Jong H (1991) Inheritance of anthocyanin pigmentation in the cultivated potato: A critical review. Am J Potato Res 68: 585–593.

De Jong WS, De Jong DM, Bodis M (2003a) A fluorogenic 5′ nuclease (TaqMan) assay to assess dosage of a marker tightly linked to red skin color in autotetraploid potato. Theor Appl Genet 107: 1384–1390.

De Jong WS, De Jong DM, De Jong H, Kalazich J, Bodis M (2003b) An allele of dihydroflavonol 4-reductase associated with the ability to produce red anthocyanin pigments in potato (*Solanum tuberosum* L.). Theor Appl Genet 107: 1375–1383.

De Koeyer D, Douglass K, Murphy A, Whitney S, Nolan L, Song Y, De Jong W (2010) Application of high-resolution DNA melting for genotyping and variant scanning of diploid and autotetraploid potato. Mol Breed 25(1): 67–90.

de Vetten N, Wolters AM, Raemakers K, van der Meer I, ter Stege R, Heeres E, Heeres P, Visser R (2003) A transformation method for obtaining marker-free plants of a cross-pollinating and vegetatively propagated crop. Nat Biotechnol 21: 439–442.

Demeke T, Lynch DR, Kawchuk LM, Kozub GC, Armstrong JD (1996) Genetic diversity of potato determined by random amplified polymorphic DNA analysis. Plant Cell Rep 15: 662–667.

Ducreux LJM, Morris WL, Taylor MA, Millam S (2005) Agrobacterium-mediated transformation of *Solanum phureja*. Plant Cell Rep 24: 10–14.

Edwards A, Borthakur A, Bornemann S, Venail J, Denyer K, Waite D, Fulton D, Smith A, Martin C (1999) Specificity of starch synthase isoforms from potato. Eur J Biochem 266: 724–736.

Erali M, Voelkerding KV, Wittwer CT (2008) High resolution melting applications for clinical laboratory medicine. Exp Mol Pathol 85: 50–58.

Finkers-Tomczak A, Danan S, van Dijk T, Beyene A, Bouwman L, Overmars H, van Eck H, Goverse A, Bakker J, Bakker E (2009) A high-resolution map of the *Grp1* locus on chromosome V of potato harbouring broad-spectrum resistance to the cyst nematode species *Globodera pallida* and *Globodera rostochiensis*. Theor Appl Genet 119: 165–173.

Flis B, Hennig J, Strzelczyk-Zyta D, Gebhardt C, Marezewski W (2005) The *Ry-f*$_{sto}$ gene from *Solanum stoloniferum* for extreme resistant to Potato virus Y maps to potato chromosome XII and is diagnosed by PCR marker GP122–718 in PVY resistant potato cultivars. Mol Breed 15: 95–101.

Ford R, Taylor PWJ (1997) The application of RAPD markers for potato cultivar identification. Aust J Agri Res 48: 1213–1217.

Fu YB, Peterson GW, Richards KW, Tarn TR, Percy JE (2009) Genetic diversity of Canadian and exotic potato germplasm revealed by simple sequence repeat markers. Am J Potato Res 86: 38–48.

Gallais A (2003) Quantitative genetics and breeding methods in autopolyploid plants. Institut National de la Recherche Agronomique, Paris, France.

Gebhardt C (2005) Potato genetics: Molecular maps and more. Biotechnol Agri For 55: 215–224.

Gebhardt C, Valkonen JP (2001) Organization of genes controlling disease resistance in the potato genome. Annu Rev Phytopathol 39: 79–102.

Gebhardt C, Ballvora A, Walkemeier B, Oberhagemann P, Schuler K (2004) Assessing genetic potential in germplasm collections of crop plants by marker-trait association: a case study for potatoes with quantitative variation of resistance to late blight and maturity type. Mol Breed 13: 93–102.

Gebhardt C, Bellin D, Henselewski H, Lehmann W, Schwarzfischer J, Valkonen JPT (2006) Marker-assisted combination of major genes for pathogen resistance in potato. Theor Appl Genet 112: 1458–1464.

Ghislain M, Zhang D, Fajardo D, Huaman Z, Hijmans RJ (1999) Marker-assisted sampling of the cultivated Andean potato *Solanum phureja* collection using RAPD markers. Genet Resour Crop Evol 46: 547–555.

Ghislain M, Spooner DM, Rodriguez F, Villamon F, Nunez J, Vasquez C, Waugh R, Bonierbale M (2004) Selection of highly informative and user-friendly microsatellites (SSRs) for genotyping of cultivated potato. Theor Appl Genet 108: 881–890.

Ghislain M, Andrade D, Rodriguez F, Hijmans RJ, Spooner DM (2006) Genetic analysis of the cultivated potato *Solanum tuberosum* L. Phureja Group using RAPDs and nuclear SSRs. Theor Appl Genet 113: 1515–1527.

Ghislain M, Nunez J, del Rosario Herrera M, Pignataro J, Guzman F, Bonierbale M, Spooner DM (2009a) Robust and highly informative microsatellite-based genetic identity kit for potato. Mol Breed 23: 377–388.

Ghislain M, Nunez J, del Rosario Herrera M, Spooner DM (2009b) The single Andigenum origin of Neo-Tuberosum potato materials is not supported by microsatellite and plastid marker analyses. Theor Appl Genet 118: 963–969.

Gidoni D, Srivastava V, Carmi N (2008) Site-specific excisional recombination strategies for elimination of undesirable transgenes from crop plants. In Vitro Cell Dev Biol-Plant 44: 457–467.

Hale AL, Miller Jr JC, Renganayaki K, Fritz AK, Coombs JJ, Frank LM, Douches DS (2005) Suitability of AFLP and microsatellite marker analysis for discriminating intraclonal variants of the potato cultivar Russet Norkotah. J Am Soc Hort Sci 130: 624–630.

Halterman DA, Kramer LC, Wielgus S, Jiang J (2008) Performance of transgenic potato containing the late blight resistance gene *RB*. Plant Dis 92: 339–343.

Hämäläinen JH, Watanabe KN, Valkonen JPT, Arihara A, Plaisted RL, Pehu E, Miller L, Slack SA (1997) Mapping and marker-assisted selection for a gene for extreme resistance to potato virus Y. Theor Appl Genet 94: 192–197.

Hawkes JG (1990) The Potato: Evolution, Biodiversity and Genetic Resources. Belhaven, London, UK.

Hellwege EM, Czapla S, Jahnke A, Willmitzer L, Heyer AG (2000) Transgenic potato (*Solanum tuberosum*) tubers synthesize the full spectrum of inulin molecules naturally occurring in globe artichoke (*Cynara scolymus*) roots. Proc Natl Acad Sci USA 97: 8699–8704.

Hijmans RJ, Gavrilenko T, Stephenson S, Bamberg J, Salas A, Spooner DM (2007) Geographical and environmental range expansion through polyploidy in wild potatoes (*Solanum* section *Petota*). Glob Ecol Biogeogr 16: 485–495.

Hofinger BJ, Jing HC, Hammond-Kosack KE, Kanyuka K (2009) High-resolution melting analysis of cDNA-derived PCR amplicons for rapid and cost-effective identification of novel alleles in barley. Theor Appl Genet 119: 851–865.

Iovene M, Barone A, Frusciante L, Monti L, Carputo D (2004) Selection for aneuploid potato hybrids combining a low wild genome content and resistance traits from *Solanum commersonii*. Theor Appl Genet 109: 1139–1146.

Isenegger DA, Taylor PWJ, Ford R, Franz P, McGregor GR, Hutchinson JF (2001) DNA fingerprinting and genetic relationships of potato cultivars (*Solanum tuberosum* L.) commercially grown in Australia. Aust J Agri Res 52: 911–918.

Ispizúa VN, Guma IR, Feingold S, Clausen AM (2007) Genetic diversity of potato landraces from northwestern Argentina assessed with simple sequence repeats (SSRs). Genet Resour Crop Evol 54: 1833–1848.

Jacobs MM, van den Berg RG, Vleeshouwers VG, Visser M, Mank R, Sengers M, Hoekstra R, Vosman B (2008) AFLP analysis reveals a lack of phylogenetic structure within *Solanum* section *Petota*. BMC Evol Biol 8: 145.

Jacobsen E (2007) The Canon of Potato Science: 6. Genetic Modification and Cis- and Transgenesis. Potato Res 50: 227–230.

Jansky S (2006) Overcoming hybridization barriers in potato. Plant Breed 125: 1–12.

Jansky SH, Simon R, Spooner DM (2006) A test of taxonomic predictivity: resistance to white mold in wild relatives of cultivated potato. Crop Sci 46: 2561–2570.

Jansky SH, Simon R, Spooner DM (2008) A test of taxonomic predictivity: resistance to early blight in wild relatives of cultivated potato. Phytopathology 98: 680–687.

Jansky SH, Simon R, Spooner DM (2009) A test of taxonomic predictivity: resistance to the Colorado potato beetle in wild relatives of cultivated potato. J Econ Entomol 102: 422–431.

Jiménez JP, Brenes A, Fajardo D, Salas A, Spooner DM (2008) The use and limits of AFLP data in the taxonomy of polyploid wild potato species in *Solanum* series *Conicibaccata*. Conserv Genet 9: 381–387.

Joung YH, Youm JW, Jeon JH, Lee BC, Ryu CJ, Hong HJ, Kim HC, Joung H, Kim HS (2004) Expression of the hepatitis B surface S and preS2 antigens in tubers of *Solanum tuberosum*. Plant Cell Rep 22: 925–930.

Jung CS, Griffiths HM, De Jong DM, Cheng S, Bodis M, De Jong WS (2005) The potato *P* locus codes for flavonoid 3′,5′-hydroxylase. Theor Appl Genet 110: 269–275.

Kalushkov P, Batchvarova R (2005) Effectiveness of Bt Newleaf® potato to control *Leptinotarsa decemlineata* (Say) (Coleoptera: Chrysomelidae) in Bulgaria. Biotechnol Biotechnol Equip 19: 28–34.

Kaniewski WK, Thomas PE (2004) The potato story. Agbioforum 7: 1–2.

Kasai K, Morikawa Y, Sorri VA, Valkonen JP, Gebhardt C, Watanabe KN (2000) Development of SCAR markers to the PVY resistance gene Ry_{adg} based on a common feature of plant disease resistance genes. Genome 43: 1–8.

Kawchuk LM, Lynch DR, Yada RY, Bizimungu B, Lynn J (2008) Marker assisted selection of potato clones that process with light chip color. Am J Potato Res 85: 227–231.

Kim TG, Langridge WHR (2003) Assembly of cholera toxin B subunit full-length rotavirus NSP4 fusion protein oligomers in transgenic potato. Plant Cell Rep 21: 884–890.

Kramer LC, Choudoir MJ, Wielgus SM, Bhaskar PB, Jiang J (2009) Correlation between transcript abundance of the *RB* gene and the level of the *RB*-mediated late blight resistance in potato. Mol Plant-Microbe Interact 22: 447–455.

Lande R, Thompson R (1990) Efficiency of marker-assisted selection in the improvement of quantiative traits. Genetics 124: 743–756.

Lara-Cabrera SI, Spooner DM (2004) Taxonomy of North and Central American diploid wild potato (*Solanum* sect. *Petota*) species: AFLP data. Plant Syst Evol 248: 129–142.

Lara-Cabrera SI, Spooner DM (2005) Taxonomy of Mexican diploid wild potatoes (*Solanum* sect. *Petota*): morphological and microsatellite data. In: RC Keating, VC Hollowell, TB Croat (eds) A Festschrift for William G D'arcy: The Legacy of a Taxonomist. Missouri Botanical Garden, St. Louis, MO, USA, pp 199–225.

Lauterslager TGM, Florack DEA, van der Wal TJ, Molthoff JW, Langeveld JPM, Bosch D, Boersma WJA, Hilgers LAT (2001) Oral immunisation of naive and primed animals with transgenic potato tubers expressing LT-B. Vaccine 19: 2749–2755.

Lawson EC, Weiss JD, Thomas PE, Kaniewski WK (2001) NewLeaf PlusReg. Russet Burbank potatoes: replicase-mediated resistance to potato leafroll virus. Mol Breed 7: 1–12.

Li L, Strahwald J, Hofferbert HR, Lubeck J, Tacke E, Junghans H, Wunder J, Gebhardt C (2005) DNA variation at the invertase locus *inv GE/GF* is associated with tuber quality traits in populations of potato breeding clones. Genetics 170: 813–821.

Li L, Paulo MJ, Strahwald J, Lubeck J, Hofferbert HR, Tacke E, Junghans H, Wunder J, Draffehn A, van Eeuwijk F, Gebhardt C (2008a) Natural DNA variation at candidate loci is associated with potato chip color, tuber starch content, yield and starch yield. Theor Appl Genet 116: 1167–1181.

Li X, Haroon M, Coleman SE, Sullivan A, Mathuresh S, Ward L, Boer SHd, Zhang T, Donnelly DJ (2008b) A simplified procedure for verifying and identifying potato cultivars using multiplex PCR. Can J Plant Sci 88: 583–592.

Lu H, Zhou X, Gong Z, Upadhyaya NM (2001) Generation of selectable marker-free transgenic rice using double right-border (DRB) binary vectors. Aust J Plant Physiol 28: 241–248.

Malosetti M, van der Linden CG, Vosman B, van Eeuwijk FA (2007) A mixed-model approach to association mapping using pedigree information with an illustration of resistance to *Phytophthora infestans* in potato. Genetics 175: 879–889.

Marczewski W, Talarczyk A, Hennig J (2001) Development of SCAR markers linked to the *Ns* locus in potato. Plant Breed 120: 88–90.

Matsubayashi M (1991) Phylogenetic relationships in the potato and its related species. In: T Tsuchiya, PK Gupta (eds) Chromosome Engineering in Plants: Genetics, Breeding, Evolution, part B. Elsevier Science, Amsterdam, The Netherlands, pp 93–118.

McGregor CE, Lambert CA, Greyling MM, Louw JH, Warnich L (2000) A comparative assessment of DNA fingerprinting techniques (RAPD, ISSR, AFLP and SSR) in tetraploid potato (*Solanum tuberosum* L.) germplasm. Euphytica 113: 135–144.

McGregor CE, van Treuren R, Hoekstra R, van Hintum TJL (2002) Analysis of the wild potato germplasm of the series *Acaulia* with AFLPs: implications for ex situ conservation. Theor Appl Genet 104: 146–156.

Meiyalaghan S, Takla MF, Jaimess O, Yongjin S, Davidson MM, Cooper PA, Barrell PJ, Jacobs ME, Wratten SD, Conner AJ (2005) Evaluation of transgenic approaches for controlling tuber moth in potatoes. Commun Agri Appl Biol Sci 70: 641–650.

Melander M (2006) Potato transformed with a 57-kDa readthrough portion of the Tobacco rattle virus replicase gene displays reduced tuber symptoms when challenged by viruliferous nematodes. Euphytica 150: 123–130.

Milbourne D, Meyer R, Bradshaw JE, Baird E, Bonar N, Provan J, Powell W, Waugh R (1997) Comparison of PCR-based marker systems for the analysis of genetic relationships in cultivated potato. Mol Breed 3: 127–136.

Nanto K, Ebinuma H (2008) Marker-free site-specific integration plants. Transgen Res 17: 337–344.

Niewöhner J, Salamini F, Gebhardt C (1995) Development of PCR assays diagnostic for RFLP marker alleles closely linked to alleles *Gro1* and *H1*, conferring resistance to the root cyst nematode *Globodera rostochiensis* in potato. Mol Breed 1: 65–78.

Norero N, Malleville J, Huarte M, Feingold S (2002) Cost efficient potato (*Solanum tuberosum* L.) cultivar identification by microsatellite amplification. Potato Res 45: 131–138.

Oberacher H, Parson W, Hölzl G, Oefner PJ, Huber CG (2004) Optimized suppression of adducts in polymerase chain reaction products for semi-quantitative SNP genotyping by liquid chromatography-mass spectrometry. J Am Soc Mass Spectrom 15: 1897–1906.

Oberhagemann P, Chatot-Balandras C, Schafer-Pregl R, Wegener D, Palomino C, Salamini F, Bonnel E, Gebhardt C (1999) A genetic analysis of quantitative resistance to late blight in potato: towards marker-assisted selection. Mol Breed 5: 399–415.

Osusky M, Osuska L, Kay W, Misra S (2005) Genetic modification of potato against microbial diseases: *in vitro* and *in planta* activity of a dermaseptin B1 derivative, MsrA2. Theor Appl Genet 111: 711–722.

Ottoman RJ, Hane DC, Brown CR, Yilma S, James SR, Mosley AR, Crosslin JM, Vales MI (2009) Validation and implementation of marker-assisted selection (MAS) for PVY resistance (Ry_{adg} gene) in a tetraploid potato breeding program. Am J Potato Res 86: 304–314.

Paal J, Henselewski H, Muth J, Meksem K, Menendez CM, Salamini F, Ballvora A, Gebhardt C (2004) Molecular cloning of the potato *Gro1-4* gene conferring resistance to pathotype Ro1 of the root cyst nematode *Globodera rostochiensis*, based on a candidate gene approach. Plant J 38: 285–297.

Pajerowska-Mukhtar K, Stich B, Achenbach U, Ballvora A, Lubeck J, Strahwald J, Tacke E, Hofferbert H-R, Ilarionova E, Bellin D, Walkemeier B, Basekow R, Kersten B, Gebhardt C (2009) Single nucleotide polymorphisms in the *allene oxide synthase 2* gene are associated with field resistance to late blight in populations of tetraploid potato cultivars. Genetics 181: 1115–1127.

Pattanayak D, Chakrabarti SK, Naik PS (2002) Genetic diversity of late blight resistant and susceptible Indian potato cultivars revealed by RAPD markers. Euphytica 128: 183–189.

Paz MM, Veilleux RE (1997) Genetic diversity based on randomly amplified polymorphic DNA (RAPD) and its relationship with the performance of diploid potato hybrids. J Am Soc Hort Sci 122: 740–747.

Pribylova R, Pavlik I, Bartos M (2006) Genetically modified potato plants in nutrition and prevention of diseases in humans and animals: a review. Vet Med 51: 212–223.

Raker CM, Spooner DM (2002) Chilean tetraploid cultivated potato, *Solanum tuberosum*, is distinct from the Andean populations: microsatellite data. Crop Sci 42: 1451–1458.

Rickert AM, Premstaller A, Gebhardt C, Oefner PJ (2002) Genotyping of SNPs in a polyploid genome by pyrosequencing™. Biotechniques 32: 592–603.

Rickert AM, Kim JH, Meyer S, Nagel A, Ballvora A, Oefner PJ, Gebhardt C (2003) First-generation SNP/InDel markers tagging loci for pathogen resistance in the potato genome. Plant Biotechnol J 1: 399–410.

Rickert AM, Ballvora A, Matzner U, Klemm M, Gebhardt C (2005) Quantitative genotyping of single-nucleotide polymorphisms by allele-specific oligonucleotide hybridization on DNA microarrays. Biotechnol Appl Biochem 42: 93–96.

Rios D, Ghislain M, Rodriguez F, Spooner DM (2007) What is the origin of the European potato? Evidence from Canary Island landraces. Crop Sci 47: 1271–1280.

Rodriguez F, Spooner DM (2009) Nitrate reductase phylogeny of potato (*Solanum* sect. *Petota*) genomes with emphasis on the origins of the polyploid species. Syst Bot 34: 207–219.

Romano A, van der Plas LHW, Witholt B, Eggink G, Mooibroek H (2005) Expression of poly-3-(R)-hydroxyalkanoate (PHA) polymerase and acyl-CoA-transacylase in plastids of transgenic potato leads to the synthesis of a hydrophobic polymer, presumably medium-chain-length PHAs. Planta 220: 455–464.

Rommens CM, Humara JM, Ye JS, Yan H, Richael C, Zhang L, Perry R, Swords K (2004) Crop improvement through modification of the plant's own genome. Plant Physiol 135: 421–431.

Rommens CM, Bougri O, Yan H, Humara JM, Owen J, Swords K, Ye J (2005) Plant-Derived Transfer DNAs. Plant Physiol 139: 1338–1349.

Sattarzadeh A, Achenbach U, Lubeck J, Strahwald J, Tacke E, Hofferbert HR, Rothsteyn T, Gebhardt C (2006) Single nucleotide polymorphism (SNP) genotyping as basis for developing a PCR-based marker highly diagnostic for potato varieties with high resistance to *Globodera pallida* pathotype Pa2/3. Mol Breed 18: 301–312.

Schaart JG, Krens FA, Pelgrom KT, Mendes O, Rouwendal GJ (2004) Effective production of marker-free transgenic strawberry plants using inducible site-specific recombination and a bifunctional selectable marker gene. Plant Biotechnol J 2: 233–240.

Sevenier R, van der Meer IM, Bino R, Koops AJ (2002) Increased production of nutriments by genetically engineered crops. J Am Coll Nutr 21: 199S–204.

Sharma JR (ed) (1994) Principles and Practice of Plant Breeding. Tata McGraw-Hill, New Delhi, India.

Simko I, Haynes KG, Ewing EE, Costanzo S, Christ BJ, Jones RW (2004a) Mapping genes for resistance to *Verticillium albo-atrum* in tetraploid and diploid potato populations using haplotype association tests and genetic linkage analysis. Mol Genet Genom 271: 522–531.

Simko I, Haynes KG, Jones RW (2004b) Mining data from potato pedigrees: tracking the origin of susceptibility and resistance to *Verticillium dahliae* in North American cultivars through molecular marker analysis. Theor Appl Genet 108: 225–230.

Solano Solis J, Morales Ulloa D, Anabalon Rodriguez L (2007) Molecular description and similarity relationships among native germplasm potatoes (*Solanum tuberosum* ssp. *tuberosum* L.) using morphological data and AFLP markers. Electron J Biotechnol 10: 436–443.

Song J, Bradeen JM, Naess SK, Raasch JA, Wielgus SM, Haberlach GT, Liu J, Kuang H, Austin-Phillips S, Buell CR, Helgeson JP, Jiang J (2003) Gene *RB* cloned from *Solanum bulbocastanum* confers broad spectrum resistance to potato late blight. Proc Natl Acad Sci USA 100: 9128–9133.

Song Y-S, Schwarzfischer A (2008) Development of STS markers for selection of extreme resistance (Ry_{sto}) to PVY and maternal pedigree analysis of extremely resistant cultivars. Am J Potato Res 85: 159–170.

Song Y-S, Hepting L, Schweizer G, Hartl L, Wenzel G, Schwarzfischer A (2005) Mapping of extreme resistance to PVY (*Ry (sto)*) on chromosome XII using anther-culture-derived primary dihaploid potato lines. Theor Appl Genet 111: 879–887.

Sorri VA, Watanabe KN, Valkonen JPT (1999) Predicted kinase-3a motif of a resistance gene analogue as a unique marker for virus resistance. Theor Appl Genet 99: 164–170.

Spooner DM, Salas A (2006) Structure, biosystematics, and genetic resources. In: J Gopal, SMP Khurana (eds) Handbook of Potato Production, Improvement, and Postharvest Management. Haworth Press, Binghampton, NY, USA, pp 1–39.

Spooner DM, McLean K, Ramsay G, Waugh R, Bryan GJ (2005a) A single domestication for potato based on multilocus amplified fragment length polymorphism genotyping. Proc Natl Acad Sci USA 102: 14694–14699.

Spooner DM, Nunez J, Rodriguez F, Naik PS, Ghislain M (2005b) Nuclear and chloroplast DNA reassessment of the origin of Indian potato varieties and its implications for the origin of the early European potato. Theor Appl Genet 110: 1020–1026.

Spooner DM, Nunez J, Trujillo G, del Rosario Herrera M, Guzman F, Ghislain M (2007) Extensive simple sequence repeat genotyping of potato landraces supports a major reevaluation of their gene pool structure and classification. Proc Natl Acad Sci USA 104: 19398–19403.

Spooner DM, Rodriguez F, Polgar Z, Ballard HEJ, Jansky SH (2008) Genomic origins of potato polyploids: GBSSI gene sequencing data. Crop Sci 48: S27–S36.

Sukhotu T, Kamijima O, Hosaka K (2005) Genetic diversity of the Andean tetraploid cultivated potato (*Solanum tuberosum* L. subsp. *andigena* Hawkes) evaluated by chloroplast and nuclear DNA markers. Genome 48: 55–64.

Sun G, Wang-Pruski G, Mayich M, de Jong H (2003) RAPD and pedigree-based genetic diversity estimates in cultivated diploid potato hybrids. Theor Appl Genet 107: 110–115.

Tai GCC, Murphy AM, Xiong X (2009) Investigation of long-term field experiments on response of breeding lines to common scab in a potato breeding program. Euphytica 167: 69–76.

Tang L, Kim M, Yang K-S, Kwon S-Y, Kim S-H, Kim J-S, Yun D-J, Kwak S-S, Lee H-S (2008) Enhanced tolerance of transgenic potato plants overexpressing nucleoside diphosphate kinase 2 against multiple environmental stresses. Transgen Res 17: 705–715.

Trethewey RN, Geigenberger P, Riedel K, Hajirezaei MR, Sonnewald U, Stitt M, Riesmeier JW, Willmitzer L (1998) Combined expression of glucokinase and invertase in potato tubers leads to a dramatic reduction in starch accumulation and a stimulation of glycolysis. Plant J 15: 109–118.

Valkonen JPT, Wiegmann K, Hamalainen JH, Marczewski W, Watanabe KN (2008) Evidence for utility of the same PCR-based markers for selection of extreme resistance to Potato virus Y controlled by *Ry(sto)* of *Solanum stoloniferum* derived from different sources. Ann Appl Biol 152: 121–130.

van den Berg RG, Bryan GJ, Del Rio A, Spooner DM (2002) Reduction of species in the wild potato *Solanum* section *Petota* series *Longipedicellata*: AFLP, RAPD and chloroplast SSR data. Theor Appl Genet 105: 1109–1114.

van der Voort JNAMR, Janssen GJW, Overmars H, van Zandvoort PM, van Norel A, Scholten OE, Janssen R, Bakker J (1999) Development of a PCR-based selection assay for root-knot nematode resistance (R_{mc1}) by a comparative analysis of the *Solanum bulbocastanum* and *S. tuberosum* genome. Euphytica 106: 187–195.

van Treuren R, Magda A, Hoekstra R, van Hintum TJL (2004) Genetic and economic aspects of marker-assisted reduction of redundancy from a wild potato germplasm collection. Genet Resour Crop Evol 51: 277–290.

Veteläinen M, Gammelgård E, Valkonen JPT (2005) Diversity of Nordic landrace potatoes (*Solanum tuberosum* L.) revealed by AFLPs and morphological characters. Genet Resour Crop Evol 52: 999–1010.

Visser RGF, Bachem CWB, de Boer JM, Bryan GJ, Chakrabati SK, Feingold S, Gromadka R, van Ham RCHJ, Huang S, Jacobs JME, Kuznetsov B, de Melo PE, Milbourne D, Orjeda G, Sagredo B, Tang X (2009) Sequencing the potato genome: Outline and first results to come from the elucidation of the sequence of the world's third most important food crop. Am J Potato Res 86: 417–429.

Walker JT (1969) Selection and quantitative characters in field crops. Biol Rev 44: 207–243.

Witek K, Strzelczyk-Zyta D, Hennig J, Marczewski W (2006) A multiplex PCR approach to simultaneously genotype potato towards the resistance alleles *Ry-f(sto)* and *Ns*. Mol Breed 18: 273–275.

Xu H, Khalilian H, Eweida M, Squire S, Abouhaidar MG (1995) Genetically engineered resistance to potato virus X in four commercial potato cultivars. Plant Cell Rep 15: 91–96.

Yasmin S, Shahidul Islam M, Nasiruddin KM, Samsul Alam M (2006) Molecular characterization of potato germplasm by random amplified polymorphic DNA markers. Biotechnology 5: 27–31.

Yeo ET, Kwon HB, Han SE, Lee JT, Ryu JC, Byu MO (2000) Genetic engineering of drought resistant potato plants by introduction of the trehalose-6-phosphate synthase (TPS1) gene from *Saccharomyces cerevisiae*. Mol Cells 10: 263–268.

Yu J, Pressoir G, Briggs WH, Bi IV, Yamasaki M, Doebley JF, McMullen MD, Gaut BS, Nielsen DM, Holland JB, Kresovich S, Buckler ES (2006) A unified mixed-model method for association mapping that accounts for multiple levels of relatedness. Nat Genet 38: 203–208.

Zhang LH, Mojtahedi H, Kuang H, Baker B, Brown CR (2007) Marker-assisted selection of Columbia root-knot nematode resistance introgressed from *Solanum bulbocastanum*. Crop Sci 47: 2021–2026.

Zhang Y, Jung CS, De Jong WS (2009) Genetic analysis of pigmented tuber flesh in potato. Theor Appl Genet 119: 143–150.

Molecular Linkage Maps: Strategies, Resources and Achievements

Harpartap Mann,[1,] Massimo Iorizzo,[1,2] Liangliang Gao,[1] Nunzio D'Agostino,[2] Domenico Carputo,[2] Maria Luisa Chiusano[2]* and *James M. Bradeen[1]*

ABSTRACT

Modern linkage mapping has its roots, in essence, in the genetic segregation studies of Mendel. Although the autotetraploid nature of potato slowed early linkage mapping efforts, extensive linkage maps have been developed. Early maps, based primarily on RFLP markers, revealed a high degree of conserved marker order along homologous potato and tomato chromosomes. RFLP markers were subsequently augmented by a bevy of PCR-based markers including SSRs and AFLPs. A linkage map for potato based on more than 10,000 AFLP markers is today the densest linkage map based on genetic recombination ever generated for any organism. This map was integrated with a BAC-based physical map of the potato genome—a resource that supported efforts to sequence the potato genome. Gene and transcript mapping efforts are common in potato. Marker-assisted selection, the use of molecular markers linked to specific genes as a proxy in genotype selection by breeders, holds promise for potato, but its potential has not been fully realized to date. Molecular marker types continue to evolve with the single nucleotide polymorphism representing the ultimate in DNA-based molecular markers. Mapping in potato can be done either at the diploid or tetraploid level. Potato linkage mapping frequently

[1]University of Minnesota, Department of Plant Pathology, 495 Borlaug Hall/1991 Upper Buford Circle, St. Paul, MN 55108, USA.
[2]University of Naples, Federico II, Department of Soil, Plant, Environmental and Animal Production Sciences, Via Università 100, 80055 Portici, Italy.
*Corresponding author: *mann0188@umn.edu*

employs the pseudo-testcross strategy, whereby linkage maps are created separately for each parent using dominant marker data and the maps are subsequently integrated into a single unified map using co-dominant data. Most of the approximately 200 wild potato species carry genes useful for potato improvement. The development of comparative genomics resources in the form of linkage maps suitable for cross-species analyses are underway and should speed efforts to map and clone genes from wild species. DArT marker mapping in the disease resistant wild potato *Solanum bulbocastanum* yielded a genome-wide linkage map comprised of 439 markers and spanning 403 cM. Sequencing of mapped DArT markers was followed by in silico comparisons to emerging potato and tomato genome sequences. From these analyses emerged an understanding of genome structure in *S. bulbocastanum* relative to cultivated potato and tomato. Further, the developed *S. bulbocastanum* linkage map serves as a scaffold upon which genes of interest (e.g., disease resistance genes) can be anchored.

Keywords: comparative genomics, linkage mapping, molecular markers, potato, pseudo-testcross

4.1 Introduction

Gregor Mendel's experiments on the pea are all the more remarkable in hindsight for the absence of linkage among the phenotypic traits that were studied. Ever since, analysis of co-inheritance of traits (linkage) from one generation to the next has formed the basis of classical and modern genetics. Linkage analysis has progressed from studies on linkage among phenotypic traits to analysis of linkage among specific DNA sequences, resulting in increasingly dense linkage maps. Linkage mapping based on molecular markers has subsequently enabled elucidation of the complete genome sequence of multiple plant species, and genome sequencing of many more, including potato is currently underway.

4.2 Brief History of Mapping Efforts in Potato

Despite being the most important food crop of the Solanaceae, genetic linkage analysis in potato has lagged behind that of the tomato, another member of the family. The highly heterozygous, tetraploid nature of potato and lack of useful genetic markers have been cited as reasons for the slower progress of genetic linkage analysis in potato (Bonierbale et al. 1988). In retrospect, the development of the first linkage map in potato was a result of convergence of three independent lines of research: development of diploid potato lines that preceded linkage analysis in potato by several years (Hougas 1957; Hougas and Peloquin 1958), development of DNA-based restriction fragment length polymorphism (RFLP) markers (Botstein et al. 1980) and an RFLP marker-based genetic linkage map of the tomato genome (Bernatzky and Tanksley 1986).

The first potato map was developed using an interspecific cross involving a diploid potato line, [*S. phureja* Juz. et Buk. x (*S. tuberosum* ($2n = 2x = 24$) x *S. chacoense* Bitt)] and RFLP markers previously mapped in tomato (Bernatzky and Tanksley 1986; Bonierbale et al. 1988). This first potato map consisted of 134 RFLP and isozyme markers, which were assigned to 12 linkage groups for a total map length of 606 cM. When compared to the tomato RFLP map, it was revealed that marker order on nine out of 12 linkage groups was highly conserved between potato and tomato. Only potato linkage groups 5, 9 and 10 showed inversions compared to the homologous linkage groups in tomato (inversions on linkage groups 11 and 12 were added to this list in later studies). In addition, the successful use of tomato RFLP probes for mapping the potato genome suggested a high level of sequence similarity among the tomato and potato genomes.

The first intraspecific linkage map for potato was developed with RFLP markers in a mapping population derived from a cross between two diploid *S. tuberosum* lines (Gebhardt et al. 1989). Markers were developed from a set of random potato cDNA clones, potato genomic clones, and potato cDNA clones for specific genes. A total of 141 markers were assigned to 12 linkage groups with a total map distance of 690 cM. Subsequently, more potato- and tomato-derived RFLP markers were added to the potato linkage map (Gebhardt et al. 1991). This extended the total map length to 1,034 cM with 304 mapped loci. Linkage analysis was also conducted on a second intraspecific mapping population (Gebhardt et al. 1991). Similar marker order was found for this mapping population, allowing integration of potato maps constructed in different genetic backgrounds. In addition, a tomato linkage map was constructed with RFLP markers derived from potato DNA (Gebhardt et al. 1991). Based on intraspecific marker analysis between potato and tomato linkage maps, a uniform nomenclature was developed for tomato and potato chromosomes. Tanksley et al. (1992) further enriched the potato genetic map with RFLP, isozyme, and cleaved amplified polymorphic sequence (CAPS) markers, thus generating a dense genetic map on which a marker can be found, on average, every 0.7 cM distance. This higher map density also revealed two additional chromosomal inversions in potato compared to tomato in linkage groups 11 and 12 of potato (Tanksley et al. 1992).

A second wave of linkage mapping studies appeared when Jacobs et al. (1995) added another 92 molecular markers to the potato linkage map. Molecular markers were developed from a potato leaf cDNA library and from genomic sequences flanking integration sites for T-DNA or the *Ac* transposable element. The same mapping population was then used to analyze linkage among 264 amplified fragment length polymorphism (AFLP) markers (van Eck et al. 1995). This was the first reported use of AFLP markers in potato. The AFLP map was integrated with the RFLP map of Jacobs et al. (1995).

PCR-based simple sequence repeat (SSR) markers were utilized for map construction in potato by Milbourne et al. (1998). Fifty-five SSR markers were placed on the potato linkage map relative to previously mapped RFLP markers. SSR markers were developed based on either published potato sequences in GenBank or EMBL databases, or from a potato cDNA library screened with tri- or tetra-nucleotide repeats. At the time, only 451 potato sequences were available for use by Milbourne et al. (1998). Seven years later, Feingold et al. (2005) reported development of SSR markers from a collection of more than 150,000 publicly available potato expressed sequence tag (EST) sequences. In total, 57 SSR markers were mapped in two diploid *S. tuberosum* mapping populations (Feingold et al. 2005).

Linkage mapping efforts in potato culminated in an ultra-high density (UHD) genetic map of 10,365 AFLP markers (van Os et al. 2006). Marker analysis was performed on a mapping population obtained from diploid *S. tuberosum* lines. A bin mapping strategy was used to obtain the UHD linkage maps. Each bin was described as a group of markers that was separated by a single recombination event from the adjacent bin. Such high marker density meant that average distance between markers was smaller than the average insert size of the potato bacterial artificial chromosome (BAC) library. In practice, this would facilitate chromosomal landing for cloning genes of interest. Furthermore the UHD map was subsequently integrated with a BAC-based potato physical map. This integrated resource is currently being employed to facilitate whole-genome sequencing of potato (van Os et al. 2006; *www.potatogenome.net*).

Ever since the publication of the first potato map, various mapping studies have attempted to map genes of known function (Bonierbale et al. 1988; Jacobs et al. 1995; Chen et al. 2001). Continuing this trend, a potato transcriptome map with 700 cDNA-AFLP markers directly established genomic location of 700 DNA fragments with putative biological function (Ritter et al. 2008). Transcriptome mapping in combination with the UHD map can potentially facilitate positional cloning of genes of interest.

Development of whole genome sequence for potato is likely to significantly change the nature of research in potato biology. A primary focus of research might shift more towards functional genomics, with extensive genetic screens in mutagenized populations for finding gene function, and proteomic and metabolomic analysis. In this changing research environment, linkage mapping would have a role in phenotype to genotype association, quantitative trait loci (QTL) discovery, genetic and metabolomic network determination (systems biology), and for allele mining in cultivated and wild germplasm (Jansen and Nap 2001; Nordborg and Weigel 2008; Keurentjes 2009).

Linkage mapping in potato, although essential to understanding basic potato biology, is not an end in itself. At a biological level, linkage mapping in potato has revealed genome structure of potato and has documented synteny with tomato and other *Solanum* species (Bonierbale et al. 1988; *www.sgn.cornell.edu/*). But, the true utility of linkage mapping is its facilitation of potato improvement through its contributions to marker-assisted selection and transgenic approaches. To this effect, linkage mapping efforts to date have provided molecular markers linked to specific traits, especially disease resistance traits (see Chapter 5), but utilization of those markers for marker-assisted selection still remains a work in progress (Rickert et al. 2003; Bormann et al. 2004; Achenbach et al. 2009) (see Chapter 3). Linkage mapping has facilitated positional cloning of genes for underlying traits of interest (Song et al. 2003; Foster et al. 2009) (see Chapter 8), but deployment of transgenic cultivars is still challenged by public opinion and regulatory hurdles (see Chapter 12).

4.3 Evolution of Marker Types: RFLPs to SNPs

Molecular marker systems are based on variation in protein (allozymes) or DNA [RFLPs, SSRs, single nucleotide polymorphisms (SNPs)] sequence. Variation in DNA sequence can be detected by hybridization (RFLPs), PCR (SSRs, AFLPs), or by direct sequencing (SNPs). RFLP markers were the pioneer DNA-based marker technology and were used to develop early potato maps (Bonierbale et al. 1988; Gebhardt et al. 1989). In fact, most of the mapping studies in potato have used RFLP as the marker system of choice, but more recent studies have employed PCR-based markers, for example AFLPs (van Os et al. 2006; Ritter et al. 2008; Table 4-1).

4.3.1 RFLP Markers

RFLP markers are co-dominant markers that detect sequence differences among specific DNA fragments (Botstein et al. 1980). DNA probes of specific sequence are labeled with radioactive isotopes or phosphorescent dyes and hybridized to genomic DNA that has been cut with specific restriction enzymes, separated on an agarose gel and immobilized on a cellulose or nylon filter. Sequence differences detected by hybridization are the result of loss or gain of restriction sites and insertion or deletion of blocks of DNA within a fragment. RFLP marker technology has its limitations, namely, requirement of large amounts of DNA for Southern blotting and labeling of probes with radioactive isotopes or costly phosphorescent labels.

Table 4-1 Summary of significant linkage maps for potato and its wild relatives.

Mapping population type	Parent	Number of progeny	Marker type and number	LOD	Linkage groups	Map length (cM)	Reference
F_1	*S. phureja* x [diploid *S. tuberosum* x *S. chacoense*]	65	RFLP (134)		12	606	Bonierbale et al. 1988
BC_1	Diploid *S. tuberosum*	67	RFLP (263)		12	690	Gebhardt et al. 1989
BC_1 F_1	diploid *S. tuberosum*, same as Gebhardt et al. 1989 *S. tuberosum* ssp. *tuberosum x* (*S. tuberosum x S. spegazzinii*)	67 100	RFLP (230) RFLP (83)		12	1,034	Gebhardt et al. 1991
BC_1	(Diploid *S. tuberosum* x *S. berthaulii*) x *S. berthaulii*	155	RFLP (228) CAPS, Isozyme, and morphological (32)	3.0	12	684	Tanksley et al. 1992
BC_1	Diploid *S. tuberosum* clones	67	RFLP (175) Isozyme (8) Morphological (10)	≥3.0	12	1,120	Jacobs et al. 1995
BC_1	Diploid *S. tuberosum* clones	68	AFLP (264), mapped relative to 193 markers in Jacobs et al. 1995	≥3.0	12	1,170	Van Eck et al. 1995
BC_1 F_1	Same as Gebhardt et al. 1991 Diploid *S. tuberosum*	67 91	SSR (55)	5.0	12	879	Milbourne et al. 1998
BC_1 F_1	Same as Gebhardt et al. 1991 Diploid *S. tuberosum* clones Diploid *S. tuberosum* x *S. chacoense*	67 90-150 49	RFLP (26) SCAR and CAPS (50)		12	--	Chen et al. 2001
F_1	*S. phureja (phu) x* diploid *S. tuberosum* (dih-*tbr*)	246	RAPD (170) AFLP (456) SSR (31)	>3.0	12	*phu* 987.4 dih-*tbr* 773.7	Ghislain et al. 2001 Ghislain et al. 2004

Table 4-1 **contd....**

Table 4-1 **contd....**

Mapping population type	Parent	Number of progeny	Marker type and number	LOD	Linkage groups	Map length (cM)	Reference
BC_1	(Diploid *S. tuberosum* x *S. berthaulii*) x *S. berthaulii* (Diploid *S. tuberosum* x *S. berthaulii*) x *S. tuberosum*		SSR (56)	≥3.0	12		Feingold et al. 2005
F_1	Diploid *S. tuberosum* lines	136	AFLP (10,365)	>4.0	12	Maternal 751 Paternal 773	van Os et al. 2006
F_1	Diploid *S. tuberosum* lines	90	cDNA – AFLP (700)		12	795	Ritter et al. 2008
BC_2	(*S. bulbocastanum* + *S. tuberosum*) somatic hybrid backcrossed to *S. tuberosum*	62	RFLP (51)		12	--	Brown et al. 1996
BC_1	(*S. pinnatisectum* x *S. cardiophyllum*) x *S. cardiophyllum*	115	RFLP (99)	3.0	12	683.9	Kuhl et al. 2001
F_1	*S. bulbocastanum* accessions	92	DArT (439)	3.0	12	403	Bradeen et al. (unpublished)

4.3.2 PCR-based Markers

Subsequently, PCR-based marker technologies such as SSR and AFLP were developed, thus significantly lowering the amount of DNA required for genetic analysis and omitting or minimizing the use of radioactive label. SSR markers are based on DNA polymorphisms among tandem repeats of specific DNA motifs. SSR markers are highly polymorphic and generally evenly distributed in the genome (Schlötterer 2004). For AFLP markers, genomic DNA is digested with restriction enzymes and restriction fragments are ligated to short DNA fragments of known sequence referred to as adapters. Multiple rounds of PCR are performed on the ligated DNA using primers homologous to the adapter. This results in reduced complexity of the DNA mixture. Ultimately, PCR products are separated on polyacrylamide gels or capillaries and scored for presence/absence of polymorphisms using either associated radioactive or phosphorescent labels or direct DNA staining. SSR markers, like RFLPs, are generally co-dominant in nature, while AFLP markers are mostly dominant in nature. While SSR markers require knowledge of genome sequence for PCR primer design, AFLP markers can be applied to any genome without a priori sequence information. Analysis of a genome with a single AFLP primer combination can yield 100 or more amplicons, a varying percentage of which might be polymorphic and thus useful for linkage mapping. Both SSR and AFLP markers have been used for mapping in potato (Jacobs et al. 1995; van Eck et al. 1995; van Os et al. 2006; Ritter et al. 2008).

4.3.3 DArT Markers

Diversity Array Technology (DArT) (Jaccoud et al. 2001; *www.diversityarrays.com/*) is a high-throughput and community accessible marker technology that appears to be gaining popularity among potato researchers. A DArT array consists of a microarray slide spotted with DNA fragments from a group of related species or a group of genotypes from within a species. DNA fragments are in fact a representation of the entire genome (a genomic representation) of each member of the group, obtained in a specific manner through restriction digestion and cloning. For linkage mapping, genomic representations from genotypes of a mapping population are hybridized to the DArT array. Marker segregation data are deduced from the presence or absence of hybridization at each microarray spot.

DArT markers offer several advantages relative to RFLP, SSR, and AFLP. Like AFLP, DArT markers require no a priori genome knowledge; the hybridization-based detection of DArT markers guarantees marker identity across populations or species, like RFLP; and DArT markers are high-throughput, reproducible, and cost effective. The current potato DArT array enables researchers to simultaneously survey approximately 16,000 potential markers at a cost as low as a few cents per data point, depending on the level of polymorphism present within a population. In addition, each DArT marker corresponds to a cloned DNA fragment that can be sequenced for comparative in silico mapping or development of PCR-based diagnostic markers. DArT markers have been used for linkage mapping and genetic diversity analysis in wheat, barley, *Arabidopsis*, rice, cassava, pigeonpea, and eucalyptus (Lezar et al. 2004; Wittenberg et al. 2005; Xia et al. 2005; Semagn et al. 2006; Wenzl et al. 2006; Xie et al. 2006; Yang et al. 2006). On going collaborative efforts between the James Bradeen (University of Minnesota, USA) and Domenico Carputo (University of Naples, Italy) laboratories are expected to yield DArT-based linkage maps for the disease resistant wild potatoes *S. bulbocastanum*, *S. commersonii*, and *S. pinnatisectum* in the near future (see below).

4.3.4 SNP Markers

SNP markers represent another high-throughput marker technology. De novo generated or existing genomic sequence information from multiple genotypes is analyzed to establish nucleotide variation for a species. Once SNP markers are identified and validated, more genotypes can then be scanned for SNPs using microarray based methods. SNP markers can be used for generating genetic maps, genome-wide association mapping, analyzing genetic diversity, phylogenetic analysis, and marker assisted selection (Rickert et al. 2003; Simko et al. 2004; Batley and Edwards 2007). Currently large-scale SNP marker assays are under development for potato (*solcap.msu.edu/objectives.shtml*). Limited locus-specific SNPs are also available for mapping in potato (Rickert et al. 2003; Simko et al. 2004).

4.4 Potato Mapping Populations: Structures and Strategies

Solanum tuberosum is a tetraploid ($2n = 4x = 48$), highly heterozygous, outcrossing plant species (see Chapter 2). Although linkage analysis is possible in tetraploid potato (Luo et al. 2001), linkage mapping efforts have largely focused on haploid (diploid $2n = 2x = 24$) populations owing to simpler marker and gene segregation patterns. But a significant hurdle remains, as haploids are generally male-sterile and most hybrids between haploid *S. tuberosum* and diploid potato species are self-incompatible. This hinders the development of true breeding lines, useful for downstream utilization of linkage data, and near-isogenic lines, which are generally more ideally suited for some genetic linkage analyses.

Initial mapping in potato was done in the F_1 or BC_1 generations using diploid *S. tuberosum* mapping populations (Bonierbale et al. 1988; Gebhardt et al. 1989, 1991). Importantly, Gebhardt et al. (1991) established that for an allogamous species like *S. tuberosum*, sufficient heterozygosity exists among parents to allow efficient linkage mapping in F_1 populations. Use of F_1 populations, however, necessitated development of suitable statistical treatments (Ritter et al. 1990). For example, in RFLP analysis, restriction fragments can be identified as dominant or co-dominant. Dominant restriction fragments are scored as present or absent. If presence of a fragment is coded as A then its absence can be coded as O. Progeny from the cross between parental genotypes AO and OO would be expected, based on Mendelian genetics, to segregate 1:1 for the presence:absence of a dominant RFLP marker (Fig. 4-1). In contrast, progeny for the cross between heterozygous parental genotypes AO and AO would be expected to segregate 3:1 for the presence:absence of a dominant RFLP marker. Ritter et al. (1990) described the theoretical framework for mapping such single fragment dominant markers along with co-dominant allelic markers. However,

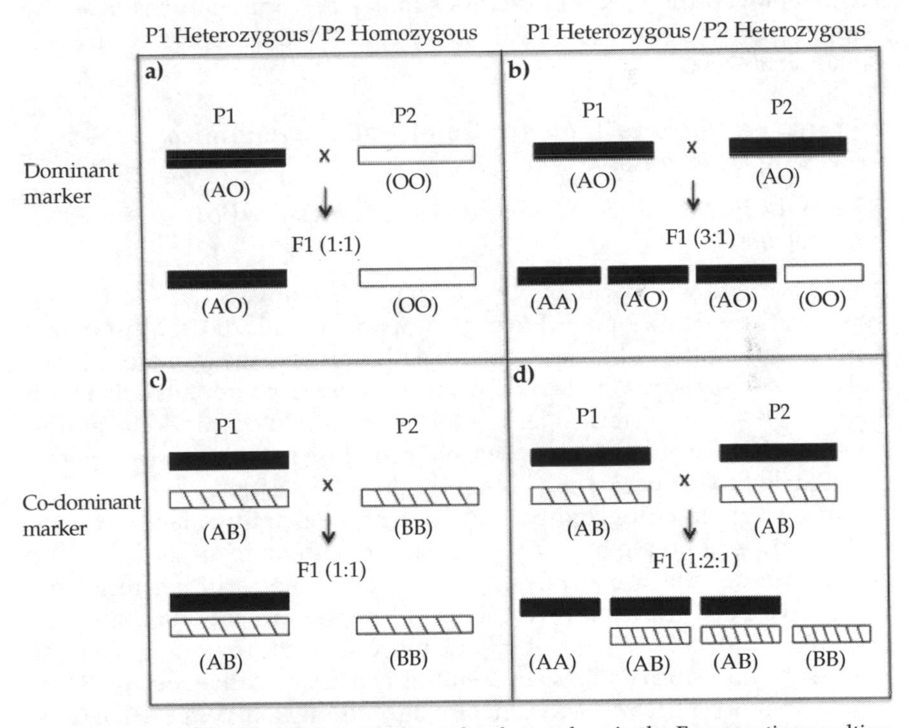

Figure 4-1 Expected segregation ratios for molecular markers in the F_1 generation resulting from a cross between two diploid parents. (a) Parent P_1 is heterozygous (AO, where A is the presence and O is the absence of allele A); parent P_2 is homozygous (OO); marker is dominant in nature. (b) Both parents are heterozygous (AO); marker is dominant in nature. (c) Parent P_1 is heterozygous for alleles A and B (AB); parent P_2 is homozygous (BB); marker is co-dominant in nature. (d) Both parents are heterozygous (AB); marker is co-dominant in nature.

mapping software available at the time (e.g., MAPMAKER) was unable to combine 1:1 and 3:1 markers into one data set and thus special software was developed for generating linkage groups using such markers (Ritter et al. 1990; Gebhardt et al. 1991). This approach to mapping in F_1 populations obtained from heterozygous parents later matured into the pseudo-testcross strategy and was instrumental for linkage analysis in not only potato but also many other outcrossing species (Grattapaglia and Sederoff 1994; Maliepaard et al. 1997). The pseudo-testcross strategy involves generating single parent maps using dominant markers. Each maternal and paternal linkage group is then integrated or aligned using allelic or co-dominant markers, yielding a unified map comprised of all markers.

The pseudo-test cross mapping strategy was employed by Jacobs et al. (1995) in constructing two parental maps and aligning those maps with the help of co-dominant markers. Analysis was done using the software JoinMap (Stam 1993). Thereafter most linkage mapping studies in potato have employed the pseudo-testcross strategy and JoinMap has been the mapping software of choice (Milbourne et al. 1998; Feingold et al. 2005; van Os et al. 2006).

4.5 Linkage Maps as Tools for Comparative Genomics: Accessing Biodiversity

4.5.1 Wild Potato Species as Sources of Genes for Potato Improvement

Potato is grown worldwide, but modern potato cultivars have originated from a narrow genetic pool (Love 1999; Spooner et al. 2005). Most potato cultivars lack sufficient genetic potential to overcome the numerous biotic and abiotic stresses that they encounter in the diverse environments in which they are grown. A historical and notorious example of this is the almost complete absence of resistance to late blight disease in the entire Irish potato crop that led to the great Irish Potato Famine of the 1840s.

In contrast to cultivated potato, some of the primitive landraces and wild relatives of potato have evolved to cope with many of the biotic and abiotic stresses that potato faces. A number of the resulting resistance traits have been transferred to potato through conventional breeding, beginning with the transfer of late blight resistance from *S. demissum* in the early 20th century (Hawkes 1990). It is estimated that up to 50% of modern potato cultivars have genetic contribution from *S. demissum* (Ross 1986). But merely 15 out of approximately 200 wild potato species have been used in the improvement of potato cultivars worldwide (Ross 1986). While potential exists to utilize wild germplasm more broadly for potato improvement, difficulty in obtaining hybrids between wild species and *S. tuberosum* and the requirement of several years of backcrossing to remove undesirable wild genetic material has limited the number of wild species used by potato breeders.

Among promising donor species of genes for potato improvement are species comprising the potato tertiary genepool. According to the Genepool Concept of Harlan and de Wet (1971), the tertiary genepool encompasses wild species distantly related to a crop species. Gene flow between the crop species and the tertiary genepool species is restricted by genetic incongruity. Thus, genes harbored by tertiary genepool species cannot be accessed for crop improvement using traditional breeding strategies (Chapter 1). For potato, tertiary genepool species include the diploid, 1EBN (Johnston et al. 1980) tuber-bearing species *S. bulbocastanum* and *S. commersonii.*

Solanum bulbocastanum has been reported as a source of disease resistance genes against late blight (Naess et al. 2000), nematodes (Austin et al. 1993; Di Vito et al. 2003), blackleg (Lojkowska and Kelman 1989; Chen et al. 2003), bacterial wilt, early blight, PLRV, PVM, ring rot, Verticillium, and wart (USDA GRIN database, *www.ars-grin.gov/*). *Solanum commersonii* has remarkable cold tolerance (Palta and Li 1979) and is a reported source of resistance to nematodes (Di Vito et al. 2003), early blight (Jansky et al. 2008), Verticillium (Bastia et al. 2000), TEV (Valkonen 1997), PVX (Tozzini et al. 1991), bacterial wilt, blackleg, and ring rot (USDA GRIN database, *www.ars-grin.gov/*).

Like other tertiary genepool species, neither *S. bulbocastanum* nor *S. commersonii* can be directly crossed with cultivated potato. Resistance genes from *S. bulbocastanum* and *S. commersonii* have been accessed for potato improvement using complicated, multi-species bridge crosses (Hermsen and Ramanna 1973; Carputo et al. 1997) and somatic hybridization (Cardi et al. 1993; Helgeson et al. 1998). Both of these strategies are time consuming and require multiple rounds of selection to recover the desirable phenotypic attributes of the cultivated potato. Cloning genes from wild potato species and transforming into commercially prominent potato cultivars is a more straightforward strategy. In recent years, four late blight resistance genes have been cloned from *S. bulbocastanum* (Song et al. 2003; van der Vossen et al. 2003, 2005; Lokossou et al. 2009; Oosumi et al. in press) and transformed potato cultivars have been tested for late blight resistance (Bradeen et al. 2009) and yield attributes (Halterman et al. 2008). While successful, gene cloning itself has historically been a time consuming, costly endeavor.

4.5.2 Documenting and Capitalizing upon Genome Synteny

There is documented and extensive genome synteny between potato and tomato (Bonierbale et al. 1988). Grube et al. (2000) demonstrated that phenotypic resistance is conditioned by corresponding genome locations in tomato and potato. More recently, Mazourek et al. (2009) demonstrated that homologs of disease resistance genes also occupy corresponding genome locations among Solanaceous species. Capitalizing upon this observation, late blight resistance gene *R3a* was cloned from potato using a comparative genomics approach and genome information from tomato (Huang et al. 2005).

Under the premise of comparative genomics, information about genome location of candidate genes in wild potato species can be deduced from genomics resources being developed for the cultivated potato and tomato. Given that complete genome sequence of cultivated potato and tomato will soon be publicly available, the potential for employing comparative genomics approaches to access genes harbored by wild relatives of potato remains high. Of course, applying comparative genomics approaches in this manner is predicated upon the assumption of knowledge of the differences between the genome structure of wild donor species (e.g., *S. bulbocastanum* or *S. commersonii*) and reference species (e.g., cultivated potato or tomato).

The potato tertiary genepool species comprise a clade that is phylogenetically distinct from cultivated potato or tomato. To date, the only whole genome linkage map developed for any tertiary genepool species is that of Brown et al. (1996). These researchers mapped in *S. bulbocastanum* 48 RFLP markers that had been previously mapped in cultivated potato. Although, based upon marker order, Brown et al. (1996) concluded a significant degree of synteny exists between *S. bulbocastanum* and cultivated potato, four of the mapped RFLP markers were assigned to different linkage groups in these two species. Brown et al. (1996) speculate this may be due to disruption of genome synteny. For the efficient application of comparative genomics approaches to gene mapping and cloning, a more robust understanding of genome structure for *S. bulbocastanum* and *S. commersonii* vis-à-vis cultivated potato and tomato is required.

In an effort to widen the scope of comparative genomics for improvement of potato, the Bradeen and Carputo laboratories have developed linkage maps of *S. bulbocastanum* and *S. commersonii* with plans to map in *S. pinnatisectum* as well. These maps will serve as scaffolds upon which candidate disease resistance genes will be located. These maps are also starting points for in silico genome structure comparisons with cultivated potato and tomato, utilizing burgeoning genome sequences.

4.5.3 Genome Mapping in S. bulbocastanum

First, a mapping population of 110 *S. bulbocastanum* F_1 individuals was analyzed with AFLP marker technology (Syverson 2007). Fingerprinting of parental genotypes provided 155 polymorphic fragments. Thirty-six of the polymorphic fragments did not segregate amongst the progeny. Of the remaining fragments, some were discarded, as their parental origin was unclear or their amplification was unreliable. An additional 23 fragments did not fit the expected 1:1 (present:absent) Mendelian segregation ratio. Ultimately, 58 AFLP fragments segregating at a 1:1 ratio were mapped. The resulting linkage map was of low density and was further limited by segregation distortion and marker clustering. In the end, only one third of the polymorphic AFLP fragments could be mapped. The relatively low proportion of mappable markers was attributed to the dominant nature of AFLP markers and a highly heterozygous F_1 population (Syverson 2007).

Results from AFLP mapping led to the development of the potato DArT array described above. Like AFLP, DArT technology yields data that are mostly dominant in nature. However, the high throughput potential for DArT suggested that mapping in an F_1 potato population could nevertheless be successful and efficient. Additionally, the potential to sequence mapped DArT clones for in silico analyses makes this marker technology ideal for comparative genomics applications. Here we detail our efforts to create a DArT-based linkage map for *S. bulbocastanum* and to compare genome structure of this wild species to those of cultivated potato and tomato.

4.5.3.1 A DArT-based Linkage Map for S. bulbocastanum

An F_1 *S. bulbocastanum* mapping population (PT29 x G15) comprised of 92 individuals was genotyped using the ~16,000 DArT element microarray in collaboration with Andrzej Kilian and DArT Pty. Ltd. (Canberra, Australia). Up to 1,332 arrayed elements hybridized differentially to genome representations generated from the progeny. Due to the dominant nature of DArT markers, two Mendelian marker segregation ratios are possible in an F_1 population generated from highly heterozygous parents (1:1 or 3:1; Fig. 4-1). Of the 1,332 arrayed elements, more than 800 segregated at either 1:1 or 3:1 ($P > 0.05$) and were considered for subsequent linkage analysis.

Linkage amongst markers was analyzed with JoinMap 4.0 (Kyazma B. V., Wageningen, The Netherlands) using the pseudo-testcross mapping strategy described above. First, two parental maps were generated for *S. bulbocastanum* genotypes PT29 and G15 at LOD 3.0 using only data segregating at a 1:1 ratio. The linkage map for parent PT29 consists of 366 markers distributed on 10 linkage groups with a total map length of 451 cM. The linkage map for parent G15 consists of 37 markers distributed on 7 linkage groups with total map length of 187 cM.

Markers with a 3:1 segregation ratio carry genetic information from both parents and are thus able to integrate the two parental maps. In the next stage of mapping, 73 markers segregating at a 3:1 ratio were included in the dataset. Subsequent mapping generated 12 linkage groups consisting of 439 markers with a total map length of 403 cM (Fig. 4-2). The final map represents 303 distinct loci after accounting for redundant markers. We now have a medium-density genetic linkage map for *S. bulbocastanum* with an average distance of 1.3 cM between adjacent loci.

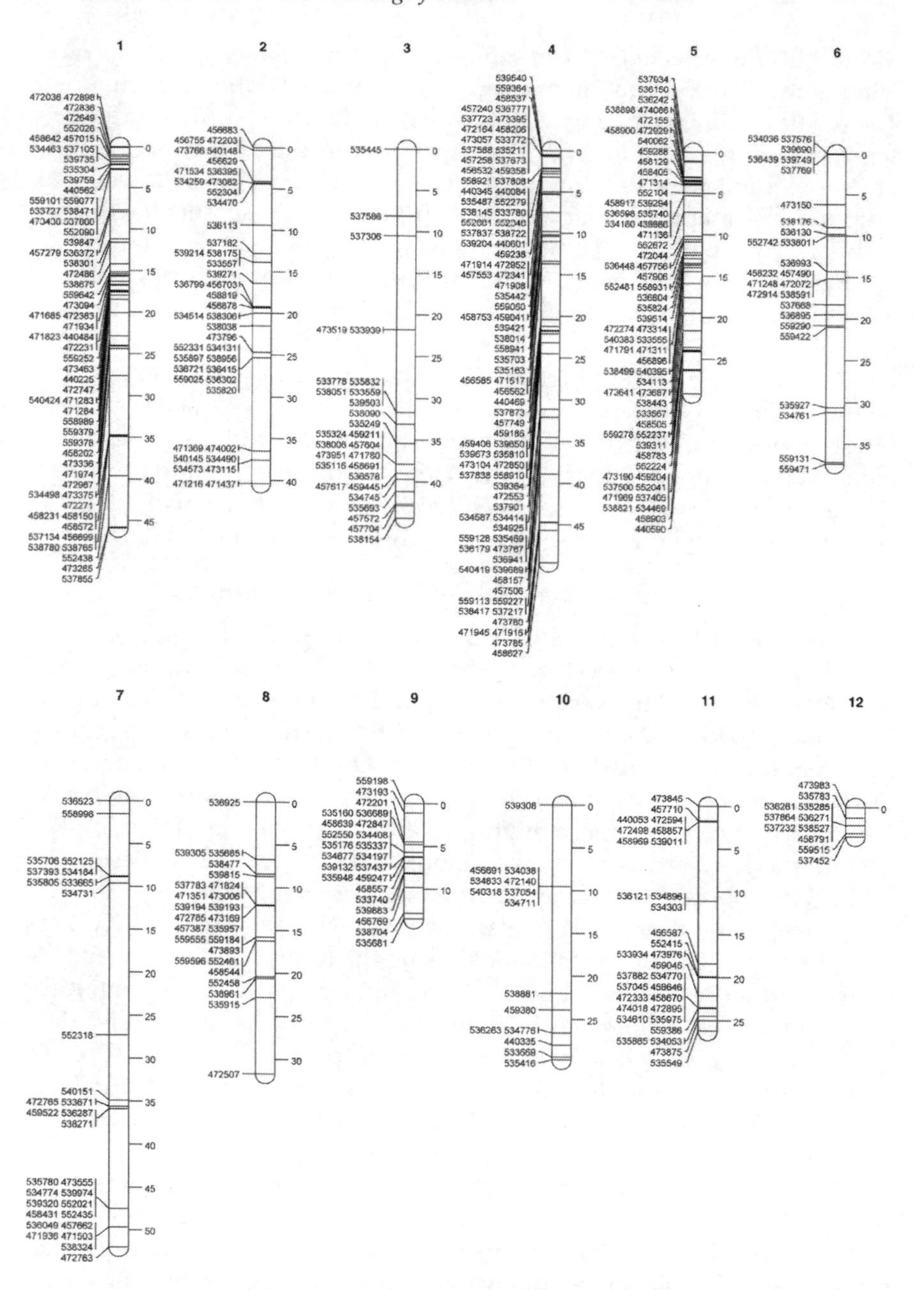

Figure 4-2 A genetic linkage map of *Solanum bulbocastanum* based on DArT markers. Marker labels appear to the left of each linkage group. Distance between markers in cM is shown to the right of each linkage group. Linkage groups are numbered from 1 through 12; labels do not necessarily correspond to chromosome numbers.

4.5.3.2 Multi-species Comparison of Genome Structure Based on Mapped DArT Marker Sequences

Ongoing genome sequencing efforts for potato (*www.potatogenome.net*) and tomato (Mueller et al. 2009) have yielded 91 Mb and 133 Mb of sequence information, respectively, as of the writing of this chapter (Table 4-2). *Solanum tuberosum* is being sequenced by an international consortium (the Potato Genome Sequencing Consortium; PGSC). The PGSC is sequencing two genotypes: RH89-039-16 (RH), a diploid, heterozygous potato variety, and DM1-3 516R44 (DM), a doubled monoploid. Initially, efforts of the PGSC focused exclusively on sequencing RH by applying a BAC-by-BAC sequencing strategy. With changing sequencing technologies, these efforts have been more recently complemented by whole genome shotgun sequencing using both 454 GS FLX and Illumina GA2 technologies.

The genome of tomato (*S. lycopersicum* L.) is being sequenced by an international consortium of 10 countries as part of the "International Solanaceae Genome Project (SOL): Systems Approach to Diversity and Adaptation" initiative. The tomato genome project is based on a BAC-by-BAC approach that aims to determine a high-quality genome sequence from the euchromatic, gene-rich portion of the genome (van der Hoeven 2002). This effort is expected to be near completion by 2010 (Mueller et al. 2009) and will provide a reference for Solanaceae and euasterid comparative genomics. In the meantime, a tomato whole-genome shotgun effort has also been undertaken to complete the genome sequencing. The current status of potato and tomato sequencing projects is summarized in Table 4-2.

Table 4-2 Current status of genome sequencing efforts for tomato and potato (as of September 2009).

Chromosome	*Solanum lycopersicum*		*Solanum tuberosum*	
	Sequenced BACs	Total nucleotides (Mb)	Sequenced BACs	Total nucleotides (Mb)
0	70	8.20	24	2.16
1	19	2.52	243	32.13
2	196	21.98	3	0.41
3	33	3.35	0	0.00
4	141	14.57	56	7.25
5	69	6.95	174	21.90
6	157	17.28	138	18.66
7	160	15.65	0	0.00
8	190	21.43	0	0.00
9	94	9.27	55	7.26
10	4	0.49	0	0.00
11	24	2.95	14	1.85
12	81	8.88	1	0.13
Total	1238	133.53	708	91.75

Although not yet complete, the ongoing genome sequencing efforts in potato and tomato can support validation of linkage analysis in *S. bulbocastanum*. An in silico comparative mapping approach was adopted to assign *S. bulbocastanum* linkage groups to specific chromosomes and to explore genome structure between this species and cultivated potato and tomato. DArT clones associated with each mapped marker were sequenced and sequences were compared to burgeoning whole genome (potato) or euchromatic (tomato) reference sequences. For this comparison, identity and coverage thresholds were set at 70% and were used to define the location of DArT sequences on the reference genomes. Importantly, our analyses confirmed a high degree of synteny between *S. bulbocastanum* and both potato and tomato (Fig. 4-3). Although our current analysis is limited by the amount of potato and tomato genome sequence that is publicly available, we anticipate completion of the reference genome sequencing projects in the near future will enable more detailed comparisons. An ongoing project that complements the *S. bulbocastanum* mapping efforts described here will result in the localization of approximately 100 candidate disease resistance genes onto the scaffold *S. bulbocastanum* map. This effort will allow extension of our cross species in silico comparisons to gene locations.

We find that comparative mapping combined with DArT marker technology is a powerful tool for generating structural genomics resources, especially for species lacking prior genome structure characterization. For *S. bulbocastanum*, mapping resources have progressed from a sparse RFLP map to a marker-rich map with an average distance of 1.3 cM between marker loci. We are also generating DArT marker-based genetic maps for other tertiary genepool species (*S. commersonii* and *S. pinnatisectum*) and expect similar marker densities. Making use of tomato and cultivated potato reference genome sequence, detailed comparison of genome structure across these three wild species will be completed and eventual map merging may be possible. We predict that these efforts will ultimately provide the tools and knowledge needed for efficient access to agriculturally useful genes harbored by wild species.

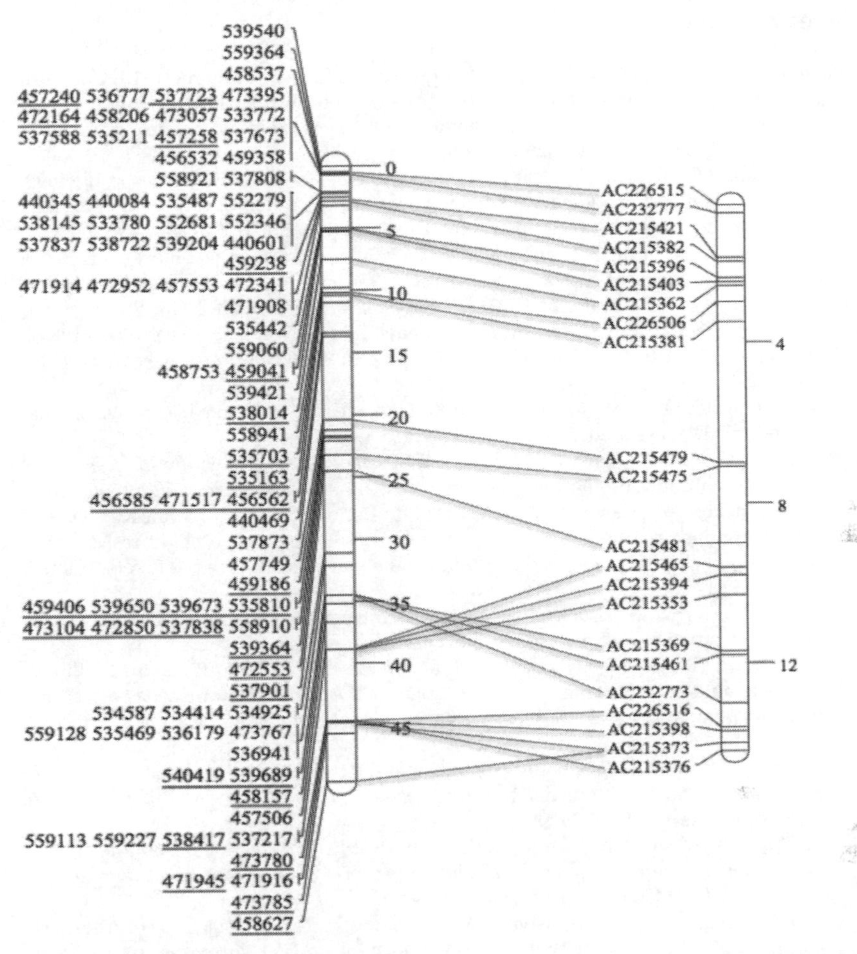

Figure 4-3 Synteny between *Solanum bulbocastanum* linkage group 4 and chromosome 2 of *Solanum lycopersicum*. Underlined DArT marker labels represent markers mapped in *S. bulbocastanum* with best sequence match to tomato genomic (BAC) sequences. Numbers to the right of the *S. bulbocastanum* map represent genetic map distance in cM. Numbers to the right of the *S. lycopersicum* map represent physical distance in Mb.

Acknowledgements

The development of potato DArT resources, construction of the DArT-based *S. bulbocastanum* linkage map, and comparison of genome structure between *S. bulbocastanum* and cultivated potato and tomato was supported by the National Research Initiative of USDA's National Institute of Food and Agriculture.

References

Achenbach U, Paulo J, Ilarionova E, Lübeck J, Strahwald J, Tacke E, Hofferbert HR, Gebhardt C (2009) Using SNP markers to dissect linkage disequilibrium at a major quantitative trait locus for resistance to the potato cyst nematode *Globodera pallida* on potato chromosome V. Theor Appl Genet 118: 619–629.

Austin S, Pohlman JD, Brown CR, Mojtahedi H, Santo GS, Douches DS, Helgeson JP (1993) Interspecific somatic hybridization between *Solanum tuberosum* L. and *S. bulbocastanum* Dun. as a means of transferring nematode resistance. Am Potato J 70: 485–495.

Bastia T, Carotenuto N, Basile B, Zoina A, Cardi T (2000) Induction of novel organelle DNA variation and transfer of resistance to frost and Verticillium wilt in *Solanum tuberosum* through somatic hybridization with 1EBN *S. commersonii.* Euphytica 116: 1–10.

Bately J, Edwards D (2007) SNP applications in plants. In: NC Oraguzie, EHA Rikkerink, SE Gardiner, NH De Silva (eds) Association Mapping in Plants. Springer, New York, USA, pp 95–102.

Bernatzky R, Tanksley SD (1986) Toward a saturated linkage map in tomato based on isozymes and random cDNA sequences. Genetics 112: 887–898.

Bonierbale MW, Plaisted RL, Tanksley SD (1988) RFLP Maps based on a common set of clones reveal modes of chromosomal evolution in potato and tomato. Genetics 120: 1095–1103.

Bormann CA, Rickert AM, Castillo Ruiz RA, Paal J, Lubeck J, Strahwald J, Buhr K, Gebhardt C (2004) Tagging quantitative trait loci for maturity-corrected late blight resistance in tetraploid potato with PCR-based candidate gene markers. Mol Plant-Microbe Interact 17: 1126–1138.

Botstein D, White RL, Skolnick M, Davis RW (1980) Construction of a genetic linkage map in man using restriction fragment length polymorphisms. Am J Hum Genet 32: 314–331.

Bradeen JM, Iorizzo M, Mollov DS, Raasch J, Colton Kramer L, Millett BP, Austin-Phillips S, Jiang J, Carputo D (2009) Higher copy numbers of the potato RB transgene correspond to enhanced transcript and late blight resistance levels. Mol Plant-Microbe Interact 22: 437–446.

Brown CR, Yang CP, Mojtahedi H, Santo GS, Masuelli R (1996) RFLP analysis of resistance to Columbia root-knot nematode derived from *Solanum bulbocastanum* in a BC2 population. Theor Appl Genet 92: 572–576.

Cardi T, D'Ambrosio F, Consoli D, Puite KJ, Ramulu KS (1993) Production of somatic hybrids between frost tolerant *Solanum commersonii* and *S. tuberosum*: characterization of hybrid plants. Theor Appl Genet 87: 193–200.

Carputo D, Barone A, Cardi T, Sebastiano A, Frusciante L, Peloquin SJ (1997) Endosperm Balance Number manipulation for direct *in vivo* germplasm introgression to potato from a sexually isolated relative (*Solanum commersonii* Dun.). Proc Natl Acad Sci USA 94: 12013–12017.

Chen Q, Kawchuck LM, Lynch DR, Goettel MS, Fujimoto DK (2003) Identification of late blight, Colorado potato beetle, and Blackleg resistance in three Mexican and two South American wild 2x (1EBN) *Solanum* species. Am J Potato Res 80: 9–19.

Chen X, Salamini F, Gebhardt C (2001) A potato molecular-function map for carbohydrate metabolism and transport. Theor Appl Genet 102: 284–295.

Di Vito M, Greco N, Carputo D, Frusciante L (2003) Response of wild and cultivated potato clones to Italian populations of root knot nematodes *Meloidogyne* spp. Nematropica 33: 65–72.

Feingold S, Lloyd J, Norero N, Bonierbale M, Lorenzen J (2005) Mapping and characterization of new EST-derived microsatellites for potato (*Solanum tuberosum* L.). Theor Appl Genet 111: 456–466.

Foster SJ, Park T, Pel M, Brigneti G, Sliwka J, Jagger L, van der Vossen E, Jones JDG (2009) *Rpi-vnt1. 1*, a *Tm-2*^{2} Homolog from *Solanum venturii*, confers resistance to potato late blight. Mol Plant-Microbe Interact 22: 589–600.

Gebhardt C, Ritter E, Debener T, Schachtschabel U, Walkemeier B, Uhrig H, Salamini F (1989) RFLP analysis and linkage mapping in *Solanum tuberosum*. Theor Appl Genet 78: 65–75.

Gebhardt C, Ritter E, Barone A, Debener T, Walkemeier B, Schachtschabel U, Kaufmann H, Thompson RD, Bonierbale MW, Ganal MW, Tanksley SD, Salamini F (1991) RFLP Maps of potato and their alignment with the homeologous tomato genome. Theor Appl Genet 83: 49–57.

Ghislain M, Trognitz B, del R. Herrera M, Solis J, Casallo G, Vásquez C, Hurtado O, Castillo R, Portal L, Orrillo M (2001) Genetic loci associated with field resistance to late blight in offspring of *Solanum phureja* and *S. tuberosum* grown under short-day conditions. Theor Appl Genet 103: 433–442.

Ghislain M, Spooner DM, Rodriguez F, Villamon F, Nunez J, Vásquez C, Waugh R, Bonierbale M (2004) Selection of highly informative and user-friendly microsatellites (SSRs) for genotyping of cultivated potato. Theor Appl Genet 108: 881–890.

Grattapaglia D, Sederoff R (1994) Genetic linkage maps of *Eucalyptus grandis* and *Eucalyptus urophylla* using a pseudo-testcross: mapping strategy and RAPD markers. Genetics 137: 1121–1137.

Grube RC, Radwanski ER, Jahn M (2000) Comparative genetics of disease resistance within the Solanaceae. Genetics 155: 873–887.

Halterman D, Kramer LC, Weilgus S, Jiang J (2008) Performance of transgenic potato containing the late blight resistance gene *RB*. Plant Dis 92: 339–343.

Harlan JR, de Wet JMJ (1971) Toward a rational classification of cultivated plants. Taxon 20: 509–517.

Hawkes JG (1990) The Potato: Evolution, biodiversity and genetic resources. Belhaven Press, London, UKHelgeson JP, Pohlman JD, Austin S, Haberlach GT, Wielgus SM, Ronis D, Zambolim L, Tooley P, McGrath JM, James RV, Stevenson WR (1998) Somatic hybrids between *Solanum bulbocastanum* and potato: a new source of resistance to late blight. Theor Appl Genet 96: 738–742.

Hermsen JGT, Ramanna MS (1973) Double-bridge hybrids of *Solanum bulbocastanum* and cultivars of *Solanum tuberosum*. Euphytica 22: 457–466.

Hougas RW (1957) A haploid plant of the potato variety Katahdin. Nature 180: 1209–1210.

Hougas RW, Peloquin SJ (1958) The potential of potato haploids in breeding and genetic research. Am Potato J 35: 701–707.

Huang S, van der Vossen EAG, Kuang H, Vleeshouwers VGAA, Zhang N, Borm TJA, van Eck HJ, Baker B, Jacobsen E, Visser RGF (2005) Comparative genomics enabled the isolation of the *R3a* late blight resistance gene in potato. Plant J 42: 251–261.

Jaccoud D, Peng K, Feinstein D, Kilian A (2001) Diversity Arrays: a solid state technology for sequence information independent genotyping. Nucl Acids Res 29: e25.

Jacobs JME, van Eck HJ, Arens P, Verkerk-Bakker B, te Lintel Hekkert B, Bastiaanssen HJM, El Kharbotly A, Pereira A, Jacobsen E, Stiekema WJ (1995) A genetic map of potato (*Solanum tuberosum*) integrating molecular markers, including transposons, and classical markers. Theor Appl Genet 91: 289–300.

Jansen RC, Nap JP (2001) Genetical genomics: the added value from segregation. Trends Genet 17: 388–391.

Jansky SH, Simon R, Spooner DM (2008) A test of taxonomic predictivity: Resistance to Early Blight in wild relatives of cultivated potato. Phytopathology 98: 680–687.

Johnston SA, den Nijs TPM, Peloquin SJ, Hanneman RE (1980) The significance of genic balance to endosperm development in interspecific crosses. Theor Appl Genet 57: 5–9.

Keurentjes JJB (2009) Genetical metabolomics: closing in on phenotypes. Curr Opin Plant Biol 12: 223–230.

Kuhl JC, Hanneman RE, Havey MJ (2001) Characterization and mapping of *Rpi1*, a late blight resistance locus from diploid (1EBN) Mexican *Solanum pinnatisectum*. Mol Genet Genom 265: 977–985.

Lezar S, Myburg AA, Berger DK, Wingfield MJ, Wingfield BD (2004) Development and assessment of microarray-based DNA fingerprinting in *Eucalyptus grandis*. Theor Appl Genet 109: 1329–1336.

Lojkowska E, Kelman A (1989) Screening of seedlings of wild *Solanum* species for resistance to Bacterial Stem Rot caused by Soft Rot Erwinias. Am Potato J 66: 379–390.

Lokossou AA, Park T, van Arkel G, Arens M, Ruyter-Spira C, Morales J, Whisson SC, Birch PRJ, Visser RGF, Jacobsen E, van der Vossen EAG (2009) Exploiting knowledge of R/Avr genes to rapidly clone a new LZ-NBS-LRR family of late blight resistance genes from potato linkage group IV. Mol Plant-Microbe Interact 22: 630–641.

Love SL (1999) Founding clones, major contributing ancestors, and exotic progenitors of prominent North American potato cultivars. Am J Potato Res 76: 263–272.

Luo ZW, Hackett CA, Bradshaw JE, McNicol JW, Milbourne D (2001) Construction of a genetic linkage map in tetraploid species using molecular markers. Genetics 157: 1369–1385.

Maliepaard C, Jansen J, van Ooijen JW (1997) Linkage analysis in a full-sip family of an outbreeding plant species: Overview and consequences of applications. Genet Res 70: 237–250.

Mazourek M, Cirulli ET, Collier SM, Landry LG, Kang BC, Quirin EA, Bradeen JM, Moffett P, Jahn M (2009) The fractionated orthology of *Bs2* and *Rx/Gpa2* and the comparative genomics model of disease resistance in the Solanaceae. Genetics 182: 1351–1364.

Milbourne D, Meyer RC, Collins AJ, Ramsay LD, Gebhardt C, Waugh R (1998) Isolation, characterization and mapping of simple sequence repeat loci in potato. Mol Gen Genet 259: 233–245.

Mueller LA, Lankhorst KR, Tanksley ST, Giovannoni JJ, White R, Vrebalov J, Fei Z, van Eck J, Buels R, Mills AA, et al. (2009) A snapshot of the emerging tomato genome sequence. Plant Genome 2: 78–92.

Naess SK, Bradeen JM, Wielgus SM, Haberlach GT, McGrath JM, Helgeson JP (2000) Resistance to late blight in *Solanum bulbocastanum* is mapped to chromosome 8. Theor Appl Genet 101: 697–704.

Nordborg M, Weigel D (2008) Next-generation genetics in plants. Nature 456–720–723.

Oosumi T, Rockhold DR, Maccree MM, Deahl KL, McCue KF, Belknap WR (in press) Gene *Rpi-bt1* from *Solanum bulbocastanum* confers resistance to late blight in transgenic potatoes. Am J Pot Res.

Palta JP, Li PH (1979) Frost hardiness in relation to leaf anatomy and natural distribution of several *Solanum* species. Crop Sci 19: 665–671.

Rickert AM, Kim JH, Meyer S, Nagel A, Ballvora A, Oefner PJ, Gebhardt C (2003) First-generation SNP/InDel markers tagging loci for pathogen resistance in the potato genome. Plant Biotechnol J 1: 399–410.

Ritter E, Gebhardt C, Salamini F (1990) Estimation of recombination frequencies and construction of RFLP linkage maps in plants from crosses between heterozygous parents. Genetics 125: 645–654.

Ritter E, de Galarreta JIR, van Eck HJ, Sanchez I (2008) Construction of a potato transcriptome map based on the cDNA-AFLP technique. Theor Appl Genet 116: 1003–1013.

Ross H (1986) Potato Breeding—problems and perspectives. Adv Plant Breed 13: 11–13.

Schlötterer C (2004) The evolution of molecular markers—just a matter of fashion? Nat Rev Genet 5: 63–69.

Semagn K, Bjornstad A, Skinnes H, Maroy AG, Tarkegne Y, William M (2006) Distribution of DArT, AFLP, and SSR markers in a genetic linkage map of a doubled-haploid hexaploid wheat population. Genome 49: 545–55.

Simko I, Haynes KG, Ewing EE, Costanzo S, Christ BJ, Jones RW (2004) Mapping genes for resistance to *Verticillium albo-atrum* in tetraploid and diploid potato populations using haplotype association tests and genetic linkage analysis. Mol Genet Genom 271: 522–531.

Song J, Bradeen JM, Naess SK, Raasch JA, Haberlach GT, Wielgus SM, Liu J, Kuang H, Michelmore RW, Austin-Phillips S, Buell CR, Helgeson JP, Jiang J (2003) Gene *RB* from

Solanum bulbocastanum confers broad spectrum resistance against potato late blight pathogen *Phytophthora infestans*. Proc Natl Acad Sci USA 100: 9128–9133.

Spooner DM, McLean K, Ramsay G, Waugh R, Bryan GJ (2005) A single domestication for potato based on multilocus amplified fragment length polymorphism genotyping. Proc Natl Acad Sci USA 102: 14694–14699.

Stam P (1993) Construction of integrated genetic linkage maps by means of a new computer package: JoinMap. Plant J 3: 739–744.

Syverson RL (2007) Towards development of molecular resources in disease resistant wild potato *Solanum bulbocastanum*, including a linkage map and SNP based markers. MS Thesis, Univ of Minnesota, St. Paul, MN, USA.

Tanksley SD, Ganal MW, Prince JP, de Vicente MC, Bonierbale MW, Broun P, Fulton TM, Giovannoni JJ, Grandillo S, Martin GB (1992) High density molecular linkage maps of the tomato and potato genomes. Genetics 132: 1141–1160.

Tozzini AC, Ceriani MF, Saladrigas MV, Hopp HE (1991) Extreme resistance to infection by potato virus X in genotypes of wild tuber-bearing *Solanum* species. Potato Res 34: 317–324.

Valkonen JPT (1997) Novel resistances to four potyviruses in tuber-bearing potato species, and temperature sensitive expression of hypersensitive resistance to potato virus Y. Ann Appl Biol 130: 91–104.

van der Hoeven R, Ronning C, Giovannoni J, Martin G, Tanksley S (2002) Deductions about the number, organization, and evolution of genes in the tomato genome based on analysis of a large expressed sequence tag collection and selective genomic sequencing. Plant Cell 14: 1441–1456.

van der Vossen EA, Sikkema A, te Lintel-Hekkert B, Gros J, Stevens P, Muskens M, Wouters D, Pereira A, Stiekema W, Allefs S (2003) An ancient R gene from the wild potato species *Solanum bulbocastanum* confers broad-spectrum resistance to *Phytophthora infestans* in cultivated potato and tomato. Plant J 36: 867–882.

van der Vossen EA, Gros J, Sikkema A, Muskens M, Wouters D, Wolters P, Pereira A, Allefs S (2005) The *Rpi-blb2* gene from *Solanum bulbocastanum* is an *Mi-1* gene homolog conferring broad-spectrum late blight resistance in potato. Plant J 44: 208–222.

van Eck HJ, Voort JR, Draaistra J, van Zandvoort P, van Enckevort E, Segers B, Peleman J, Jacobsen E, Helder J, Bakker J (1995) The inheritance and chromosomal localization of AFLP markers in a non-inbred potato offspring. Mol Breed 1: 397–410.

van Os H, Andrzejewski S, Bakker E, Barrena I, Bryan G, Caromel B, Ghareeb B, Isidore E, de Jong W, van Koert P, Lefebvre V, Milbourne D, Ritter E, van der Voort JN, Rousselle-Bourgeois F, van Vliet J, Waugh R, Visser RGF, Bakker J, van Eck HJ (2006) Construction of a 10,000-marker ultradense genetic recombination map of potato: providing a framework for accelerated gene isolation and a genomewide physical map. Genetics 73: 1075–1087.

Wenzl P, Li H, Carling J, Zhou M, Raman H, Paul E, Hearnden P, Maier C, Xia L, Caig V, Ovesna J, Cakir M, Poulsen D, Wang J, Raman R, Smith KP, Muehlbauer GJ, Chalmers KJ, Kleinhofs A, Huttner E, Kilian A (2006) A high-density consensus map of barley linking DArT markers to SSR, RFLP and STS loci and phenotypic traits. BMC Genom 7: 206.

Wittenberg AHJ, van der Lee T, Cayla C, Kilian A, Visser RGF, Schouten HJ (2005) Validation of the high-throughput marker technology DArT using the model plant *Arabidopsis thaliana*. Mol Genet Genom 274: 30–39.

Xia L, Peng K, Yang S, Wenzl P, de Vicente MC, Fregene M, Kilian A (2005) DArT for high-throughput genotyping of cassava (*Manihot esculenta*) and its wild relatives. Theor Appl Genet 110: 1092–1098.

Xie Y, McNally K, Li CY, Leung H, Zhu YY (2006) A high-throughput genomic tool: diversity array technology complementary for rice genotyping. J Integr Plant Biol 48: 1069–1076.

Yang S, Pang W, Ash G, Harper J, Carling J, Wenzl P, Huttner E, Zong X, Kilian A (2006) Low level of genetic diversity in cultivated pigeonpea compared to its wild relatives is revealed by Diversity Arrays Technology (DArT). Theor Appl Genet 113: 585–595.

5

Mapping and Tagging of Simply Inherited Traits

Joseph C. Kuhl

ABSTRACT

The rediscovery of Gregor Mendel's 1866 work, *Experiments in Plant Hybridization*, in the early 1900s established the standard for simply inherited traits. One requirement of a simply inherited trait is the presence of a single gene or tightly grouped cluster of linked genes inherited as a single unit. For any given population structure, simply inherited traits segregate among progeny at expected Mendelian segregation ratios, assuming minimal influence by biological or environmental variables, and typically involve a completely dominant phenotype that can be qualitatively scored. This chapter reviews studies involving simply inherited traits that precede subsequent tagging with molecular markers, with particular focus on disease resistance to viruses, nematodes, and late blight. Disease related phenotypes, such as immunity and the hypersensitive response, tend to provide clear, qualitative evaluation facilitating mapping of simply inherited traits. Additionally, the gene-for-gene hypothesis, although somewhat dated, provides guidance for setting up simply segregating populations by utilizing appropriate resistant host genotypes while challenging with a pathogen carrying the cognate avirulence gene. A wide range of techniques has been utilized to tag resistance loci. One technique, bulked segregant analysis, has been extensively used to streamline marker identification and facilitate the tagging of disease resistant loci. Emerging molecular data will need to be mined and new techniques applied to continue the identification of molecular tags to valuable traits.

Keywords: disease resistance, inheritance, mapping, Mendelian, molecular markers

---

Department of Plant, Soil, and Entomological Sciences, University of Idaho, P.O. Box 442339, Moscow, ID 83844, USA; e-mail: *jkuhl@uidaho.edu*

5.1 Introduction

The current era of gene inheritance analysis began with the rediscovery in the early 1900s of Gregor Mendel's 1866 work, *Experiments in Plant Hybridization* (Mendel 1866). Since then a plethora of plant inheritance studies has occurred, in some of which the trait of interest was simply inherited. In general, for a trait to be defined as simply inherited, a single gene or a tightly grouped cluster of linked genes, inherited as a unit, must be responsible. At a whole plant level, this is revealed by phenotypic segregation ratios among progeny. For any given population structure, simply inherited traits segregate among progeny at expected Mendelian segregation ratios, assuming minimal influence by biological or environmental variables. Simply inherited traits typically include those in which a completely dominant phenotype can be qualitatively scored. Potato has many examples of simply inherited traits.

It is not intented in this chapter to document all studies in potato that describe simply inherited traits. The focus of this chapter is inheritance and tagging, therefore particular attention is paid to inheritance studies that precede subsequent tagging/mapping studies. Furthermore, the scope of traits reviewed is limited to host resistance evaluations. This is in part to narrow the focus of this review, but it also reflects the underlying complexity of inheritance studies. Analysis of inheritance in segregating progeny requires clearly defined and reproducible phenotypes (whether qualitative or quantitative), and resistance traits have a propensity to meet this requirement (Fig. 5-1). Obviously, accurate resistance phenotype determination is critical in establishing simple inheritance ratios—complex or multifaceted phenotypes will invariably complicate the study of genetic segregation ratios. In some cases resistance traits aid the observer by providing clear differentiation between resistant and susceptible qualitative phenotypes. For example, in the case of immunity, a plant cannot be infected by a given pathogen (Agrios 1988), resulting in little or no visible response from the host when challenged by the pathogen. In the case of virus infections, this might be called extreme resistance when the plant does not become infected under any circumstances (Matthews 1992). The complementary susceptible response shows signs and symptoms of successful infection. At the simplest, immunity produces a black and white qualitative phenotype. In other cases, many plants have evolved a resistance mechanism which initiates cell death when a pathogen is detected, resulting in necrosis at the site of infection and frequently resulting in visible necrotic spots—the so-called hypersensitive response. Both immunity and the hypersensitive response result in clear and distinct resistance responses that differ from susceptible responses. Of course, resistance phenotypes may be less obvious, and phenotypes can be very complex and continuous

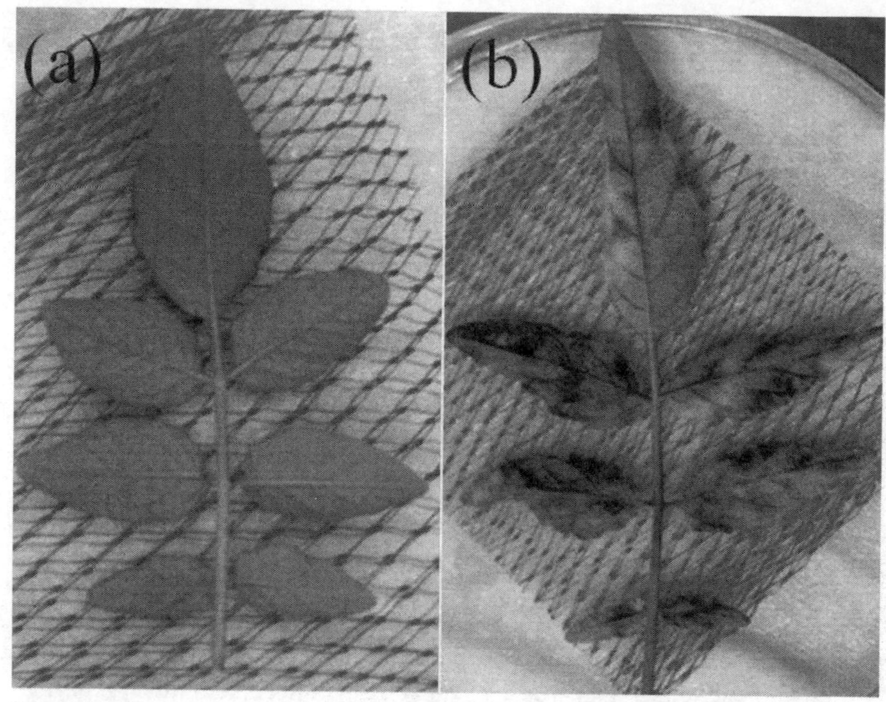

Figure 5-1 Backcross 1 progeny from a cross between resistant clone *S. pinnatisectum* (PI# 253214) and susceptible clone *S. cardiophyllum* (PI# 347759), backcrossed to the susceptible parent, 6 days post inoculation with *P. infestans* isolate MSU96 (US-8, A2 mating type), incubated at 18°C under 12 hour light/dark cycles, (a) resistant and (b) susceptible phenotypes.

Color image of this figure appears in the color plate section at the end of the book.

in nature, in which case quantitative analysis becomes appropriate (see Chapter 6). In some cases, quantitative trait loci (QTL) analysis identifies a single major contributing locus that ultimately results in simple segregation ratios. Simple resistance phenotypes are also suited to bulked segregant analysis (BSA). Although many variations of BSA have been used (see below for specific examples), the basic concept involves pooling progeny with similar phenotypes (resistance or susceptibility) for combined analysis. This might be done to streamline molecular marker screening to identify polymorphic or linked markers.

The plethora of simply inherited resistance traits reported in potato literature reflects plant research priorities resulting from the gene-for-gene hypothesis, which describes interactions between flax and the flax rust fungus, *Melampsora lini* (Flor 1956, 1971). This concept hypothesizes that for a plant resistance gene, there is a corresponding pathogen avirulence gene that leads to an incompatible reaction (resistance). In general, the host resistance allele is dominant, while the pathogen virulence allele is recessive. Therefore, through careful selection of germplasm and pathogen strains, a one-to-one genetic relationship can be established. Further exploitation of the model can be made through controlled crosses such that only a single host resistance gene segregates amongst progeny. Although the molecular/biochemical basis of the gene-for-gene hypothesis has since been refined (van der Biezen and Jones 1998; Jones and Dangl 2006), from a classical genetics standpoint the model still provides constructive guidance.

The cultivated potato, *Solanum tuberosum*, is tetraploid ($2n = 4x = 48$) with tetrasomic inheritance. Asexual propagation allows for high levels of heterozygosity to be maintained, in part due to severe inbreeding depression after repeated selfing. Each locus may have one to four different alleles. A locus can be homozygous or quadruplex (nulliplex) (e.g., R1R1R1R1), or one of four heterozygous genotypes: simplex/triplex (e.g., R1R2R2R2), duplex/duplex (e.g., R1R1R2R2), duplex/simplex/simplex (e.g., R1R1R2R3), or 4-times simplex (e.g., R1R2R3R4). The simplest research circumstance for studying a single dominant resistant gene results from a cross between a heterozygous, simplex resistant parent (Rrrr) and a nulliplex susceptible parent (rrrr). In this case, segregation ratios among progeny remain relatively simple. However, ratios become more complex for almost any deviations from this situation (e.g., duplex or triplex genotypes, incomplete dominance, or multiple-gene systems).

It is possible to conduct genetic analysis of the potato at the diploid level, in part because many wild *Solanum* species are diploid. Alternatively, anther culture or parthenogenesis can be used to generate diploid lines from tetraploids (Hougas et al. 1958; Rokka et al. 1996); from here on in this chapter, these will be called haploids. Diploid genetic analysis in the potato is no different than in other diploid species in which partially heterozygous parents generate segregating F_1 offspring from controlled crosses. In research directed at mapping specific traits, some studies have started with one highly heterozygous parent carrying the dominant allele and a second parent homozygous for the recessive allele. The simplest genetic model for the inheritance of disease resistance is based on the monogenic inheritance of a single completely dominant resistance allele. In diploid systems, a resistant, heterozygous parent (Rr) crossed to a susceptible, homozygous recessive (rr) will yield progeny in which resistance segregates at a ratio of 1: 1 (resistant to susceptible). In the case of haploids derived from tetraploid cultivars, a segregation ratio of 1: 1 is expected for one resistance allele that is in simplex and 5: 1 when in duplex, assuming unbiased segregation. A chi-square (χ^2) statistical test is frequently performed to determine how well the observed ratio fits the expected ratio. If the chi-square value has a corresponding probability (P) > 0.05 it is considered a good fit. The following sections will provide examples of simple inheritance and tagging in potatoes focusing on virus, nematode and late blight resistance. Single resistance genes tagged with molecular markers are listed in Table 5-1.

Table 5-1 Selected single resistance genes in potato tagged with molecular markers.

Gene[a]	Phenotype	Species of Origin	Chromosome Location	Marker Name (Marker Type)	Reference
Virus Resistance					
Ry_{sto}[b]	Extreme Resistance to PVY^{N}	*S. stoloniferum*	XI	M45, M5 (AFLP)	Brigneti et al. 1997
Ny-1	Hypersensitive Resistance to PVY^{O}, PVY^{N}, PVY^{NIN}	cultivar Rywal	IX	$SC895_{1139}$(SCAR)	Szajko et al. 2008
Ry_{che}	Extreme Resistance to PVY^{N}	*S. chacoense*	IX	38–530 (RAPD)	Hosaka et al. 2001
Ry_{che}	Extreme Resistance to PVY^{O}, PVY^{N}	*S.chacoense*	IX	CT220 (RFLP)	Sato et al. 2006
Ry_{adg}	Extreme Resistance to PVY^{O}	*S. tuberosum* ssp. *andigena*	XI	TG508 (RFLP)	Hamalainen et al. 1997
Ry_{adg}	Extreme Resistance to PVY^{O}, PVY^{N}	*S. tuberosum* ssp. *andigena*	XI	ADG1, ADG2 (RFLP)	Hamalainen et al. 1998
Ry_{adg}	Extreme Resistance to PVY^{O}	*S. tuberosum* sp. *andigena*	XI	ADG2/Bbv1 (CAPS)	Sorri et al. 1999
Ry_{adg}	Extreme Resistance to PVY	*S. tuberosum* ssp. *andigena*	XI	RYSC3, RYSC4 (SCAR)	Kasai et al. 2000
Ny_{tbr}	Hypersensitive Resistance to PVY^{O}	*S. tuberosum* ssp. *tuberosum*	IV	TG506 (RFLP)	Celebi-Toprak et al. 2002
Ry_{sto}	Extreme Resistance to PVY^{O}, PVY^{C}, PVY^{N}, PVY^{NIN}	*S. stoloniferum*	XII	E+ATC/M+CGC-110, STM0003 (AFLP, SSR)	Song et al. 2005
Ry_{sto}	Extreme Resistance to PVY	*S. stoloniferum*	XII	YES3-3A, YES3-3B (STS)	Song & Schwarfischer 2008
$Ry\text{-}f_{sto}$	Extreme Resistance to PVY^{N}	*S. stoloniferum*	XII	$GP122_{718}$ (CAPS)	Flis et al. 2005
$Ry\text{-}f_{sto}$	Extreme Resistance to PVY	*S. stoloniferum*	XII	$GP122_{564}$ (CAPS)	Witek et al. 2006
Ns	Hypersensitive Resistance to PVY	*S. tuberosum* ssp. *andigena*	VIII	$SC811_{260}$ (CAPS)	Witek et al. 2006
Ry_{sto}	Extreme Resistance to PVY^{NIN}	*S. stoloniferum*	XII	M1 (RAPD)	Cernak et al. 2008a
Ry_{sto}	Extreme Resistance to PVY^{NIN}	*S. stoloniferum*	XII	$SCAR_{YSTO4}$ (SCAR)	Cernak et al. 2008b
Rx1	Extreme Resistance to PVX	*S. tuberosum* ssp. *andigena*	XII	CP60 (RFLP)	Ritter et al. 1991
Rx2	Extreme Resistance to PVX	*S. acaule*	V	GP21 (RFLP)	Ritter et al. 1991
Nb	Hypersensitive Resistance to PVX	*S. tuberosum* ssp. *tuberosum*	V	SPUD237 (CAPS)	De Jong et al. 1997
Ns	Hypersensitive Resistance to PVS	*S. tuberosum* ssp. *andigena*	VIII	$OPE15_{550}$ (RAPD)	Marezewski et al. 1998
Ns	Hypersensitive Resistance to PVS	*S. tuberosum* ssp. *andigena*	VIII	$UBC811_{660}$ (ISSR)	Marezewski et al. 2001
Ns	Hypersensitive Resistance to PVS	*S. tuberosum* ssp. *andigena*	VIII	$SCG17_{321}$ (SCAR)	Marezewski et al. 2001
Ns	Hypersensitive Resistance to PVS	*S. tuberosum* ssp. *andigena*	VIII	CP16 (STS)	Marezewski et al. 2002

Nematode	Resistance				
H1	Resistance to *G. rostochiensis*, Ro1	*S. tuberosum* ssp. *andigena*	V	CP113 (RFLP)	Gebhardt et al. 1993
H1	Resistance to *G. rostochiensis*, Ro1	*S. tuberosum* ssp. *andigena*	V	CD78 (RFLP)	Pineda et al. 1993
H1	Resistance to *G. rostochiensis*, Ro1	*S. tuberosum* ssp. *andigena*	V	CM1, EM15 (AFLP)	Bakker et al. 2004
Gro1	Resistance to *G. rostochiensis*, Ro1	*S. spegazzinii*	VII	CP51(c) (RFLP)	Barone et al. 1990
Gro1	Resistance to *G. rostochiensis*, Ro1	*S. spegazzinii*	VII	CP56, CP51 (c), GP516 (c) (RFLP)	Ballvora et al. 1995
GroV1	Resistance to *G. rostochiensis*, Ro1	*S. vernei*	V	SCAR-U14, SCAR-XO2 (SCAR)	Jacobs et al. 1996
Gpa2	Resistance to *G. pallida*, Pa2	*S. tuberosum* ssp. *andigena*	XII	GP34 (RFLP)	Rouppe van der Voort et al. 1997
Gpa2	Resistance to *G. pallida*, Pa2	*S. tuberosum* ssp. *andigena*	XII	77R, 45L, 221R, IPM4a (CAPS)	Rouppe van der Voort et al. 1999b
Rmc1	Resistance to *M. chitwoodi, M. fallax, M. hapla*	*S. bulbocastanum*	XI	M39b, CT182 (CAPS)	Rouppe van der Voort et al. 1999a
R_{mc1} (*blb*)	Resistance to *M. chitwoodi*	*S. bulbocastanum*	XI	19319, 56F6, 39E18, 524F16, 406L19 (STS)	Zhang et al. 2007
NR	Resistance to *M. fallax*	*S. sparsipilum*	XII	IPM4 (CAPS)	Bakari et al. 2006
Late Blight	**Resistance**				
R1	Resistance to *P. infestans*	*S. demissum*	V	GP21 (RFLP)	Leonards-Schippers et al. 1992
R1	Resistance to *P. infestans*	*S. demissum*	V	AFLP, AFLP2 (AFLP)	Meksem et al. 1995
R3	Resistance to *P. infestans*	*S. demissum*	XI	TG105a, GP185, GP250(a) (RFLP)	El-Kharbotly et al. 1994
R1	Resistance to *P. infestans*	*S. demissum*	V	TD_s259 (RFLP)	El-Kharbotly et al. 1996a
R5, R7	Resistance to *P. infestans*	*S. demissum*	XI	185 (a), GP250(a) (RFLP)	El-Kharbotly et al. 1996b
R2	Resistance to *P. infestans*	*S. demissum*	IV	ACC/CAT-535, ACT/CAC-189, AGC/CCA-369 (AFLP)	Li et al. 1998
RB	Resistance to *P. infestans*	*S. bulbocastanum*	VIII	CT88, OPG02-625 (RFLP, RAPD)	Naess et al. 2000
	Resistance to *P. infestans*	*S. pinnatisectum*	VII	TG20A (RFLP)	Kuhl et al. 2001
Rpil	Resistance to *P. infestans*	*S. tuberosum* ssp.	X	TG63 (RFLP)	Ewing et al. 2000
Rpl-ber	Resistance to *P. infestans*	*tuberosum*	X	mCT240 (STS)	Rauscher et al. 2006
Rpl-ber	Resistance to *P. infestans*	*S. berthaultii*	X	CT214 (STS)	Park et al. 2009
Rpi-ber1, Rpi-ber2	Resistance to *P. infestans*	*S. berthaultii*	IX	TG328 (CAPS)	Smilde et al. 2005
		S. mochiquense			
Rpi-moc1	Resistance to *P. infestans*		IV	TH21 (SCAR)	Park et al. 2005b
Rpi-blb3		*S. bulbocastanum*			

[a]Genes listed order they appear in text.
[b]Probably Ry_{adg} (Valkonen et al. 2008).

5.2 Virus Resistance

Cockerham's 1970 article *Genetical Studies on Resistance to Potato Viruses X and Y* (Cockerham 1970) is a model example of using classical genetic evaluation to identify resistance loci and forms the basis of numerous future studies investigating the inheritance of virus resistance—in some cases resulting in mapping and tagging of Cockerham's original loci. Cockerham (1970) reported on resistance to potato viruses X, Y, and A (PVX, PVY, and PVA). Data were reported at the tetraploid level from cultivars of *S. tuberosum* ssp. *tuberosum* as well as from hybrids of *S. tuberosum* ssp. *andigena* and *S. acaule*. Inheritance was primarily tetrasomic, however disomic inheritance is reported for materials including *S. stoloniferum*, *S. demissum*, and *S. hougasii*. Certainly not all segregation ratios were simple, however several examples were presented in which a genotype carrying a dominant resistance allele in simplex crossed with a nulliplex susceptible resulted in 1:1 segregation ratios in the progeny. Different factors were compared in test crosses to determine linkage and allelic relationships, resulting in double and triple factor segregation ratios. Simply inherited genes in diploid material were reported in *S. chacoense* and *S. microdontum*. For example, Nx_{chc}, conditioning a hypersensitive reaction in response to PVX in *S. chacoense*, was identified, as was a similar or identical gene/allele in *S. microdontum*. Ny_{chc}, causing hypersensitive reactions against PVY and PVA with linkage to Nx_{chc} and Nx_{tbr}^{spl}, was reported in *S. chacoense* and *S. microdontum*. Cockerham (1970) reports on many more gene and linkage relationships than mentioned here.

Numerous studies have evaluated PVY resistance at the tetraploid level. A few are highlighted here that present simple segregation ratios and subsequently enable mapping of resistance loci using linked molecular markers. Brigneti et al. (1997) evaluated a tetraploid population in which the resistant parent carried Ry_{sto} in the simplex condition. F_1 progeny segregated 1:1, extreme resistance to susceptible, when challenged with PVY^N. The authors subsequently mapped Ry_{sto} to chromosome XI using amplified fragment length polymorphisms (AFLPs) [this result is questionable due to unreliable pedigree information (Gebhardt and Valkonen 2001), and Ry_{sto} is probably identical to Ry_{adg} (Valkonen et al. 2008)]. AFLP markers M5 and M45 co-segregated with resistance. Szajko et al. (2008) mapped a hypersensitive response to PVY^N and PVY^{NTN} strains using 200 F_1 tetraploids. This population segregated 93:107, hypersensitive resistant to susceptible, fitting a 1:1 genetic model ($\chi^2 = 0.98$, $P = 0.32$) for a single dominant gene, *Ny-1*. Utilizing BSA, an inter-simple sequence repeat (ISSR) marker was identified with linkage to resistance. This fragment was sequenced and sequence characterized amplified region (SCAR) marker $SC895_{1139}$ developed, linked to *Ny-1* at 0.5 cM on chromosome IX.

Hosaka et al. (2001) evaluated 92 F_1 plants from a cultivar by cultivar cross with extreme resistance to PVY originating from *S. chacoense* in the cultivar Konafubuki. Segregation within progeny was 48 resistant and 44 susceptible to PVY^N, indicating a single dominant gene, Ry_{chc}. Although random amplified polymorphic DNA (RAPD) marker 38–530 was shown to be loosely linked to resistance (16.3% recombination frequency), it could still be used to detect resistance from cultivar Konafubuki. Continuing this work, Sato et al. (2006) generated haploids from "Konafubuki". Seven resistant and five susceptible haploids were generated, consistent with a single dominant locus. Progeny from a cross between resistant and susceptible haploid genotypes segregated 50 resistant to 64 susceptible, not significantly different from 1:1 ($P > 0.10$). Using restriction fragment length polymorphism (RFLP) and RAPD markers, Ry_{chc} was mapped to the distal end of chromosome IX.

Valkonen et al. (1994), working at the diploid level, conducted graft inoculations to distinguish between necrotic and extreme resistance to PVY^o. This approach allowed them to identify Ry_{adg} and Ny_{adg}, extreme resistance and hypersensitive loci, respectively, in a cross between an *S. tuberosum* ssp. *andigena* resistant haploid and a susceptible *S. tuberosum* ssp. *andigena* hybrid haploid. F_1 progeny segregated 34:8:20, extreme resistance: necrotic resistance: susceptible, respectively ($0.05 < P < 0.1$), consistent with a two gene model with extreme resistance and hypersensitive response segregating independently, and extreme resistance epistatic to the hypersensitive response. The responsible loci were dubbed *Ry* and *Ny*, respectively. The progeny resulting from a cross of the resistant parent (*Ry/ry Ny/ny*) to the susceptible parent (*ry/ry ny/ny*) would therefore be expected to segregate 2:1:1 for extreme resistance: hypersensitive response: susceptible. The authors also presented results from a resistant haploid *S. stoloniferum* crossed with the above mentioned susceptible parent. Progeny from that cross segregated 1:1, resistant: susceptible, suggesting an *Ry/ry ny/ny* composition for the resistant parent.

Working with 54 of the 62 F_1 progeny studied by Valkonen et al. (1994), Hämäläinen et al. (1997) identified linkage between four RFLP markers from chromosome XI and resistance locus Ry_{adg}. These four markers were not associated with the hypersensitive response, providing further evidence that Ry_{adg} is inherited independently of Ny_{adg}. Tomato RFLP marker TG508 was closest to Ry_{adg} with a maximum estimated map distance of 2.0 cM. A test for independent assortment for TG508 resulted in $\chi^2 = 51.00$, $P < 0.005$. A screen of unrelated diploid and tetraploid potato lines accurately associated extreme resistance with the TG508 fragment except when derived from *S. phureja* or *S. brevidens*.

Hämäläinen et al. (1998) added 23 F_1 progeny to the 54 progeny reported by Hämäläinen et al. (1997). F_1 progeny segregated 51:14:12, extreme resistance: hypersensitive: susceptible when challenged with PVY^O. The 51 extreme resistance progeny were also resistant to PVY^N. The 14 hypersensitive and 12 susceptible progeny were susceptible to PVY^N. Fifty-six progeny were resistant to PVA. Of these, two displayed a hypersensitive reaction to PVY^O and three were susceptible to PVY^O. Of 21 progeny susceptible to PVA, 12 and nine displayed hypersensitivity and susceptiblity to PVY^O. The data suggested that PVA resistance is controlled by a single dominant locus, Ra_{adg}, linked to Ry_{adg} at a distance of approximately 6.8 cM, on chromosome XI. Two resistance gene-like (RGL) DNA fragments (ADG1 and ADG2), generated from polymerase chain reaction (PCR) primers developed by Leister et al. (1996), were identified that co-segregated both with RFLP markers from potato chromosome XI and extreme resistance to PVY. The primers of Leister et al. (1996) were designed to amplify RGLs by targeting gene regions conserved between the *N* gene of *Nicotiana glutinosa* and *RPS2* from *Arabidopsis thaliana*. Further analysis of ADG2 was conducted by Sorri et al. (1999). Sequence analysis of the ADG2 fragment from PVY-resistant and PVY-susceptible genotypes revealed 12 nucleotide differences. One of these differences was located in a predicted kinase-3a motif and resulted in the presence/absence of a *Bbv*I restriction endonuclease recognition site, allowing development of a cleaved amplified polymorphic sequence (CAPS) marker. Use of the ADG2/*Bbv*I marker accurately identified Ry_{adg} and is among the first examples of a PCR-based marker for selection of an important trait in potato. Kasai et al. (2000) used sequence data from ADG2 to design SCAR markers linked Ry_{adg}. Two SCARs, RYSC3 and RYSC4, were highly accurate in identifying Ry_{adg} among 103 genotypes tested (diploid and tetraploid lines), 24 of which were extremely resistant to PVY. RYSC3 was exclusively present in genotypes with Ry_{adg}. RYSC4 was also always present in genotypes carrying Ry_{adg}, however it was also present in four PVY-susceptible genotypes. Both SCAR markers only detected *Ry* from *S. tuberosum* ssp. *andigena* and not from other *Solanum* species.

Continuing their earlier work (Hämäläinen et al. 1998), Hämäläinen et al. (2000) next focused on PVA resistance. Working with the 78 diploid F_1 progeny detailed above, the authors sap-inoculated with two strains of PVA. The data suggested a recessive resistance gene, designated ra_{adg}, controlling/blocking vascular transport of PVA and not defeated by graft inoculation. A second, independently segregating dominant locus, linked to or possibly allelic to Ry_{adg}, confers a hypersensitive reaction to PVA, but did not prevent vascular movement. This was designated Na_{adg}. This two gene model replaced the formerly identified Ra_{adg} (Hämäläinen et al. 1998).

Celebi-Toprak et al. (2002) evaluated hypersensitivity to PVY^O in two backcross populations derived from an F_1 cross between a susceptible diploid *S. berthaultii* genotype and a resistant haploid of potato cultivar Saco. In one population, BCB, 73 progeny displayed mosaic symptoms with clearly positive ELISA values while 64 progeny developed necrotic local lesions on inoculated leaves and, frequently, top necrosis, fitting a 1:1 segregation ratio ($P = 0.44$) for a single dominant gene. The BCT population, however, was skewed toward the necrotic phenotype. The authors concluded that altogether the data are most consistent with a single dominant gene, Ny_{tbr}, from *S. tuberosum* controlling hypersensitivity to PVY. Ny_{tbr} was located on chromosome IV, between RFLP markers TG316 and TG208.

Song et al. (2005) evaluated extreme resistance to PVY originating from *S. stoloniferum* in the cultivar Assia. Fifty-seven haploid lines were generated from "Assia" and inoculated with a mixture of PVY^{O}, PVY^{C}, PVY^{N}, and PVY^{NTN}. Twenty-eight resistant and 29 susceptible lines were detected, consistent with a 1:1 ratio when tested with chi-square ($P = 0.05$). AFLP and simple sequenced repeat (SSR) markers mapped the single dominant gene, Ry_{sto}, to chromosome XII. Extreme resistance derived from *S. stoloniferum* was accurately identified in 106 potato varieties using AFLP (E+ATC/M+CGC-110) and SSR (STM0003) markers linked to Ry_{sto}. Song and Schwarzfischer (2008) developed two sequence tagged site (STS) markers for detecting Ry_{sto} based on AFLP markers identified by Song et al. (2005). Both markers, YES3-3A and YES3-3B, successfully identified extreme resistance to PVY in 188 European cultivars.

Analysis of a tetraploid hybrid F_1 population by Flis et al. (2005) for extreme resistance to isolate PVY^{N} Wi identified 87 resistant and 82 susceptible progeny, not significantly different from a 1:1 segregation ratio of a single dominant locus, $Ry\text{-}f_{sto}$. ISSR markers mapped extreme resistance to chromosome XII, with CAPS marker $GP122_{718}$ 1.2 cM from $Ry\text{-}f_{sto}$. A comparison of linked markers suggests that Ry_{sto} and $Ry\text{-}f_{sto}$, both of which impart extreme resistance to PVY, could be independent (Song et al. 2005). Witek et al. (2006) developed a multiplex protocol incorporating CAPS markers, $GP122_{564}$, linked to $Ry\text{-}f_{sto}$, and $SC811_{260}$ linked to *Ns* (see discussion below); resistance was reliably identified in 55 Polish cultivars.

Valkonen et al. (2008) evaluated a diploid population of 112 F_1 genotypes for extreme resistance to PVY originating from *S. stoloniferum*, using a *S. stoloniferum* accession different from that used by Song et al. (2005) and Flis et al. (2005). The F_1 population segregated 57:7:48 (extreme resistance: hypersensitive response: susceptible), fitting a single dominant gene model for extreme resistance, Ry_{sto}, with heterozygosity in the resistant parent. As with earlier results, extreme resistance is epistatic to the hypersensitive response, also controlled by a single dominant gene. CAPS and SSR markers from Song et al. (2005), Flis et al. (2005), and Witek et al. (2006) co-segregated with Ry_{sto}, and new CAPS markers were added. Marker distance from Ry_{sto} was estimated at approximately 15 cM.

Cernák et al. (2008a) mapped extreme resistance to PVY originating from *S. stoloniferum* in a tetraploid F_1 population from a cross between resistant cultivar White Lady and susceptible breeding line S440. A total of 195 progeny segregated 95:100, resistant and susceptible, fitting a 1:1 genetic model for a single dominant simplex allele crossed to a nulliplex. The Ry_{sto} locus was mapped to chromosome XII. Three RAPD markers were identified with linkage to Ry_{sto}; these were shown to accurately detect extreme resistance to PVY in cultivars when it originated from *S. stoloniferum*. Later work developed SCAR markers based on these linked RAPDs. *SCARysto4* reliably detected the presence of Ry_{sto} in 21 cultivars tested (Cernák et al. 2008b).

Working with two different diploid populations, including one reported by Barone et al. (1990), Ritter et al. (1991) evaluated extreme resistance to PVX. One population segregated 53 resistant to 64 susceptible progeny, fitting a 1:1 segregation ratio for a single dominant resistance gene, *Rx1*, which may have originated from *S. tuberosum* ssp. *andigena*. *Rx1* was mapped to chromosome XII using RFLP markers. The second population segregated 31 resistant to 79 susceptible progeny, significantly different from a 1:1 segregation ratio. Relying on markers specific for the resistant parent of this population, Ritter et al. (1991) found linkage (4.5% recombination) between RFLP locus GP21 and a second putative resistance gene, *Rx2*, which likely originated from *S. acaule*. With the addition of two other markers, *Rx2* was located on chromosome V. The skewed segregation ratio observed amongst progeny was explained by the authors as reduced recombination and segregation distortion on chromosome V.

Unlike extreme resistance genes *Rx1* and *Rx2*, the *Nb* gene confers hypersensitive resistance against PVX (Cockerham 1970). De Jong et al. (1997) selfed the tetraploid cultivar Pentland Ivory, which carries *Nb*. They observed 84:25 resistant to susceptible segregation ratio, fitting the 3:1 ratio predicted if "Pentland Ivory" carries *Nb* in the simplex condition. Using BSA, these researchers identified eight AFLP fragments specific to the resistant pool. Two of these markers when converted to CAPS, SPUD237 and SPUD128, were revealed to be polymorphic between *Solanum lycopersicum* and *S. pennellii*. Segregation of these markers in a tomato mapping population suggested a potato chromosome V location for *Nb*, in the same region as the *R1* late blight resistance locus and the *Rx2* locus.

Marczewski et al. (1998) evaluated hypersensitive resistance to PVS, most likely from *S. tuberosum* ssp. *andigena*, in F_1 diploid progeny. Progeny segregated 75 resistant and 77 susceptible, fitting the expected segregation for a single dominant gene, *Ns* ($\chi^2 = 0.026$, $P = 0.87$). RAPD markers were identified with linkage to *Ns*. The closest, $OPE15_{550}$, was linked at 2.6 cM. Linked RAPD markers failed to identify resistant tetraploid clones with resistance originating from a related *S. tuberosum* ssp. *andigena* clone, raising the possibility of different resistance loci existing in *S. tuberosum* ssp. *andigena*. Additional work identified SCAR marker $SCG17_{321}$ (Marczewski et al. 2001) and ISSR marker $UBC811_{660}$ (Marczewski 2001) linked to *Ns*. These markers have been used for marker-assisted selection (Marczewski et al. 2002) and were mapped to chromosome VIII by Marczewski et al. (2002) in a population lacking *Ns*. Chromosomal location of *Ns* was confirmed by showing linkage with chromosome VIII markers.

5.3 Nematode Resistance

Similar to Cockerham's 1970 report on PVY and PVX resistance, early nematode resistance studies have contributed to later mapping and tagging research. One of those is Toxopeus and Huijsman's 1953 study titled *Breeding for Resistance to Potato Root Eelworm* (Toxopeus and Huijsman 1953). The authors evaluated resistance to root cyst nematode, *Globodera rostochiensis*, among progeny from a self of Commonwealth Potato Collection (CPC) accession 1673 (among others), *S. tuberosum* ssp. *andigena*. Genetic analysis identified a single dominant gene named *H1* (Toxopeus and Huijsman 1953; Huijsman 1955). Clone CPC 1673 has been widely used as a source of resistance to *G. rostochiensis* and *G. pallida*, and was later discovered to carry *Rx1*.

Working with *G. rostochiensis* resistance derived from CPC 1673, Gebhardt et al. (1993) evaluated progeny from a cross between resistant haploid line Amaryl H5 and a susceptible *S. phureja* line. Fifty-eight F_1 progeny were found to be resistant and 53 susceptible, fitting a 1:1 ratio for a single dominant gene ($P > 0.5$), *H1*, in which the resistant parent Amaryl H5 is heterozygous for *H1*. Six RFLP markers from chromosome V showed linkage to *H1*, with RFLP CP113 linked to resistance without recombination. In related work, the cultivar Atlantic, resistant to *G. rostochiensis* and simplex for *H1* gene, was used to generate 101 haploid progeny, of which 51 were resistant to pathotype Ro1, 38 were susceptible, and 13 were not tested (Pineda et al. 1993). Observed data fit a 1:1 ratio for a single dominant gene and a simplex maternal parent ($\chi^2 = 1.899$, $p < 0.25$). RFLP markers were applied to resistant and susceptible bulked pools and linkage was detected for eight markers on chromosome V. Analysis of individual plants located RFLP marker CD78 2.7 cM from *H1*. Bakker et al. (2004) continued this work with a high-resolution map of the *H1* region. A diploid population of 120 F_1 progeny segregated 41 resistant to 79 susceptible. Applying 704 AFLP primer combinations resulted in six markers linked to *H1*. The authors then utilized 136 F_1 genotypes from the same population used by Rouppe van der Voort et al. (1997) to map *Gpa2* to chromosome XII (see below), which also segregated for resistance to *G. rostochiensis*. This population was then expanded to 1,209 F_1 progeny for high-resolution mapping. The additional progeny enabled identification of two flanking markers, EM1 and EM14 at 0.2 cM and 0.8 cM, respectively, and two co-segregating markers, CM1 and EM15.

Huijsman (1960) continued the study of resistance to *G. rostochiensis* by analyzing segregation of diploid progeny from crosses between resistant *S. kurtzianum* and susceptible *S. goniocalyx* and *S. stenotomum*. Progeny segregation ratios of 1:1 and 3:1 (resistant: susceptible) indicated one and two dominant genes, respectively. The level of resistance from *S. kurtzianum* was similar to that observed in *S. tuberosum* ssp. *andigena* clone CPC 1673.

Barone et al. (1990) utilized two diploid, highly heterozygous parents, one, an interspecific hybrid between *S. spegazzinii* and *S. tuberosum* ssp. *tuberosum*, heterozygous for resistance to root cyst nematode (*G. rostchiensis*) and the other a susceptible *S. tuberosum* ssp. *tuberosum* clone. Eighty-three F_1 progeny were evaluated for resistance to pathotype Ro1 and segregated 40 resistant and 43 susceptible, fitting a 1:1 ratio ($\chi^2 = 0.108$) for a single dominant gene, *Gro1*. Linked RFLP markers assigned *Gro1* to chromosome VII, with TG51(c) and TG20(a) flanking *Gro1*. Ballvora et al. (1995) continued this work by generating a high-resolution map from a total of 1,105 F_1 plants using RFLP, RAPD and AFLP markers. Thirty-nine F_1 plants from Barone et al. (1990) were included in four DNA pools (A-D), representing the four possible *Gro1* allelic combinations based on linked markers. Eight RAPD markers (out of 700) were linked to resistance, four of which were linked without recombination to *Gro1* when screening the 39 F_1s. Using two markers linked to *Gro1*, 1,006 new F_1 plants (533 resistant and 473 susceptible) were screened for recombinant resistant (RR) and recombinant susceptible (RS) individuals. DNA from recombinant individuals was used to construct RR (nine individuals) and RS (ten individuals) pools. These were screened with AFLPs in a search for fragments originating from the resistant parent and present in the RR pool while absent from the RS pool. Two AFLP markers flanked *Gro1* at 0.6 cM and 0.8 cM. Using a candidate gene approach Paal et al. (2004) ultimately cloned *Gro1*.

Jacobs et al. (1996) evaluated resistance to *G. rostochiensis* pathotype Ro1 in a diploid F_1 population with resistance originating from *S. vernei*. A total of 108 F_1 genotypes were evaluated with pathotype Ro1. Forty-six individuals were resistant and 56 were susceptible, fitting a model for a single dominant gene, *GroV1*, with an expected ratio of 1:1 ($\chi^2 = 0.98$, $P >$ 0.5). A smaller set of progeny was evaluated with pathotype Ro4. Resistance to Ro4 segregated 16: 18 (resistant: susceptible; $\chi^2 = 0.12$, $P > 0.25$), fitting a 1:1 ratio and showing correlation with Ro1 resistance. However, subsequent testing on other genotypes failed to confirm this relationship between Ro1 and Ro4 resistance. TG69, an RFLP marker from the region of *H1* (resistance locus from *S. tuberosum* ssp. *andigena*) on chromosome V co-segregated with *GroV1*. Fine mapping utilized BSA of pooled DNA from resistant and susceptible individuals. Five RAPDs were identified that co-segregated with resistance phenotypes among F_1 individuals, the closest at 4 cM. SCARs were designed based on sequences from cloned RAPD markers.

Rouppe van der Voort et al. (1997) evaluated a diploid mapping population with resistance to *G. pallida* (population D383) originating from CPC 1673. A total of 181 F_1 progeny segregated 77 resistant: 78 susceptible, with 26 genotypes not classified. These data fit a 1:1 ratio ($\chi^2 = 0.01$, $P > 0.95$) for a single dominant locus, *Gpa2*, conferring resistance for *G. pallida*. AFLPs mapped *Gpa2* to chromosome XII in the proximity to *Rx1* (extreme resistance to PVX) and this was confirmed by chromosome XII specific RFLPs. Fine mapping using tetraploid and diploid populations showed that *Gpa2* and *Rx1* were 0.02 cM apart and that both genes had been introgressed from *S. tuberosum* ssp. *andigena* CPC 1673 (Rouppe van der Voort et al. 1999b). van der Vossen et al. (2000) utilized these results to isolate *Gpa2* and the previously isolated *Rx1* (Bendahmane et al. 1999).

Resistance to Columbia root-knot nematode (*Meloidogyne chitwoodi*) derived from *S. bulbocastanum* was first mapped in a somatic hybrid-derived BC_2 tetraploid/aneuploid population to a major quantitative locus on chromosome XI (Brown et al. 1996), R_{Mc1}, later called *Rmc1* (Rouppe van der Voort et al. 1999a) and $R_{Mc1(blb)}$ (Zhang et al. 2007). This work was pursued by Rouppe van der Voort et al. (1999a) working with a diploid F_1 population derived from *S. bulbocastanum* and a tetraploid/aneuploid BC_3 population, derived from the BC_2 population of Brown et al. (1996). Multiple diploid populations were evaluated for *M. chitwoodi* and *M. fallax* resistance. Segregation ratios fit either 1:0 or 1:1 depending on whether the resistant parent was homozygous or heterozygous for *Rmc1*, and results indicated *Rmc1* was highly effective against *M. chitwoodi* and *M. fallax*. AFLP primer combinations previously mapped in *S. tuberosum* were applied to the *S. bulbocastanum* genome, however marker transfer between these two species proved unsuccessful due to fragment dissimilarity between the two genomes. CAPS markers specific to chromosome XI, originally designed for fine mapping Ry_{sto} (Brigneti et al. 1997), successfully located *Rmc1* to chromosome XI. No recombinants, among 29 BC_3 clones, were identified between *Rmc1* and the CAPS markers, however flanking CAPS markers were each 2 cM away. Pursuing closer markers, Zhang et al. (2007) utilized a diploid population derived from *S. bulbocastanum* and BC_5 tetraploid/aneuploid lines. For tagging *Rmc1*, BSA of the diploid population was used to determine closely linked markers on chromosome XI. STS markers were derived from bacterial artificial chromosome (BAC) end sequences from 240 BAC clones positively identified from hybridization with a *S. demissum* probe with homology to N-like (*Nl*) resistance genes. BSA located two previously mapped CAPS markers, M33 and M39, to the *Rmc1* region. Additionally, one AFLP marker and five STS markers were closely linked with *Rmc1*. The five STS markers were applied to 180 plants of

18 BC_5 families. All five STS markers were 100% accurate in detecting resistant phenotypes and *Rmc1*.

Bakari et al. (2006) took a different approach to identify resistance to *M. fallax*. Working with F_1 progeny from a cross between resistant *S. sparsipilum* and a susceptible haploid clone of *S. tuberosum* ssp. *tuberosum*, two different resistance phenotypes were evaluated. A necrotic reaction at the feeding site was detected in 68 progeny, while 60 progeny showed no reaction, fitting a 1:1 ratio ($\chi^2 = 0.47$, $P = 0.488$) for a single dominant gene, *NR*, heterozygous in the resistant parent. A second phenotype, number of females developed on roots, proved to have a continuous distribution. Using AFLPs, the necrotic phenotype was mapped as a qualitative trait to chromosome XII, in the same region as tomato resistance genes *Mi3*/*Mi5*. Quantitative trait loci analysis for the number of female nematodes identified a single major locus (explaining 94.5% of observed phenotypic variation), *MfaXIIspl*, in the same region as *NR* on chromosome XII.

5.4 Late Blight Resistance

Identification of resistance to late blight of potato, caused by the oomycete *Phytophthora infestans*, has a long history with extensive publications . Much of the history of late blight resistance has focused on *R* genes, particularly major dominant genes originating from *S. demissum*. Results of segregation analysis of crosses among tetraploid potato clones demonstrated the inheritance of *R* genes as single dominant factors (Masternbroek 1953; Malcolmson and Black 1966). Resistance loci were classified as *R1*-*R11* on a range of tetraploid differential potato clones (Malcolmson and Black 1966; Ewing et al. 2000). This early work laid the basis for many more recent studies which will now be discussed, in particular those genetic studies in diploid populations which map and/or tag late blight resistance loci. For the purpose of this section, *R* gene resistance refers to the incompatible reaction observed when testing with *P. infestans* strains carrying the cognate avirulence factor(s).

Leonards-Schippers et al. (1992) crossed a resistant haploid *S. tuberosum* line carrying the *R1* gene to a susceptible line. Inoculating 92 F_1 plants with *P. infestans* race 0, 54 plants were hypersensitive resistant and 38 plants were susceptible ($\chi^2 = 2.78$, $P > 0.05$), fitting a 1:1 ratio for a single dominant gene in the heterozygous state in the resistant parent. Application of RFLP markers located *R1* to chromosome V, flanked by GP21 and GP179, with GP21 closest at 2.5 cM. Meksem et al. (1995) added an additional 375 plants, for a total of 461 F_1 plants, from which 17 recombinants were identified in the GP21 to GP179 interval. These recombinants were used to generate DNA pools similar to those of Barone et al. (1990), as described above. AFLP technology was applied to these pools and 29 markers linked to

R1 were identified, eight of which were within the GP21-GP179 interval. Two of these AFLP markers, AFLP1 and AFLP2, co-segregated with *R1*. A combination of positional cloning and candidate gene approach cloned *R1* (Ballvora et al. 2002).

El-Kharbotly et al. (1994) generated haploids from 10 different cultivars with different *R* gene compositions. Numbers were limited, so haploids were crossed and segregations followed. F_1 progeny from *R1*- and *R3*-containing haploids segregated 1:1 resistant to susceptible. RFLP marker GP21 was used to confirm the position of *R1* on chromosome V. *R3* was located to chromosome XI based on co-segregation with three chromosome XI markers. El-Kharbotly et al. (1996a) utilized a different tagging strategy by introducing the *Dissociation* transposable element (*Ds*) of maize into a diploid potato clone through *Agrobacterium tumefaciens* transformation. Flanking sequence was amplified from T-DNA insertions in 312 transformants and used as probes for RFLP analysis. *R1*, heterozygous in the original transformed line, was linked in repulsion to T-DNA loci in two of the transformants. Crosses with a susceptible haploid indicated that the T-DNA insertion in one of the lines was 18 cM from *R1*.

El-Kharbotly et al. (1996b) generated haploids from tetraploid differentials carrying *R6* and *R7*. These were crossed with susceptible lines to generate multiple segregating diploid populations. Four *R6* F_1 populations segregated at 1:1 (resistant: susceptible) ratios, while one population had an excess of susceptible individuals. The single *R7* population also had an excess of susceptible individuals. RFLP markers located both *R6* and *R7* on the distal end of chromosome XI in the vicinity of the *R3* locus.

Li et al. (1998) generated a tetraploid mapping population from a cross between resistant parent BET95-4200-3 (*R_2rrr*) and susceptible parent DJ93-6707-10 (*rrrr*). Resistance in BET95-4200-3 originated from *S. demissum* accession CPC 2127, which was used to characterize *R* gene specificity (Black et al. 1953). Progeny segregated 44 resistant to 40 susceptible, fitting a 1:1 segregation ratio indicating a simplex inheritance of the resistance allele. Two DNA pools, B_R and B_S, were generated from eight resistant and eight susceptible genotypes, respectively. BSA identified 12 AFLP markers detected in the resistant parent and B_R, and absent in the susceptible parent and B_S. Eleven of these were confirmed to be linked to *R2*, three of which, ACC/CAT-535, ACT/CAC-189, and AGC/CCA-369, were found to co-segregate with *R2*. AFLP markers linked to *R2* were applied to a reference population and located on chromosome IV.

Naess et al. (2000) utilized backcross progeny originating from somatic fusion products (Helgeson et al. 1998) between resistant *S. bulbocastanum* (PI 243510) and susceptible *S. tuberosum* (PI 23900) to characterize broad-spectrum late blight resistance derived from *S. bulbocastanum*. Preliminary field data on the BC_1 population suggested resistance was caused by a single dominant gene. Highly resistant BC_1 individuals were used to generate three BC_2 populations. Two of the BC_2 populations produced 58% and 59% resistant individuals, not significantly different from a 1:1 segregation ratio. The third BC_2 population was skewed to susceptible individuals and did not fit a 1:1 segregation ratio. RAPD marker G02-0625 was shown to be linked to late blight resistance in all three BC_2 populations and was located on chromosome VIII. RFLP markers were added to 64 individuals of one of the BC_2 populations, 1K6. The resistance locus, later named *RB* (Song et al. 2003), was flanked by RFLP markers CP53 and CT64, and co-segregated with RFLP marker CT88 and RAPD marker OPG02-625.

Kuhl et al. (2001) utilized an interspecific cross between late blight resistant *S. pinnatisectum* and susceptible *S. cardiophyllum* genotypes. A selected resistant F_1 was backcrossed to the susceptible *S. cardiophyllum* parent. The backcross population segregated 42 resistant to 56 susceptible, not significantly different from a 1:1 ratio ($P > 0.1$), consistent with a single dominant locus for resistance, *Rpi1* (Fig. 1). RFLP markers localized *Rpi1* to chromosome VII. The *P. infestans* isolated used to screen for resistant was incompatible on all differentials except LB3 carrying *R9*.

Ewing et al. (2000) also utilized an interspecific population, crossing a haploid line derived from the cultivar Saco and a *S. berthaultii* clone (PI 473331), then backcrossing an F_1 to a different haploid *S. tuberosum* line to generate a diploid backcross population (BCT) (Bonierbale et al. 1994). A little more than half of the progeny were resistance when inoculated with *P. infestans*, suggesting a single dominant gene. This qualitative resistance (later named $R_{Pi\text{-}ber}$) was most closely linked (4.8 cM) to RFLP marker TG63 located chromosome X, and was possibly identical to *R8, R9* or *R11*. However, Rauscher et al. (2006) subsequently show that $R_{Pi\text{-}ber}$ is different from all 11 of the identified *R* genes from *S. demissum*. Five additional PCR-based markers were added using a technique called multiplex allele-specific polymorphism mapping to convert previously mapped RFLPs on chromosome X to allele specific markers. Flanking markers encompass 3.9 cM, with TG63 less than 1.0 cM from $R_{Pi\text{-}ber}$.

Another group utilizing diploid interspecific crosses was that of Park et al. (2009). These authors generated two different diploid interspecific populations, one resulting from a cross between late blight resistant *S. berthaultii* [PI 473331, the same accession used by Ewing et al. (2000)] and *S. stenotomum*, and a second population resulting from a cross between a resistant *S. berthaultii* (PI 265858) genotype and a susceptible hybrid of *S. chacoense* and *S. phureja*. The first population segregated 20 resistant: 24 susceptible progeny while the second population segregated 25 resistant: 25 susceptible progeny, both fitting a single dominant gene model, *Rpi-ber1* and *Rpi-ber2*, respectively. Using BSA with RFLPs, AFLPs, and other selected PCR-based markers, both loci mapped to chromosome X in the same general region as $R_{Pi\text{-}ber}$, but, somewhat surprisingly, located to two different genetic positions.

A slightly different approach using intraspecific diploid populations was taken by Smilde et al. (2005) and Park et al. (2005a). Smilde et al. (2005) crossed resistant and susceptible plants from three *S. mochiquense* accessions, to generate multiple segregating diploid populations, however only one of these populations was used for mapping. The mapping population segregated 51 resistant and 45 susceptible F_1 progeny, fitting a 1:1 genetic model for a single dominant gene, *Rpi-moc1*. Smilde et al. (2005) then used a three step mapping strategy. One, BSA was used to identify AFLP fragments linked to resistance. Linked fragments were cloned and sequenced. Two, PCR primers developed from AFLP sequences were used to amplify orthologous fragments in *S. lycopersicum* and *S. pennellii*; polymorphism between these two tomato species allowed the use of introgression lines to roughly determine the map position of *Rpi-moc1*. Three, markers in the region were identified from online databases and selected markers were amplified and sequenced to identify polymorphisms between the two *S. mochiquense* parents. These markers were used to map *Rpi-moc1* in 70 segregating individuals to chromosome IX, with linkage to RFLP marker TG328. Park et al. (2005a) also utilized intraspecific hybridization to generate a BC_1 population from a cross between resistant and susceptible plants from different *S. bulbocastanum* accessions. A 1:1 segregation ratio in the progeny suggested a single dominant gene system, *Rpi-blb3*. Markers distributed on different potato chromosomes were screened on 10 clearly resistant and 10 clearly susceptible lines. CAPS marker TG506R and SCAR marker CT229 on chromosome IV were found to flank *Rpi-blb3*. A larger BC_1 population of 1,400 genotypes was screened and 183 resistant and 133 susceptible CT229/TG506R recombinants identified. AFLP markers were identified that flanked *Rpi-blb3* and spanned a 0.93 cM region. One AFLP marker co-segregated with *Rpi-blb3* and was converted to SCAR marker TH21. *Rpi-blb3* resides in the same *R* gene cluster on chromosome IV as *Rpi-abpt*, *R2* and *R2-like* (Li et al. 1998; Park et al. 2005b, 2005c).

5.5 Conclusions

Tagging of resistance loci with molecular markers offers the possibility, at least in the early stages of selection, of eliminating the need for time-consuming and environmentally sensitive trials. Molecular markers should be closely linked to the resistance locus, be easy to apply, and function in a wide variety of genotypes when the resistance has been introgressed from the same or related sources. A number of researchers have reported on the efficacy of various markers to accurately detect a specific locus and, more importantly, a resistance phenotype. Whitworth et al. (2009) successfully used RYSC3 and RYSC4 (Kasai et al. 2000) and ADG2/*Bbv*I (Sorri et al. 1999) markers in a tetraploid population segregating for Ry_{adg}. Progeny segregated 1: 1 for PVY resistance, and all three markers co-segregated with resistance. Ottoman et al. (2009) used RYSC3 (Kasai et al. 2000) and ADG2/*Bbv*I (Sorri et al. 1999) markers in a tetraploid population and reported a high degree of association between markers and ELISA results. Utilizing marker-assisted selection Gebhardt et al. (2006) conducted a study to combine multiple major resistance genes: Ry_{adg} for extreme resistance to PVY, *Gro1* for resistance to root cyst nematode *G. rostochiensis*, *Rx1* for extreme resistance to PVX, or *Sen1* for resistance to potato wart (*Synchytrium endobioticum*). Crosses were made and PCR-based markers were applied to 110 F_1 hybrids. Thirty tetraploid plants contained the desired marker composition and were tested for the resistance traits. All selected plants carried the expected resistance phenotypes.

Regardless of whether genetic analysis is conducted at the diploid level or tetraploid level or whether the genetic control of a trait is considered to be simple or complex, potato research has been successful at identifying genetic factors related to many traits. Efforts are still needed to identify robust molecular markers tagging genetic loci. As more and more molecular data are generated, proof-of-concept studies are needed to connect molecular markers with traits. In some cases, such as *Rx1*, the tagged locus has been cloned. In other cases, such as Ry_{adg}, markers are closely linked. Tagging of a resistance locus may enable cloning but cloning is not necessary to identify useful molecular markers. A universally key element in the successful application of markers is that recombination between the marker and the locus conditioning a trait should be limited or non-existent (i.e., the marker reliably identifies the desired trait). Potato breeding will benefit from emerging technologies and research initiatives aimed at delivering reliable, high throughput markers that can be applied at a reasonable price.

References

Agrios, GN (1988) Plant Pathology. Academic Press, San Diego, New York, Boston, London, Sydney, Tokyo, Toronto.

Bakari KA, Kerlan M, Caromel B, Dantec J, Didier F, Manzanares-Dauleux M, Ellissèche D, Mugniéry D (2006) A major gene mapped on chromosome XII is the main factor of a quantitatively inherited resistance to *Meloidogyne fallax* in *Solanum sparsipilum*. Theor Appl Genet 112: 699–707.

Bakker E, Achenbach U, Bakker J, van Vliet J, Peleman J, Segers B, van der Heijden S, van der Linde P, Graveland R, Hutten R, van Eck H, Coppoolse E, van der Vossen E, Bakker J, Goverse A (2004) A high-resolution map of the *H1* locus harbouring resistance to the potato cyst nematode *Globodera rostochiensis*. Theor Appl Genet 109: 146–152.

Ballvora A, Hesselbach J, Niewöhner J, Leister D, Salamini F, Gebhardt C (1995) Marker enrichment and high-resolution map of the segment of potato chromosome VII harbouring the nematode resistance gene *Gro1*. Mol Gen Genet 249: 82–90.

Ballvora A, Ercolano MR, Weiß J, Meksem K, Bormann CA, Oberhagemann P, Salamini F, and Gebhardt C (2002) The *R1* gene for potato resistance to late blight (*Phytophthora infestans*) belongs to the leucine zipper/NBS/LRR class of plant resistance genes. Plant J 30: 361–371.

Barone A, Ritter E, Schachtschabel U, Debener T, Salamini F, Gebhardt C (1990) Localization by restriction fragment length polymorphism mapping in potato of a major dominant gene conferring resistance to the potato cyst nematode *Globodera rostochiensis*. Mol Gen Genet 224: 177–182.

Bendahmane A, Kanyuka K, Baulcombe DC (1999) The *Rx* gene from potato controls separate virus resistance and cell death responses. Plant Cell 11: 781–791.

Black W, Mastenbroek C, Mills WR, Peterson LC (1953) A proposal for an international nomenclature of races of *Phytophthora infestans* and of gene controlling immunity in *Solanum demissum* derivatives. Euphytica 2: 173–179.

Bonierbale MW, Plaisted RL, Pineda O, Tanksley SD (1994) QTL analysis of trichome-mediated insect resistance in potato. Theor Appl Genet 87: 973–987.

Brigneti G, Garcia-Mas J, Baulcombe DC (1997) Molecular mapping of the potato virus Y resistance gene Ry_{sto} in potato. Theor Appl Genet 94: 198–203.

Brown CR, Yang CP, Mojtahedi H, Santo GS, Masuelli R (1996) RFLP analysis of resistance to Columbia root-knot nematode derived from *Solanum bulbocastanum* in a BC_2 population. Theor Appl Genet 92: 572–576.

Celebi-Toprak F, Slack SA, Jahn MM (2002) A new gene, Ny_{tbr}, for hypersensitivity to *Potato virus Y* from *Solanum tuberosum* maps to chromosome IV. Theor Appl Genet 104: 669–674.

Cernák I, Taller J, Wolf I, Fehér E, Babinszky G, Alföldi Z, Csanádi G, Polgár Z (2008a) Analysis of the applicability of molecular markers linked to the PVY extreme resistance gene Ry_{sto}, and the identification of new markers. Acta Biol Hung 59: 195–203.

Cernák I, Decsi K, Nagy S, Wolf I, Polgár Z, Gulyás G, Hirata Y, Taller J (2008b) Development of a locus-specific marker and localization of the Ry_{sto} gene based on linkage to a catalase gene on chromosome XII in the tetraploid potato genome. Breed Sci 58: 309–314.

Cockerham G (1970) Genetical studies on resistance to potato viruses X and Y. Heredity 25: 309–348.

De Jong W, Forsyth A, Leister D, Gebhardt C, Baulcombe DC (1997) A potato hypersensitive resistance gene against potato virus X maps to a resistance gene cluster on chromosome 5. Theor Appl Genet 95: 246–252.

El-Kharbotly A, Leonards-Schippers C, Huigen DJ, Jacobsen E, Pereira A, Stiekema WJ, Salamini F, Gehbardt C (1994) Segregation analysis and RFLP mapping of the *R1* and *R3* alleles conferring race-specific resistance to *Phytophthora infestans* in progeny of dihaploid potato parents. Mol Gen Genet 242: 749–754.

El-Kharbotly A, Jacobs JME, te Lintel Hekkert B, Jacobsen E, Ramanna MS, Stiekema WJ, Pereira A (1996a) Localization of Ds-transposon containing T-DNA inserts in the diploid transgenic potato: linkage to the *R1* resistance gene against *Phytophthora infestans* (Mont.) de Bary. Genome 39: 249–257.

El-Kharbotly A, Palomino-Sánchez C, Salamini F, Jacobsen E, Gebhardt C (1996b) *R6* and *R7* alleles of potato conferring race-specific resistance to *Phytophthora infestans* (Mont.) de Bary identified genetic loci clustering with R3 locus on chromosome XI. Theor Appl Genet 92: 880–884.

Ewing EE, Simko I, Smart CD, Bonierbale MW, Mizubuti ESG, May GD, Fry WE (2000) Genetic mapping of qualitative and quantitative field resistance to *Phytophthora infestans* in a population derived from *Solanum tuberosum* and *Solanum berthaultii*. Mol Breed 6: 25–36.

Flis B, Hennig J, Strzelczyk-Zyta D, Gebhardt C, Marczewski W (2005) The *Ry-*f_{sto} gene from *Solanum stoloniferum* for extreme resistant to *Potato virus Y* maps to potato chromosome XII and is diagnosed by PCR marker $GP122_{718}$ in PVY resistant potato cultivars. Mol Breed 15: 95–101.

Flor HH (1956) The complementary genic systems in flax and flax rust. Adv Genet 8: 29–54

Flor HH (1971) Current status of the gene-for-gene concept. Annu Rev Phytopathol 9: 275–296.

Gebhardt C, Valkonen JPT (2001) Organization of genes controlling disease resistance in the potato genome. Annu Rev Phytopathol 39: 79–102.

Gebhardt C, Mugniery D, Ritter E, Salamini F, Bonnel E (1993) Identification of RFLP markers closely linked to the *H1* gene conferring resistance to *Globodera rostochiensis* in potato. Theor Appl Genet 85: 541–544.

Gebhardt C, Bellin D, Henselewski H, Lehmann W, Schwarzfischer J, Valkonen JPT (2006) Marker-assisted combination of major genes for pathogen resistance in potato. Theor Appl Genet 112: 1458–1464.

Hämäläinen JH, Watanabe KN, Valkonen JPT, Arihara A, Plaisted RL, Pehu E, Miller L, Slack SA (1997) Mapping and marker-assisted selection for a gene for extreme resistance to potato virus Y. Theor Appl Genet 94: 192–197.

Hämäläinen JH, Sorri VA, Watanabe KN, Gebhardt C, Valkonen JPT (1998) Molecular examination of a chromosome region that controls resistance to potato Y and A potyviruses in potato. Theor Appl Genet 96: 1036–1043.

Hämäläinen JH, Kekarainen T, Gebhardt C, Watanabe KN, Valkonen JPT (2000) Recessive and dominant genes interfere with the vascular transport of *Potato Virus A* in diploid potatoes. Mol Plant-Microbe Interact 13: 402–412.

Helgeson JP, Pohlman JD, Austin S, Haberlach GT, Wielgus SM, Ronis D, Zambolim L, Tooley P, McGrath JM, James RV, Stevenson WR (1998) Somatic hybrids between *Solanum bulbocastanum* and potato: a new source of resistance to late blight. Theor Appl Genet 96: 738–742.

Hougas RW, Peloquin SJ, Ross RW (1958) Haploids of the common potato. J Hered 49: 103–106.

Hosaka K, Hosaka Y, Mori M, Maida T, Matsunaga H (2001) Detection of a simplex RAPD marker linked to resistance to Potato Virus Y in a tetraploid potato. Am J Potato Res 78: 191–196.

Huijsman CA (1955) Breeding for resistance to the potato root eelworm II. Data on the inheritance of resistance in andigenum-tuberosum crosses obtained in 1954. Euphytica 4: 133–140.

Huijsman CA (1960) Some data on the resistance against the potato root-eelworm (*Heterodera rostochiensis* W.) in *Solanum kurtzianaum*. Euphytica 9: 185–190.

Jacobs JME, van Eck HJ, Horsman K, Arens PFP, Verkerk-Bakker B, Jacobsen E, Pereira A, Stiekema WJ (1996) Mapping of resistance to the potato cyst nematode *Globodera rostochiensis* from the wild potato species *Solanum vernei*. Mol Breed 2: 51–60.

Jones JDG, Dangl JL (2006) The plant immune system. Nature 444: 323–329.

Kasai K, Morikawa Y, Sorri VA, Valkonen JPT, Gebhardt C, Watanabe KN (2000) Development of SCAR markers to the PVY resistance gene Ry_{adg} based on a common feature of plant disease resistance genes. Genome 43: 1–8.

Kuhl JC, Hanneman RE, Havey MJ (2001) Characterization and mapping of *Rpi1*, a late-blight resistance locus from diploid (1EBN) Mexican *Solanum pinnatisectum*. Mol Genet Genom 265: 977–985.

Leonards-Schippers C, Gieffers W, Salamini F, Gebhardt C (1992) The *R1* gene conferring race-specific resistance to *Phytophtora infestans* in potato is located on potato chromosome V. Mol Gen Genet 233: 278–283.

Leister D, Ballvora A, Salamini F, Gebhardt C (1996) A PCR-based approach for isolating pathogen resistance genes from potato with potential for wide application in plants. Nat Genet 14: 421–429.

Li X, van Eck HJ, Rouppe van der Voort JNAM, Huigen DJ, Stam P, Jacobsen E (1998) Autotetraploids and genetic mapping using common AFLP markers: the *R2* allele conferring resistance to *Phytophthora infestans* mapped on potato chromosome 4. Theor Appl Genet 96: 1121–1128.

Malcolmson JF, Black W (1966) New R genes in *Solanum demissum* Lindl. and their complementary races of *Phytophthora infestans* (Mont.) de Bary. Euphytica 15: 199–203.

Marczewski W (2001) Inter-simple sequence repeat (ISSR) markers for the *Ns* resistance gene in potato (*Solanum tuberosum* L.). J Appl Genet 42: 139–144.

Marczewski W, Ostrowska K, Zimnoch-Guzowska E (1998) Identification of RAPD markers linked to the *Ns* locus in potato. Plant Breed 117: 88–90.

Marczewski W, Talarczyk A, Hennig J (2001) Development of SCAR markers linked to the *Ns* locus in potato. Plant Breed 120: 88–90.

Marczewski W, Hennig J, Gebhardt C (2002) The *Potato virus S* resistance gene *Ns* maps to potato chromosome VIII. Theor Appl Genet 105: 564–567.

Mastenbroek C (1953) Experiments on the inheritance of blight immunity in potatoes derived from *Solanum demissum* Lindl. Euphytica 2: 197–206.

Matthews REF (1992) Fundamentals of Plant Virology. Academic Press and Harcourt Brace Jovanovich, New York, USA.

Meksem K, Leister D, Peleman J, Zabeau M, Salamini F, Gebhardt C (1995) A high-resolution map of the vicinity of the *R1* locus on chromosome V of potato based on RFLP and AFLP markers. Mol Gen Genet 249: 74–81.

Mendel G (1866) Versuche über Pflanzen-Hybriden. Verh Naturforsch Ver Brünn 4: 3–47 (first English translation in 1901, J Roy Hort Soc 26: 1–32).

Naess SK, Bradeen JM, Wielgus SM, Haberlach GT, McGrath JM, Helgeson JP (2000) Resistance to late blight in *Solanum bulbocastanum* is mapped to chromosome 8. Theor Appl Genet 101: 697–704.

Ottoman RJ, Hane DC, Brown CR, Yilma S, James SR, Mosley AR, Crosslin JM, Vales MI (2009) Validation and implementation of marker-assisted selection (MAS) for PVY resistance (Ry_{adg} gene) in a tetraploid potato breeding program. Am J Potato Res 86: 304–314.

Paal J, Henselewski H, Muth J, Meksem K, Menéndez CM, Salamini F, Ballvor A, Gebhardt C (2004) Molecular cloning of the potato *Gro1-4* gene conferring resistance to pathotype Ro1 of the root cyst nematode *Globodera rostochiensis*, based on a candidate gene approach. Plant J 38: 285–297.

Park TH, Gros J, Sikkema A, Vleeshouwers VGAA, Muskens M, Allefs S, Jacobsen E, Visser RGF, van der Vossen EAG (2005a) The late blight resistance locus *Rpi-blb3* from *Solanum bulbocastanum* belongs to a major late blight *R* gene cluster on chromosome 4 of potato. Mol Plant-Microbe Interact 18: 722–729.

Park TH, Vleeshouwers VGAA, Huigen DJ, van der Vossen EAG, van Eck HJ, Visser RGF (2005b) Characterization and high-resolution mapping of a late blight resistance locus similar to *R2* in potato. Theor Appl Genet 111: 591–597.

Park TH, Vleeshouwers VGAA, Hutten RCB, van Eck HJ, van der Vossen E, Jacobsen E, Visser RGF (2005c) High-resolution mapping and analysis of the resistance locus *Rpi-adpt* against *Phytophthora infestans* in potato. Mol Breed 16: 33–43.

Park TH, Foster S, Brigneti G, Jones JDG (2009) Two distinct potato late blight resistance genes from *Solanum berthaultii* are located on chromosome 10. Euphytica 165: 269–278.

Pineda O, Bonierbale MW, Plaisted RL, Brodie BB, Tanksley SD (1993) Identification of RFLP markers linked to the *H1* gene conferring resistance to the potato cyst nematode *Globodera rostochiensis*. Genome 36: 152–156.

Rauscher GM, Smart CD, Simko I, Bonierbale M, Mayton H, Greenalnd A, Fry WE (2006) Characterization and mapping of R_{Pi-ber}, a novel potato late blight resistance gene from *Solanum berthaultii*. Theor Appl Genet 112: 674–687.

Ritter E, Debener T, Barone A, Salamini F, Gebhardt C (1991) RFLP mapping on potato chromosomes of two genes controlling extreme resistance to potato virus X (PVX). Mol Gen Genet 227: 81–85.

Rokka V-M, Pietilä L, Pehu E (1996) Enhanced production of dihaploid lines via anther culture of tetraploid potato (*Solanum tuberosum* ssp *tuberosum*) clones. Am Potato J 73: 1–12.

Rouppe van der Voort J, Wolters P, Folkertsma R, Hutten R, van Zandvoort P, Vinke H, Kanyuka K, Bendahmane A, Jacobsen E, R Janssen, Bakker J (1997) Mapping of the cyst nematode resistance locus *Gpa2* in potato using a strategy based on comigrating AFLP markers. Theor Appl Genet 95: 874–880.

Rouppe van der Voort JNAM, Janssen GJW, Overmars H, van Zandvoort PM, van Norel A, Scholten OE, Janssen R, Bakker J (1999a) Development of a PCR-based selection assay for root-knot nematode resistance (*Rmc1*) by a comparative analysis of the *Solanum bulbocastanum* and *S. tuberosum* genome. Euphytica 106: 187–195.

Rouppe van der Voort J, Kanyuka K, van der Vossen E, Bendahmane A, Mooijman P, Klein-Lankhorst R, Stiekema W, Baulcombe D, Bakker J (1999b) Tight physical linkage of the nematode resistance gene *Gpa2* and the virus resistance gene *Rx* on a single segment introgressed from the wild species *Solanum tuberosum* subsp. *andigena* CPC 1673 into cultivated potato. Mol Plant-Microbe Interact 12: 197–206.

Sato M, Nishikawa K, Komura K, Hosaka K (2006) *Potato virus Y* resistance gene, Ry_{chc}, mapped to the distal end of potato chromosome 9. Euphytica 149: 367–372.

Smilde WD, Brigneti G, Jagger L, Perkins S, Jones JDG (2005) *Solanum mochiquense* chromosome IX carries a novel late blight resistance gene *Rpi-moc1*. Theor Appl Genet 110: 252–258.

Song Y, Schwarzfischer A (2008) Development of STS markers for selection of extreme resistance (RY_{sto}) to PVY and maternal pedigree analysis of extremely resistant cultivars. Am J Potato Res 85: 159–170.

Song J, Bradeen JM, Naess SK, Raasch JA, Wielgus SM, Haberlach GT, Liu J, Kuang H, Austin-Phillips S, Buell CR, Helgeson JP, Jiang J (2003) Gene *RB* cloned from *Solanum bulbocastanum* confers broad spectrum resistance to potato late blight. Proc Natl Acad Sci USA 100: 9128–9133.

Song Y, Hepting L, Schweizer G, Hartl L, Wenzel G, Schwarzfischer A (2005) Mapping of extreme resistance to PVY (Ry_{sto}) on chromosome XII using anther-culture-derived primary dihaploid potato lines. Theor Appl Genet 111: 879–887.

Sorri VA, Watanabe KN, Valkonen JPT (1999) Predicted kinase-3a motif of a resistance gene analogue as a unique marker for virus resistance. Theor Appl Genet 99: 164–170.

Szajko K, Chrzanowska M, Witek K, Strzelczyk-Żyta D, Zagórska H, Gebhardt C, Henning J, Marczewski W (2008) The novel gene *Ny-1* on potato chromosome IX confers hypersensitive resistance to *Potato virus Y* and is an alternative to *Ry* genes in potato breeding for PVY resistance. Theor Appl Genet 116: 297–303.

Toxopeus HJ, Huijsman CA (1953) Breeding for resistance to potato root eelworm I. Preliminary data concerning the inheritance and the nature of resistance. Euphytica 2: 180–186.

Valkonen JPT, Slack SA, Plaisted RL, Watanabe KN (1994) Extreme resistance is epistatic to hypersensitive resistance to potato virus Y^{o} in a *Solanum tuberosum* subsp. *andigena*-derived potato genotype. Plant Dis 78: 1177–1180.

Valkonen JPT, Wiegmann K, Hämäläinen JH, Marczewski W, Watanabe KN (2008) Evidence for utility of the same PCR-based markers for selection of extreme resistance to *Potato virus Y* controlled by Ry_{sto} of *Solanum stoloniferum* derived from different sources. Ann Appl Biol 152: 121–130.

van der Biezen EA, Jones JDG (1998) Plant disease resistance proteins and the gene-for-gene concept. Trends Biochem Sci 23: 454–456

van der Vossen EAG, Rouppe van der Voort JNAM, Kanyuka K, Bendahmane A, Sandbrink H, Baulcombe DC, Bakker J, Stiekema WJ, Klein-Lankhorst RM (2000) Homologues of a single resistance-gene cluster in potato confer resistance to distinct pathogens: a virus and a nematode. Plant J 23: 567–576.

Witek K, Strzelczyk-Żyta D, Hennig J, Marczewski (2006) A multiplex PCR approach to simultaneously genotype potato towards the resistance alleles $Ry\text{-}f_{sto}$ and *Ns*. Mol Breed 18: 273–275.

Whitworth JL, Novy RG, Hall DG, Crosslin JM, Brown CR (2009) Characterization of broad spectrum potato virus Y resistance in a *Solanum tuberosum* ssp. *andigena*-derived population and select breeding clones using molecular markers, grafting, and field inoculations. Am J Potato Res 86: 286–296.

Zhang J-H, Mojtahedi H, Kuang H, Baker B, Brown CR (2007) Marker-assisted selection of Columbia root-knot nematode resistance introgressed from *Solanum bulbocastanum*. Crop Sci 47: 2021–2026.

Mapping Complex Potato Traits

Glenn J. Bryan

ABSTRACT

Quantitative traits are determined by the action and interaction of two or more genes. In potato, quantitative traits include multigenic pest and disease resistance, production of glycoalkaloids, tuber life-cycle traits such as dormancy and tuberization, tuber morphological traits such as tuber shape and eye depth, and various tuber quality traits. This chapter reviews significant research in each of these areas. While progress has been made towards understanding the genetic control of each of these traits, much remains to be done. In general, the genetic study of quantitative traits requires densely populated linkage maps and well-replicated sets of phenotypic data. The obligate outcrossing, autotetraploid nature of potato presents significant challenges to the study of quantitative genetics, but the ease with which potato genotypes can be asexually propagated allows robust replication for phenotypic characterization. Efforts to map quantitative traits in potato have been advanced by improvements in both marker technologies and in analytical statistical approaches. Candidate gene approaches have been employed to further mapping of disease resistance, tuberization and tuber dormancy, and cold sweetening loci. Emerging approaches such as transcript mapping show some promise for elucidating the complex genetic control of important quantitative traits. The availability of reliable whole genome sequence for the potato is likely to advance the field of quantitative genetics.

Keywords: candidate gene, disease resistance, insect resistance, leptines, morphology, quality, tuberization

Genetics Program, Scottish Crop Research Institute, Invergowrie, Dundee DD2 5DA, UK; e-mail: *glenn.bryan@scri.ac.uk*

6.1 Introduction

The previous chapter dealt with the topic of simply inherited traits in potato, focusing on pest and disease resistance mapping studies. In this chapter, the subject of complex traits in potato will be discussed. There have been a large number of published studies of potato quantitative traits, both on resistance traits, as well as on traits impinging on tuber quality and potato life-cycle components (e.g., tuber dormancy). It is not intended for this review to be completely comprehensive, but this chapter will summarize some of the key studies that, in my view, have shed the most light on the inheritance of complex traits in potato.

Cultivated potato (*Solanum tuberosum* ssp. *tuberosum*) behaves as an outbreeding tetraploid, and this somewhat hampers genetic analysis of complex traits in this important crop. This limitation has been countered by performing most complex trait analysis at the diploid level, generally using populations produced by the intercrossing of heterozygous diploid parents. However, there are also examples of complex trait analysis at the tetraploid level. Trait analysis is highly dependent on accurate linkage maps (a topic reviewed in Chapter 4), construction of which has entailed development of molecular markers of several different types. Another option for trait analysis is the use of linkage disequilibrium or association genetics, which is reviewed in Chapter 7.

This chapter is concerned with the diverse accumulated work on the genetic analysis of complex potato traits, here defined as traits determined by the action of more than a single gene. While progress in this area has been made, there is still relatively little knowledge concerning the genes underlying many polygenic potato traits. Paradoxically, such traits include many of those in which potato breeders are most interested and have a large impact on the quality of the potato crop (e.g., yield, tuberization, maturity, dormancy, sugar accumulation, etc.).

6.2 Complex Traits in Potato

A complex trait can be defined as one whose genetic component does not follow strict Mendelian inheritance and is generally supposed to involve the action of two or more genes or gene-environment interactions. This is in contrast to "simple" traits—those primarily due to the actions of a single genetic locus, such as eye color in humans or the potato resistance traits described in the previous chapter.

Many traits of primary interest to potato breeders (e.g., yield, tuberization, dormancy, nutritional traits, organoleptic traits), as will be documented later in this chapter, are genetically quite complex. Such traits provide significant analytical challenges, requiring dense linkage maps and well-replicated sets of phenotypic data. The use of highly heterozygous (diploid or tetraploid) parents for map construction can adversely affect the quality and accuracy of linkage maps, and this in turn can affect the resolution and effectiveness of the quantitative trait loci (QTL) analysis performed. The clonal nature of potato propagation offers some advantages, in that it is straight forward to generate genetically identical tuber material for replicated phenotypic assessments in the field or glasshouse. Nonetheless, the all-important tuber traits offer significant challenges in terms of sampling at different development stages, changes during storage, etc.

Potato, a tetraploid heterozygous outbreeder, suffers from a lack of described single-gene traits, such as those that are highly numerous in inbred model and crop plants (e.g., *Arabidopsis*, tomato, rice, barley). Doubtless, potato varieties harbor many recessive mutant alleles, that if homozygous would yield interesting, possibly useful phenotypes that could be analyzed in the normal manner. However, the phenotypes of such alleles remain elusive, as they are rarely if ever going to occur in the "quadriplex" configuration, making it exceedingly difficult to assess the effect of any individual allele. Finding such variant alleles for complex traits using "eco-tilling" or mutagenesis approaches is not difficult (e.g., Muth et al. 2008) but the generation of potato plants allowing examination of their phenotypic effects is much more problematic. This explains why the majority of work carried out on simply inherited traits in potato has focussed on disease resistance genes, which are usually dominant to susceptibility. Thus, a large proportion of potato trait studies that are not directed towards disease resistance have necessarily had to deal with the genetics of complex traits. These traits include more complex types of pest and disease resistance, as well as the wide array of traits that impinge on potato tuber quality, nutritional value and tuber "life history" traits (e.g., tuberization, dormancy). Table 6-1 lists the main categories of complex potato traits that are covered in this chapter. In subsequent sections of this chapter, the most significant research on genetic analysis of complex potato traits will be summarized.

6.3 Quantitative Pest and Disease Resistance

A large number of monogenic disease resistances have been mapped in potato (reviewed in Chapter 5), including the well studied 11 resistance or "*R*" genes (*R1* to *R11*) active against late blight (caused by the oomycete *Phytophthora infestans*), several more recently identified genes providing defence against late blight, genes providing resistance against the potato cyst nematode (PCN) species *Globodera pallida* and *G. rostochiensis*, and several loci conferring resistance to different viruses. Genetically more complex

Table 6-1 Important quantitative traits in potato.

Quantitative traits in potato
Disease and pest resistance
Late blight resistance
Potato cyst nematode resistance
Trichome-mediated insect resistance
Glykoalkaloid-mediated insect resistance
Tuber morphological characteristics
Tuber shape
Eye depth
Tuber size uniformity
Tuber life-cycle
Tuberization
Dormancy/sprouting
Tuber Quality traits
Dry matter
Yield and Starch content
Cold sweetening

resistances are somewhat less numerous and more difficult to analyze but are arguably more useful from a breeder's perspective. Monogenic resistances operate largely via a "race-specific", "gene-for-gene" mechanism in which each plant resistance gene has a corresponding "avirulence" gene in the relevant pest or pathogen that can mutate so as to avoid recognition by the host plant. The outcome of this is that most of the single gene potato resistance genes, while conferring complete resistance against certain pathogen isolates, have been rapidly overcome when deployed in a field situation. Complex "horizontal" or "field resistances", normally partial in nature, are seemingly less likely to be rapidly overcome by the pathogen and may be more "durable".

6.3.1 Late Blight Resistance

One of the earliest published studies of quantitative resistance to late blight was by Leonards-Schippers et al. (1994), who employed restriction fragment length polymorphism (RFLP) markers and a leaf disc bioassay for quantitative resistance. Eleven loci on nine chromosomes were reported to show differences in resistance scores between marker genotypic classes. This study was one of the first to hint at the use of a candidate gene approach, as some of markers linked to resistance QTLs were derived from cloned "defense" genes invoked as part of the plant's response to pathogen attack. Collins et al. (1999) and Oberhagemann et al. (1999) in a parallel set of studies mapped foliage and tuber resistance QTL on 11 of the 12 potato linkage groups. These studies were the first to suggest that developmental or physiological effects could contribute to foliar resistance. Although it had been suggested long before this that blight resistance and plant maturity were associated (Toxopeus 1958), this was perhaps the first time that QTLs for the two traits were shown to co-localize on chromosome V. Moreover, these studies and the earlier study by Leonards-Schippers (1994) were the first to suggest that QTLs for late blight resistance often mapped to loci shown previously to contain major gene resistance, suggesting that quantitative resistance may be due to the action of less extreme allelic variants of major genes. A prime example of this is the co-localization of a major QTL with the *R1* locus near the marker GP179 on chromosome V.

Meyer et al. (1998) first reported quantitative late blight resistance in a now very well-studied tetraploid full-sib cross, which was part of a tetraploid potato breeding program. This cross was made between parents with complementary traits: breeding clone 12601ab1 with good processing quality and nematode resistance from *S. tuberosum* ssp. *andigena* and the table cultivar Stirling, with late blight resistance in both foliage and tubers. In this study, 94 F_1 individuals were used to map late blight resistance using amplified fragment length polymorphism (AFLP) markers. This paper illustrates the many challenges presented by tetrasomic inheritance, and along with more recent studies mentioned later, how far the "state of the art" has advanced in the past 10 years. In this initial study, only 573 of 3,173 AFLP loci observed were polymorphic and segregated in accordance with certain expected (presence:absence) segregation ratios due to simplex by nulliplex (1:1), simplex by simplex (3:1), duplex by nulliplex (5:1), and duplex by simplex (11:1) configurations in the two parents. Sixteen of the 94 F_1 clones were omitted from the genetic analysis due to a shortfall of paternal markers. Map construction was a complex process, entailing entering pairwise distances and LOD scores of 249 simplex markers, 32 double-simplex, and 117 duplex markers into JOINMAP software, standard procedures then being used to obtain both parental maps. Overall 22 major linkage groups could be identified (14 and eight from the female and male parents, respectively), with 34 small linkage groups of fewer than five markers. Many markers remained unlinked (13% and 29%), and the cumulative lengths of the female and male maps were 991 and 485 cM, respectively. The distribution of resistance scores in the population was highly skewed towards resistance, with 12601ab1 scoring higher (~ 8.0) than the 4.0–4.5 expected from previous tests, using the well-established 1–9 scale. A non-parametric test (Mann-Whitney) was used to detect marker-trait associations, with 21 associations between markers and blight resistance components detected at the 5% level. However, use of a stringent permutation test left only one association involving an AFLP marker (p75m48 = 5) that accounted for ~ 32% of the phenotypic variation. In an attempt to map the small linkage group containing p75m48 – 5 to a chromosome, STS markers were developed from the sequence of the cloned AFLP and mapped onto LG VIII of a diploid reference population. However, this result illustrates the risks inherent in using such an approach, as a later study demonstrated that this large effect QTL actually resides on chromosome IV! The study in question (Bradshaw et al. 2004) employed a larger set of clones (227) from the same cross (including 58 from the reciprocal of the original cross) to analyze late blight resistance as well as maturity and plant height. Linkage and QTL analysis benefited from the availability of improved methods (Luo et al. 2000, 2001; Hackett et al. 2001, 2003) too complex to discuss here, and which subsequently led to the publication of the TetraploidMap software (Hackett et al. 2007). High correlations were observed between foliage and tuber blight, field resistance and glasshouse foliage resistance, and maturity and all traits except glasshouse foliage resistance. Fourteen markers from "Stirling" were significantly associated with at least one trait and found to be present on chromosome IV and V by virtue of linkage to simple sequence repeat (SSR) markers that were mapped on this cross along with AFLP markers. These 14 linked markers accounted for varying amounts of phenotypic variation. An interval mapping approach was used for QTL detection. For chromosome V, the QTLs co-localized with between ~18–55% of the variation explained, and for linkage group IV the corresponding figures were 11–29%.

The now well-established association between quantitative resistance to late blight with late foliage maturity has provoked much interest. It was first described by Toxopeus (1958) and has been confirmed by other studies, such as those already described. This somewhat undesirable association has been hard to break by potato breeders, suggesting that either the horizontal resistance and foliage maturity are controlled by closely linked genes or that the genes controlling foliage physiological "ageing" also affect blight resistance in a pleiotropic manner. Colon (1994) showed that under short day-induced early maturity, resistance is lost, providing further evidence for the role of physiological changes during foliage maturation. Since the emergence of molecular marker technologies, several QTLs for race-non-specific resistance to *P. infestans*, mapping to all potato chromosomes, have been reported, but the most often reported QTL is the large effect locus on chromosome V. Analysis of populations analyzed for late blight and foliage maturity QTLs has shown that each of the relatively few loci for maturity co-localize with resistance QTL, with the most common association being with a locus on chromosome V near RFLP markers GP21 and GP179. Visker et al. (2003) further analyzed this particular association using a diploid population. Foliage maturity type showed a high heritability (0.79), with late blight resistance being somewhat less heritable (0.54). Late blight QTLs were detected on chromosomes III and V, with a significant interaction between the two loci explaining 15% of the total variation in the trait. A single QTL for foliage maturity type was found near marker GP21 on chromosome V and accounted for 84% of the total trait variation. If resistance scores were adjusted for maturity type, the magnitude of the QTL for resistance on chromosome V was reduced by one half, suggesting the possibility of two distinct genes at the locus, one with a pleiotropic effect on late blight resistance and foliage maturity type and another with an effect only on resistance. In a subsequent study (Visker et al. 2005), a set of six diploid crosses was used to identify QTLs for foliage resistance against *P. infestans* and for maturity type, and to assess their genetic relationship. As expected perhaps the most important locus for both traits was found on chromosome V near marker GP21! An additional QTL with a small effect on foliage maturity type was identified on chromosome III, and smaller QTLs for late blight resistance were found on chromosomes III and X, which seem to be independent of foliage maturity type and are not affected by epistatic effects of the locus on chromosome V. A further QTL was detected on chromosome VII following correction of resistance for the effect of maturity type.

Bradshaw et al. (2006) have further analyzed cultivar Stirling's late blight resistance QTL on chromosome IV at the diploid level, using a cross involving a resistant dihaploid clone. This work and other ongoing work suggests that these resistance QTL map very close to the major gene *R2* and the *Rpi-blb3* gene characterized by Park et al. (2005), suggesting that it may indeed be manifest by the action of one or more of the large number of nucleotide binding site-leucine rich repeat (NBS-LRR) genes found at this locus.

6.3.2 Potato Cyst Nematode Resistance

Kreike et al. (1993) published one of the earliest studies on mapping quantitatively inherited resistance to the PCN species *G. rostochiensis* in the wild diploid species *Solanum spegazzinii*. These authors reported the identification and mapping of two QTLs acting against pathotype *Ro1*. RFLP markers were used to map a diploid *S. spegazzinii* x *S. tuberosum* F_1 population, which segregated for the resistance phenotype since the wild species parent of this cross was apparently heterozygous at the nematode resistance loci. Two QTLs involved in disease resistance to *G. rostochiensis* were identified and mapped to chromosomes X and XI. In a later study (Kreike et al. 1996), the locus on chromosome X was named *Gro1.2*, and a new resistance locus, *Gro1.4*, also conferring resistance to *G. rostochiensis* pathotype *Ro1*, was found on chromosome III.

The aforementioned tetraploid cross was also used to analyze quantitative resistance to the PCN species *G. pallida* introgressed from *S. tuberosum* ssp. *andigena* (Bradshaw et al. 1998). This study employed the same 78 plants as the Meyer et al. (1998) study, as well as very similar linkage and QTL mapping methodologies. Single and double "dose" markers, whose segregation ratios could be unambiguously identified in the offspring, were used to test for linkages between markers and a putative PCN resistance QTL in the 12601ab1 parent. In short, a "double dose" QTL, explaining close to 30% of the phenotypic variation for the trait (square root of female cyst count) was found in the 12601ab1 parent to be linked in coupling to two duplex AFLP markers. Use of previously mapped SSR markers enabled the QTL to be localized to chromosome IV and a duplex allele of the SSR STM3016 was subsequently found to be in coupling with the locus. Other QTLs, explaining up to 15% of the variation, were found, but these could not be localized to any particular chromosome due to incomplete map coverage, unfavorable repulsion linkages between marker and trait alleles, or gene effects that were too minor to be detected in such a small population. Importantly, these authors suggested that QTL detection at the tetraploid level was quite feasible, but recommended that population sizes of 250 be used. They also suggested that co-dominant markers such as SSRs greatly improve the efficiency of linkage analysis. In a later study, Bryan et al. (2004) expanded the analysis to 227 clones from the same tetraploid cross. These were analyzed in part using a bulking method based on both phenotypic scores (i.e., high vs. low cyst counts) as well as marker scores for STM3016, which was found previously to be linked to the resistance. This analysis used many of the enhanced mapping methods used in the Bradshaw et al. (2004) study and proved highly effective, allowing these authors to confirm the results of the earlier study and to produce a much more marker-dense map of the chromosome IV region containing the large effect QTL. Moreover, analysis of residual scores after removal of the effects due to the major QTL allowed detection of a second, smaller QTL on chromosome XI, which explained 17% of the residual variance.

A second quantitative source of resistance to PCN that has been used extensively in potato breeding is derived from the wild diploid species *Solanum vernei*. In the UK, this source of resistance was introgressed into cultivated germplasm in the 1950s but was difficult to deploy in cultivars, possibly because its genetic basis was not established at the time. Bryan et al. (2002) used both a diploid and a tetraploid cross to map this resistance source. At the tetraploid level a "bulked segregant" approach was used to identify AFLP markers linked to a large effect QTL on linkage group V. A more conventional QTL analysis of the diploid cross further narrowed down the location of the QTL on chromosome V to the resistance "hotspot" between markers GP21 and GP179. This analysis also detected a second QTL on linkage group IX. Interestingly, an independent study using a cross whose resistant parent contained introgressions from several *G. pallida* resistant potato species, including *S. vernei*, had detected resistance loci in exactly the same locations (van der Voort et al. 2000). This work also established that the two QTLs acted in a more or less additive fashion. In this study, the phenotypic variances accounted for by the QTLs were 61% (*GpaV* on LG V) and 24% (*GpaVI* on LG IX). Together these QTLs appear to offer a high likelihood of durability against *G. pallida*.

A further source of quantitative resistance to *G. pallida* is the diploid wild species *S. spegazzinii*, which was crossed to a diploid *S. tuberosum* clone, Rosa H1 (Caromel et al. 2003). The nematode resistance scores in the progeny clearly behaved as a quantitative trait, showing a continuous distribution that conformed to normality when transformed on a log scale. A linkage map was constructed with RFLP and AFLP markers and three QTLs were found on chromosomes V, VI, and XII of the wild species parental map, suggesting resistance in this particular *S. spegazzinii* genotype is conferred by a completely different suite of resistance QTLs than was described earlier (Kreike et al. 1993). The QTL on chromosome V, *GpaM1*, was of the largest effect, explaining more than 50% of the phenotypic variation. All three QTLs mapped to known resistance gene clusters. This, along with other evidence, further suggests that components of quantitative resistance may be coded for by genes residing in major *R* gene clusters and, therefore, may be classic *R* genes themselves.

Caromel et al. (2005) also reported two QTLs conferring resistance to *G. pallida*, originating from the wild species *Solanum sparsipilum*. These two QTLs together explained 89% of the phenotypic variation, with the one mapping to the resistance hotspot on chromosome V displaying the major effect on cyst counts (R^2 = 76.6%). A key aspect of resistance to PCN is the ability of the host to reduce the development of nematode juveniles into females. The two QTLs on chromosomes V and XI restricted the development into females to 2.3% and 50.8%, respectively. However, clones in the mapping population carrying both QTLs restricted female development to 0.8%. Moreover, these clones displayed a distinct necrotic reaction in roots infected by nematodes, which was absent in plants containing either QTL alone. These results suggest that to make effective use of a resistance strategy that has evolved in a wild species such as *S. sparsipilum*, it is important to introgress all of the resistance components (in this case two unlinked QTLs) into adapted material for effective deployment in potato breeding.

6.3.3 Trichome-mediated Insect Resistance

Bonierbale et al. (1994) analyzed trichome-mediated insect resistance in potato, using a wild species, *Solanum berthaultii*, which possesses two types of secretory glandular trichomes (types A and B), products of which act to provide physical and chemical obstacles to attack by insects such as Colorado potato beetle (CPB; *Leptinotarsa decemlineata*), aphids and leafhoppers. These authors used two backcross progenies generated by crossing a hybrid between *S. tuberosum* and *S. berthaultii* to the two parental clones. They performed a wide range of phenotypic assays: direct observation of trichome phenotypes (types A and B), biochemical analysis of type-A trichomes (enzymatic browning, polyphenol oxidase concentration), and properties of type-B trichomes (presence/absence of sugar droplets, glucose assays). In this study, following phenotyping of 300 of the "BCB" (backcross to *S. berthaultii*) population, a subset of 150 individuals from the "tails" of the phenotypic distributions were selected for further phenotypic investigations and genotypic analysis with RFLP markers. The other backcross, "BCT" (backcross to *S. tuberosum*), was analyzed in a similar way, although with a phenotypic focus on the type-A trichomes, as the population did not express type-B sugar droplets, presumably a recessive character. Resistance to CPB was also assessed. Linkage analysis was compromised by the relatively small numbers of RFLP markers that could be used, by significant marker skewing, and by reduced marker segregation in the recurrent parents. The two backcross populations differed significantly in their phenotypic properties, with BCB segregating for all trichome traits and BCT segregating only for type-A trichome traits. Two QTLs were identified for the type-A trichome "browning" trait (MEBA assay) in BCB. These were located on chromosomes VI and X and explained 52% and 20% of the phenotypic variation, respectively. These two QTL regions were also found to contain QTLs for the density of type-A trichomes (40% and 27% explained, respectively). The backcross to *S. tuberosum* only segregated for the browning QTL on chromosome VI, with a lower level of variation explained (34%). The same co-localization of the browning QTL with a trichome density QTL was seen. The sucrose droplet phenotype of the type-B trichomes behaves as a single recessive gene (named *bdr* in this study), mapping to chromosome V in the BCB cross. Five QTLs, explaining ~ 68% of the phenotypic variation in total, were found for sucrose ester levels. One of these was close to *bdr*. Two QTLs were detected for type-B trichome density in population BCB, the largest of which mapped on chromosome V near to the locus influencing sucrose ester levels. Several small QTLs were detected by analyzing heterozygous loci from the recurrent parents but these will not be discussed.

6.3.4 Glycoalkaloid-mediated Insect Resistance

Leptines are a class of glycoalkaloids, high levels of which, in common with presence of leaf glandular trichomes, provide natural forms of resistance to insect pests such as CPB. Leptines are related to solanidine, an alkaloid found in cultivated potatoes. The ability to produce leptines appears to be limited to rare accessions of the wild species *Solanum chacoense* Bitt., and then only in a few individuals of any accession (Sinden et al. 1986). These observations have been taken to suggest that leptine synthesis is controlled by one or a small number of genes and observations from crosses between high and low glycoalkaloid clones suggest that high levels of leptine are suppressed by dominant "suppressor" alleles (Sanford et al. 1994, 1996). The additive, recessive nature of this trait creates severe problems for deployment in a tetraploid outbreeder. Bouarte-Medina et al. (2002) attempted to shed more light on the genetic control of leptine by use of reciprocal backcross populations between high (*S. chacoense*) and low (*S. phureja*) leptine progenitor clones. Leptine in the form of acetyleptinidine (ALD) was measured in leaves using gas chromatography. ALD content was found to segregate most widely in the backcross to the low-leptine *S. phureja* parent that had *S. phureja* cytoplasm, segregating at an approximately 1:1 (high:low) ratio. Among the heterozygous backcross segregants of this cross that contained leptine, leptine levels were lower than those of the heterozygous F_1 progenitor, suggesting a pattern of genetic control more complex than that of a single gene. The observation that the backcross with cytoplasm from *S. chacoense* showed a different pattern of segregation suggests a cytoplasmic effect on leptine production. A bulked segregant approach was used to identify random amplified polymorphic DNA (RAPD) and cleaved amplified polymorphic sequence (CAPS) markers linked to QTLs for ALD production in the backcross populations as well as a small population of monoploids derived from the *S. chacoense* x *S. phureja* F_1 hybrid clone. Three RAPD primers amplified fragments exclusively in high ALD bulks, suggesting that these markers are in coupling to ALD content. The low marker density precluded a full linkage analysis, so chromosomal assignments of the putative QTLs were not achieved.

6.4 Tuber Life-cycle Traits: Dormancy and Tuberization

The process of tuberization itself, and related traits such as tuber dormancy and maturity, have been subjected to extensive quantitative genetic analysis. These traits, which are of considerable importance to potato agronomy, represent a major challenge to genetical analysis, a fact borne out by the small number of published studies and the complexity of the results. Many of these studies have been conducted using crosses between cultivated potato genotypes with different wild species or landrace material. The rationale for this is primarily to expand the range of variation in the target traits. Particularly, in contrast with cultivated potato, many wild species lack or show extreme tuber dormancy, or are classified as late tuberizing-especially under the long summer days typical of North American or European climes.

Tuber dormancy, following early attempts by Simmonds (1964) and Thompson et al. (1980), has been studied by Freyre et al. (1994), who investigated the trait using a diploid population generated by crossing an *S. tuberosum/S. chacoense* hybrid to a *S. phureja* clone. In this population, tuber dormancy varied from 10 to 90 days, with a mean of 19 days. One-way ANOVA was used to identify six putative QTLs on chromosomes II, III, IV, V, VII and VIII, the largest mapping to chromosome VII. A multilocus model, explaining ~58% of the phenotypic variation for dormancy was fitted, and there was some evidence for epistatic interactions. In a subsequent study, van den Berg et al. (1996) used the same reciprocal backcross populations between *S. tuberosum* and *S. berthaultii* parents used to study trichome mediated resistance (see above) to perform QTL analysis of tuber dormancy. *Solanum berthaultii*, a wild Bolivian species, is characterized by a short day requirement for tuberization and long tuber dormancy. QTLs for dormancy were detected on nine chromosomes, with alleles from the wild species promoting dormancy. The largest effect, which explained ~31% of the variation, resided on chromosome II. Having an *S. tuberosum* allele at this locus shortened dormancy by 29 days! In the backcross to *S. berthaultii*, the additive effects of dormancy QTLs could explain 48% of the measured phenotypic variance, with only small epistatic effects being detected. Some of the dormancy QTLs appeared to map to loci associated with tuberization in the same crosses. These authors speculated on the implications of their findings for potato breeding, the need for controlling tuber dormancy being of primary importance. Comparisons between these two studies suggests that it is possible that different potato species may have qualitatively different genetic mechanisms for controlling dormancy and that controlling tuber dormancy within *S. tuberosum* varieties is not necessarily helped by recourse to studying wild species of potato with extreme dormancy phenotypes.

The aforementioned *S. tuberosum* by *S. berthaultii* interspecific backcrosses were used to study the process of tuberization (van den Berg et al. 1996). Earliness of tuberization was measured using an innovative method developed previously by Ewing (1978). This method involves planting stem cuttings such that buds were below the surface of the potting medium and subsequent assessments of tuber initiation at the bud. Whole plant earliness tests were also performed. A selective genotyping approach, such as that used in the trichome analysis work by the same authors, was adopted. Eleven loci spread across seven potato chromosomes were identified. Most loci were of small effect except for a locus on chromosome V, which accounted for 27% of the phenotypic variance. Most of the alleles favoring early tuberization were at least partly dominant. Surprisingly, at three loci, tuberization was favored by an allele of the wild species parent. It is interesting to postulate that the QTL on chromosome V is due to the same locus as that which affects plant maturity type reported in later works. Some of the loci for trichome-mediated insect resistance were found to be associated with tuberization QTLs, a finding that has significant implications for potato breeders trying to select for earliness and insect resistance.

More recently, Ewing et al. (2004) used a "polygene mapping" approach to study the genetics and physiology of tuberization and dormancy. They suggest that both genetic and physiological knowledge should be used in breeding for particular tuberization or tuber dormancy phenotypes. These authors performed QTL analysis of tuberization and tuber dormancy in a segregating diploid population (also involving *S. berthaultii*) and also mapped QTLs for levels of hormones implicated in the control of these traits. Large numbers of QTLs were detected for tuberization and tuber dormancy. Also detected were QTLs for polyamines, abscisic acid, tuberonic acid, gibberellin A and several other compounds. Some of the hormone QTLs appeared to map close to QTLs for tuberization or dormancy. In a way, this is an early use of a candidate gene (or hormone) approach to the study of important tuber traits.

6.5 Tuber Morphological Traits: Tuber Shape and Eye Depth

Tuber shape is one of the more straightforward potato traits to score and it is a trait of considerable significance to the potato processing industry. Long tubered varieties are preferred for French fries, but varieties with round tubers are used for making potato chips. Despite being easy to score, the trait does exhibit a degree of quantitative variation, although early studies suggested it to be a monogenic trait (Masson 1985), with round (*Ro*) being dominant to long (*ro*). Van Eck et al. (1994) analyzed tuber shape as a quantitative trait in a complex backcross involving two diploid clones, an interspecific hybrid of *S. vernei*-*S. tuberosum* origin and US-W5337.3, a *S. phureja* x *S. tuberosum* hybrid. The F_1 between these clones was backcrossed to US-W5337.3. Both parents of the F_1 generation have round tubers and 102 BC_1 progeny clones were assessed for shape. Phenotypic data were scored in two ways, simple classification into round vs. long, and, as a quantitative trait, as tuber length/width ratio. The progeny scores conformed to a 3:1 expected ratio of round:long clones, and progeny genotypes were confirmed by further crosses of progeny and parents. Cosegregation of several chromosome X RFLP markers with tuber shape was observed. These authors adopted a "gametic" mapping strategy, whereby maps were constructed separately for male and female parents of the cross. The female parent showed higher levels of recombination than the male, possibly due to the higher level of wild species-derived chromosomal material in the male parent. The quantitative analysis, based on only 50 progeny clones, showed the "ratio" trait to have a high heritability (0.8) with hardly any environmental variation. However, the observed differences between clones of the same class suggested the action of other minor genetic factors. A model suggested that the "main" locus explained 75% of the variation and the minor loci 25%. A more detailed modelling analysis further revised these components to an 80/20% split between major/minor loci. Further study of the four observed genotypic classes, and slight differences in shape of clones carrying maternal and paternal *Ro* alleles, suggested that the two parents contained two different dominant *Ro* alleles.

Tuber eye depth is a major component of tuber quality, with deep eyes detracting from the visual appearance of tubers and adding significantly to the cost of peeling in processing factories. Li et al. (2005) used a cultivated diploid potato family of 107 plants to dissect this trait. The family segregated for both eye depth (deep vs. shallow) and tuber shape (round vs. long) traits. The deep eye (Eyd) phenotype was found to be associated with round tubers (*Ro*) in most progeny clones. Further evaluation of this population with molecular markers, including SSR, AFLP, and sequence characterized amplified region (SCAR) markers revealed that the primary locus for eye depth is located on chromosome X. This map location was confirmed by evaluating a second diploid family. While this trait appears to behave as a simply inherited one, other studies (see below) suggest a more complex genetic architecture.

In a very recent study, Sliwka and co-authors (2008) analyzed another population segregating for traits introgressed from the wild species *S. verrucosum* and *S. microdontum* into a cultivated background. The 98-21 population, previously analyzed for late blight resistance, was re-examined for segregation of tuber dormancy, tuber shape, regularity of tuber shape, eye depth and flesh color. Some of these traits have been analyzed previously as simple traits in other studies (e.g., tuber shape, eye depth, flesh color) and major loci have been mapped on chromosome X and III. The current study found the largest QTLs on chromosomes II (dormancy), X (eye depth), IV (flesh color), II (shape) and III (shape regularity), with several smaller QTLs for all traits detected. This study shows the presence both of seemingly "universal" QTLs for some of these traits and new QTLs in previously unseen locations. This work also demonstrates the overall complexity of the studied traits, suggesting that breeding for them may be quite difficult. However, use of a population segregating for tuber and resistance traits allows identification of clones possessing resistance as well as good tuber characters that can be used as progenitors of new cultivars.

6.6 Tuber Quality Traits

6.6.1 Dry Matter

Dry matter content in potato is very important to the processing industry and is estimated by measuring specific gravity (SG). This trait, effectively the amount of dry matter vs. water in a potato, plays a large part in determining the food value of a potato. Dry matter is known to be influenced by environmental factors, such as temperature, rainfall, and day length. Freyre and Douches (1994) performed one of the earliest analyses of this trait, evaluating a diploid population in three environments to assess the effects of G x E interactions. Ten QTLs on six different chromosomes were detected, but there was significant variation across environments. The amount of variation accounted for by individual loci, as determined by their R^2 value, ranged from 4.0 to 15.8%. The use of a multilocus model, whereby the largest QTLs detected in the different environments was included, accounted for up to 45% of the phenotypic variation for this trait. Perhaps most significantly, a version of the model using average SG value across environments explained consistent proportions of the phenotypic variation when tested in each of the different environments.

6.6.2 Cold Sweetening

One of the most elusive and commercially relevant traits in potato is cold temperature sweetening. Potato tubers, in common with other plant storage organs, accumulate both reducing sugars (glucose and fructose) and sucrose when stored in cool temperatures—this phenomenon is known as cold sweetening (Burton 1969). Cold sweetening is due to a shift in the balance between starch degradation and glycolysis, which leads to accumulation of sucrose, which in turn is converted to glucose and fructose. The storage of tubers below 10°C to delay sprouting is a nearly universal practice. Thus, the accumulation of reducing sugars in stored potatoes is a nearly universal occurrence. High reducing sugar content reduces potato processing quality since high temperatures used in frying cause a "non-enzymatic" Maillard reaction between free aldehyde groups of reducing sugars and free α-amino groups of amino acids and proteins, resulting in a dark, bitter product (Talbut and Smith 1975). Despite elucidation of many of the pathways and enzymes, there is little knowledge of individual genetic contributions to the trait.

Sugar content of tubers is a fairly highly heritable trait and so would seem to be well suited to genetic analysis. Douches and Freyre (1994) attempted to identify genetic factors using a diploid population and a visual "chip color" assay on French fries after storage of tubers at 10°C. A total of 13 molecular markers, representing six QTLs on chromosomes II, IV, V, and X were associated with the trait. Fitting of different models accounted for 43.5-50.5% of the phenotypic variation. Interestingly, all of the significant marker associations were identified in one parent, a *S. tuberosum-S. chacoense* hybrid. This may be due in part to lower levels of marker polymorphism observed in the *S. phureja* male parent.

In a later study, Menendez et al. (2002) tried to bridge the gap between conventional QTL mapping and knowledge of the genes involved in the relevant metabolic pathways. Cold sweetening is very amenable to such a "candidate gene" approach, as the metabolic pathways involved in sugar metabolism are well known and many of the genes had previously been characterized at the molecular and biochemical level. These authors had previously published a "function map" for genes involved in carbohydrate transport and metabolism (Chen et al. 2001). The later study analyzed two diploid populations and tuber sugar levels were scored after storage for 3 months at 4°C. Marker analysis involved both AFLPs (to "fill out" linkage maps) and RFLP and CAPS markers developed from 10 candidate genes. Sugar contents were found to exhibit "transgressive" segregation (i.e. exhibiting a range outside the parental values) in all six environments used. QTLs for glucose, fructose and sucrose content were detected on all chromosomes, with QTLs for glucose and fructose tending to co-localize. Larger-effect loci were found on six linkage groups. Candidate genes found to be linked to QTLs included invertases, sucrose synthase 3, sucrose phosphate synthase, a sucrose transporter gene, and a putative sucrose sensor. The authors concluded that allelic variation in genes coding for enzymes in carbohydrate metabolism could contribute to genetic variation in cold sweetening.

6.6.3 Tuber Yield and Starch Content

Schafer-Pregl et al. (1998) performed an analysis of QTLs for potato tuber yield and starch content using RFLP markers. QTLs for tuber starch-content and tuber yield were mapped in two diploid populations using both single-marker tests and interval mapping. These authors were the first to use a model specifically developed for interval QTL mapping in non-inbred plant species. Eighteen putative QTLs for tuber starch-content were identified on all 12 potato linkage groups. Eight putative QTLs for tuber yield were identified on eight different linkage groups. Twenty of the QTLs were reproducibly detected in at least two environments and/or mapping populations. Very few major QTLs for tuber starch-content were highly stable across environments and most were detected in only one of the two mapping populations analyzed. The majority of QTLs for tuber yield were linked with QTLs for tuber starch-content, suggesting that the effects on both traits are controlled by the same genetic factors.

Bradshaw et al. (2008) performed QTL analysis for 16 yield, agronomic, and quality traits on 227 clones from the previously described tetraploid full-sib family (12601ab1 x "Stirling"). Although it was not possible to identify all linkage groups, 39 QTLs were detected for a range of traits using an interval mapping approach. All of the traits scored had moderately high heritabilities with 54–92% of the variation in clone means over three years and two replicates being due to genetic differences. The largest effect detected, residing on chromosome V, was for maturity and explained 56% of the phenotypic variance. The other QTLs individually explained between ~ 6–17% of the phenotypic variance for the various characters. Only two QTLs were found for yield. Those resided on chromosomes I and VI.

6.7 How Can the Genes for Potato Quantitative Traits be Identified?

All of the aforementioned studies, while contributing greatly to an understanding of trait "architectures" for complex potato traits, have provided relatively scant information about the genes involved in their manifestation. There are exceptions to this, such as the adoption of candidate gene approaches to study traits for which there is some *a priori* knowledge of the genes involved (e.g., Menendez et al. 2002).

There are also studies that have used more advanced genetic or gene expression profiling methods to move towards isolating genes for complex traits. Fernandez-Del-Carmen et al. (2007) performed "targeted transcript mapping" using the highly multiplex cDNA-AFLP on bulks of plants from a diploid cross segregating for early and late tuberization (i.e. maturity). Thirty-seven transcripts were found to segregate and almost 50% mapped to the well-known chromosome V locus. This suggests that cDNA analysis applied to bulks tends to identify actual transcript fragments linked to the trait locus rather than candidate genes for the trait. Nonetheless, these markers were used to identify large-insert BAC clones from the locus—perhaps the first step in map-based cloning of this important gene.

For QTLs involved in resistance traits, it is clear that the majority map to locations of major resistance genes, so it is likely that the resistance QTL phenotype is imparted by the action of the same type of gene. Now that knowledge of such genes in potato is well advanced, it should be possible to clone all of the "cognate" *R* genes at any particular locus. However, the functional analysis of these remains problematical in that the effects of such

QTLs may be relatively small yet demonstrating function requires extensive resistance testing of transgenic plants, which may only be feasible for QTLs of relatively large effect.

The impending availability of a potato genome sequence (*www.potatogenome.net*) will enhance the prospects for identifying genes underlying the components of complex potato traits and for making more efficient use of them in potato breeding programs.

References

Bonierbale MW, Plaisted RL, Pineda O, Tanksley SD (1994) QTL analysis of trichome-mediated insect resistance in potato. Theor Appl Genet 87: 973–987.

Bouarte-Medina T, Fogelman E, Chani E, Miller AR, Levin I, Levy D, Veilleux RE (2002) Identification of molecular markers associated with leptine in reciprocal backcross families of diploid potato. Theor Appl Genet 105: 1010–1018.

Bradshaw JE, Meyer RC, Milbourne D, McNicol JW, Phillips MS, Waugh R (1998) Identification of AFLP and SSR markers associated with quantitative resistance to *Globodera pallida* (Stone) in tetraploid potato (*Solanum tuberosum* subsp. *tuberosum*) with a view to marker-assisted selection. Theor Appl Genet 97: 202–210.

Bradshaw JE, Pande B, Bryan GJ, Hackett CA, McLean K, Stewart HE, Waugh R (2004) Interval mapping of quantitative trait loci for resistance to late blight [*Phytophthora infestans* (Mont.) de Bary], height and maturity in a tetraploid population of potato (*Solanum tuberosum* subsp. *tuberosum*). Genetics 168: 983–995.

Bradshaw JE, Hackett CA, Lowe R, McLean K, Stewart HE, Tierney I, Vilaro MDR, Bryan GJ (2006) Detection of a quantitative trait locus for both foliage and tuber resistance to late blight [*Phytophthora infestans* (Mont.) de Bary] on chromosome 4 of a dihaploid potato clone (*Solanum tuberosum* subsp. *tuberosum*). Theor Appl Genet 113: 943–951.

Bradshaw JE, Hackett CA, Pande B, Waugh R, Bryan GJ (2008) QTL mapping of yield, agronomic and quality traits in tetraploid potato (*Solanum tuberosum* subsp. *tuberosum*). Theor Appl Genet 116: 193–211.

Bryan GJ, McLean K, Bradshaw JE, Phillips M, Castelli L, De Jong WS, Waugh R (2002) Mapping QTL for resistance to the cyst nematode *Globodera pallida* derived from the wild potato species *Solanum vernei*. Theor Appl Genet 105: 68–77.

Bryan GJ, McLean K, Pande B, Purvis A, Hackett CA, Bradshaw JE, Waugh R (2004) Genetical dissection of H3-mediated polygenic PCN resistance in a heterozygous autotetraploid potato population. Mol Breed 14: 105–116.

Burton WG (1969) The sugar balance in some British potato varieties during storage. II. The effects of tuber age, previous storage temperature, and intermittent refrigeration upon low-temperature sweetening. Eur Potato J 12: 81–95.

Caromel B, Mugniery D, Lefebvre V, Andrzejewski S, Ellisseche D, Kerlan MC, Rousselle P, Rousselle-Bourgeois F (2003) Mapping QTLs for resistance against *Globodera pallida* (Stone) Pa2/3 in a diploid potato progeny originating from *Solanum spegazzinii*. Theor Appl Genet 106: 1517–1523.

Caromel B, Mugniery D, Kerlan MC, Andrzejewski S, Palloix A, Ellisseche D, Rousselle-Bourgeois F, Lefebvre V (2005) Resistance quantitative trait loci originating from *Solanum sparsipilum* act independently on the sex ratio of *Globodera pallida* and together for developing a necrotic reaction. Mol Plant-Microbe Interact 18: 1186–1194.

Chen X, Salamini F, Gebhardt C (2001) A potato molecular-function map for carbohydrate metabolism and transport. Theor Appl Genet 102: 284–295.

Collins A, Milbourne D, Ramsay L, Meyer R, Chatot-Balandras C, Oberhagemann P, De Jong W, Gebhardt C, Bonnel E, Waugh R (1999) QTL for field resistance to late blight in potato are strongly correlated with maturity and vigour. Mol Breed 5: 387–398.

Colon LT (1994) Resistance to *Phytophthora infestans* in *Solanum tuberosum* and wild *Solanum* species. PhD thesis. Wageningen Agricultural Univ, Wageningen, The Netherlands.

Douches DS, Freyre R (1994) Identification of genetic-factors influencing chip color in diploid potato (*Solanum* spp.). Am Potato J 71: 581–590.

Ewing EE (1978) Critical photoperiods for tuberization—screening technique with potato cuttings. Am Potato J 55: 43–53.

Ewing EE, Simko I, Omer EA, Davies PJ (2004) Polygene mapping as a tool to study the physiology of potato tuberization and dormancy. Am J Potato Res 81: 281–289.

Fernandez-Del-Carmen A, Celis-Gamboa C, Visser RGF, Bachem CWB (2007) Targeted transcript mapping for agronomic traits in potato. J Exp Bot 58: 2761–2774.

Freyre R, Douches D (1993) Mapping quantitative trait loci of tuber traits in diploid potato (*Solanum* spp.). HortScience 28: 104.

Freyre R, Douches DS (1994) Development of a model for marker-assisted selection of specific-gravity in diploid potato across environments. Crop Sci 34: 1361–1368.

Freyre R, Warnke S, Sosinski B, Douches DS (1994) Quantitative trait locus analysis of tuber dormancy in diploid potato (*Solanum* spp). Theor Appl Genet 89: 474–480.

Hackett CA, Bradshaw JE, McNicol JW (2001) Interval mapping of QTLs in autotetraploid species. Genetics 159: 1819–1832.

Hackett CA, Pande B, Bryan GJ (2003) Constructing linkage maps in autotetraploid species using simulated annealing. Theor Appl Genet 106: 1107–1115.

Hackett CA, Milne I, Bradshaw JE, Luo Z (2007) Tetraploid Map for Windows: Linkage map construction and QTL mapping in autotetraploid species. J Hered 98: 727–729.

Kreike CM, Dekoning JRA, Vinke JH, Vanooijen JW, Gebhardt C, Stiekema WJ (1993) Mapping of loci involved in quantitatively inherited resistance to the potato cyst-nematode *Globodera rostochiensis* pathotype-Ro1. Theor Appl Genet 87: 464–470.

Kreike CM, KokWesteneng AA, Vinke JH, Stiekema WJ (1996) Mapping of QTLs involved in nematode resistance, tuber yield and root development in *Solanum* spp. Theor Appl Genet 92: 463–470.

Leonards-Schippers C, Gieffers W, Schaefer-Pregl R, Ritter E, Knapp SJ, Salamini F, Gebhardt C (1994) Quantitative resistance to *Phytophthora infestans* in potato: A case study for QTL mapping in an allogamous plant species. Genetics 137: 68–77.

Li X-Q, De Jong H, De Jong DM, De Jong WS (2005) Inheritance and genetic mapping of tuber eye depth in cultivated diploid potatoes. Theor Appl Genet 110: 1068–1073.

Luo ZW, Hackett CA, Bradshaw JE, McNicol JW, Milbourne D (2000) Predicting parental genotypes and gene segregation for tetrasomic inheritance. Theor Appl Genet 100: 1067–1073.

Luo ZW, Hackett CA, Bradshaw JE, McNicol JW, Milbourne D (2001) Construction of a genetic linkage map in tetraploid species using molecular markers. Genetics 157: 1369–1385.

Masson MF (1985) Mapping combining abilities, heritabilities, and heterosis with 4x X 2x crosses in potato. PhD Thesis, Univ of Wisconsin, Madison, WI, USA.

Menendez CM, Ritter E, Schafer-Pregl R, Walkemeier B, Kalde A, Salamini F, Gebhardt C (2002) Cold sweetening in diploid potato: mapping quantitative trait loci and candidate genes. Genetics 162: 1423–1434.

Meyer RC, Milbourne D, Hackett CA, Bradshaw JE, McNichol JW, Waugh R (1998) Linkage analysis in tetraploid potato and association of markers with quantitative resistance to late blight (*Phytophthora infestans*). Mol Gen Genet 259: 150–160.

Muth J, Hartje S, Twyman RM, Hofferbert H-R, Tacke E, Prüfer D (2008) Precision breeding for novel starch variants in potato. Plant Biotechnol J 6: 576–584.

Oberhagemann P, Chatot-Balandras C, Schafer-Pregl R, Wegener D, Palomino C, Salamini F, Bonnel E, Gebhardt C (1999) A genetic analysis of quantitative resistance to late blight in potato: towards marker-assisted selection. Mol Breed 5: 399–415.

Park TH, Gros J, Sikkema A, Vleeshouwers VGAA, Muskens M, Allefs S, Jacobsen E, Visser RGF, Vossen EAGvd (2005) The late blight resistance locus *Rpi-blb3* from *Solanum bulbocastanum* belongs to a major late blight R gene cluster on chromosome 4 of potato. Mol Plant-Microbe Interact 18: 722–729.

Sanford LL, Deahl KL, Sinden SL (1994) Glycoalkaloid content in foliage of hybrid and backcross populations from a *Solanum tuberosum* x *S chacoense* cross. Am Potato J 71: 225–235.

Sanford LL, Kobayashi RS, Deahl KL, Sinden SL (1996) Segregation of leptines and other glycoalkaloids in *Solanum tuberosum* (4x) x *S. chacoense* (4x) crosses. Am Potato J 73: 21–33.

Schafer-Pregl R, Ritter E, Concilio L, Hesselbach J, Lovatti L, Walkemeier B, Thelen H, Salamini F, Gebhardt C. (1998) Analysis of quantitative trait loci (QTLs) and quantitative trait alleles (QTAs) for potato tuber yield and starch content. Theor Appl Genet 97: 834–846.

Simmonds NW (1964) Genetics of seed and tuber dormancy in cultivated potatoes. Heredity 19: 489–504.

Sinden SL, Sanford LL, Cantelo WW, Deahl KL (1986) Leptine glycoalkaloids and resistance to the Colorado Potato Beetle (Coleoptera, Chrysomelidae) in *Solanum chacoense*. Environ Entomol 15: 1057–1062.

Sliwka J, Wasilewicz-Flis I, Jakuczun H, Gebhardt C (2008) Tagging quantitative trait loci for dormancy, tuber shape, regularity of tuber shape, eye depth and flesh colour in diploid potato originated from six *Solanum* species. Plant Breed 127: 49–55.

Talbut WF, Smith O (1975) Potato Processing. 3rd edn. AVI Publ, Westport, CT, USA.

Thompson PG, Haynes FL, Moll RH (1980) Estimation of genetic variance components and heritability for tuber dormancy in diploid potatoes. Am Potato J 57: 39–46.

Toxopeus HJ (1958) Some notes on the relations between field resistance to *Phytophthora infestans* in leaves and tubers and ripening time in *Solanum tuberosum* ssp. *tuberosum*. Euphytica 7: 123–130.

van den Berg JH, Ewing EE, Plaisted RL, McMurry S, Bonierbale MW (1996) QTL analysis of potato tuber dormancy. Theor Appl Genet 93: 317–324.

van der Voort JR, van der Vossen E, Bakker E, Overmars H, van Zandroort P, Hutten R, Lankhorst RK, Bakker J (2000) Two additive QTLs conferring broad-spectrum resistance in potato to *Globodera pallida* are localized on resistance gene clusters. Theor Appl Genet 101: 1122–1130.

Van Eck HJ, Jacobs JME, Stam P, Ton J, Stiekema WJ, Jacobsen E (1994) Multiple alleles for tuber shape in diploid potato detected by qualitative and quantitative genetic-analysis using RFLPs. Genetics 137: 303–309.

Visker M, Keizer LCP, Van Eck HJ, Jacobsen E, Colon LT, Struik PC (2003) Can the QTL for late blight resistance on potato chromosome 5 be attributed to foliage maturity type? Theor Appl Genet 106: 317–325.

Visker M, Heilersig H, Kodde LP, Van de Weg WE, Voorrips RE, Struik PC, Colon LT (2005) Genetic linkage of QTLs for late blight resistance and foliage maturity type in six related potato progenies. Euphytica 143: 189–199.

Population Genetics and Association Mapping

Christiane Gebhardt

ABSTRACT

Association mapping is an experimental approach to identify DNA variants that are associated, directly or indirectly, with phenotypic variation in natural populations. In potato, the most useful for association mapping of agronomic traits are populations of tetraploid or diploid cultivars and landraces. One cornerstone of association mapping is the phenotypic evaluation of the individuals in a population, the other is the genotyping of the same individuals with DNA-based markers. Various test statistics are then used to detect marker-trait associations. In this chapter, methodology, results and lessons learned of the first generation association mapping studies performed in potato are summarized.

Keywords: Association mapping, chip color, disease resistance, molecular markers, potato (*Solanum tuberosum*), starch, yield

7.1 Introduction

The phenotypic differences observed between individuals of the same species are the consequence of both genetic and environmental variation. The cause of natural genetic variation is random mutations, which alter an individual's genomic DNA sequence at single nucleotide positions (point mutations) or over DNA stretches from a few nucleotides to whole chromosome arms (insertions, deletions, inversions, translocations). Neutral mutations do not have any detectable phenotypic effect in a given environmental context in the individuals that carry them. The majority of DNA-based markers

MPI for Plant Breeding Research, Carl von Linné Weg 10, 50829 Köln, Germany; e-mail: *gebhardt@mpiz-koeln.mpg.de* result from mutations in this category. On the other hand, non-deleterious mutations in genes that are causal for a given phenotype must be responsible for the genetic portion of the phenotypic variation among individuals. These mutations may affect coding and/or regulatory portions of a gene and give rise to alleles that modify the phenotype in one or another direction. Let's assume the DNA sequences of all gene(s) that control a particular phenotype are known. In this case, the natural molecular variation of these genes can be evaluated in populations of individuals, which have been examined for their phenotype values. The causal DNA variants will then be directly associated with the phenotypic variation (Kerem et al. 1989). In potato, there is presently no example for this ideal situation. However, not only causal DNA variants in causal gene(s) show association with the phenotype but also physically linked DNA polymorphisms within the gene(s) itself or in the chromosomal regions flanking the gene, which do not have a direct functional role in producing the phenotype. This is due to the common transmission of physically linked DNA polymorphisms through successive meiotic generations (linkage disequilibrium = LD). LD is a function of the recombination frequency between polymorphic loci and the number of meiotic generations separating the ancestral haplotype (the combination of DNA polymorphisms present on a chromosome) from the haplotypes in the population analyzed, assuming random mating among the individuals (Hartl and Clark 1997). With each generation, the common transmission of physically linked DNA polymorphisms is reduced in proportion to the recombination fraction until linkage equilibrium (LE) is reached, that is, recombinant and non-recombinant haplotypes are equally distributed in the population (Fig. 7-1). The tighter the physical linkage between two loci, the longer linkage disequilibrium persists over multiple meiotic generations in a population due to the low recombination frequency. A DNA-based marker linked, for example, to an unknown gene underlying a quantitative trait locus (QTL) for pathogen resistance, might be but is not necessarily in LD with that gene in a population of individuals descending from common ancestors (related by descent). If the marker is in LD with the gene underlying the QTL, specific marker alleles distributed at a given frequency in such a population will be associated with the variation of disease resistance. Marker-trait associations as compared to marker-trait linkages have the advantage that a marker associated with a particular phenotype is diagnostic in wide germplasm pools, whereas a linked marker is diagnostic primarily in progeny descending from a specific carrier of a specific trait allele.

Association mapping is an experimental approach to identify DNA variants associated, directly or indirectly, with phenotypic variation in natural populations. Concept and methodology have been developed for human populations, where experimental populations for genetic studies

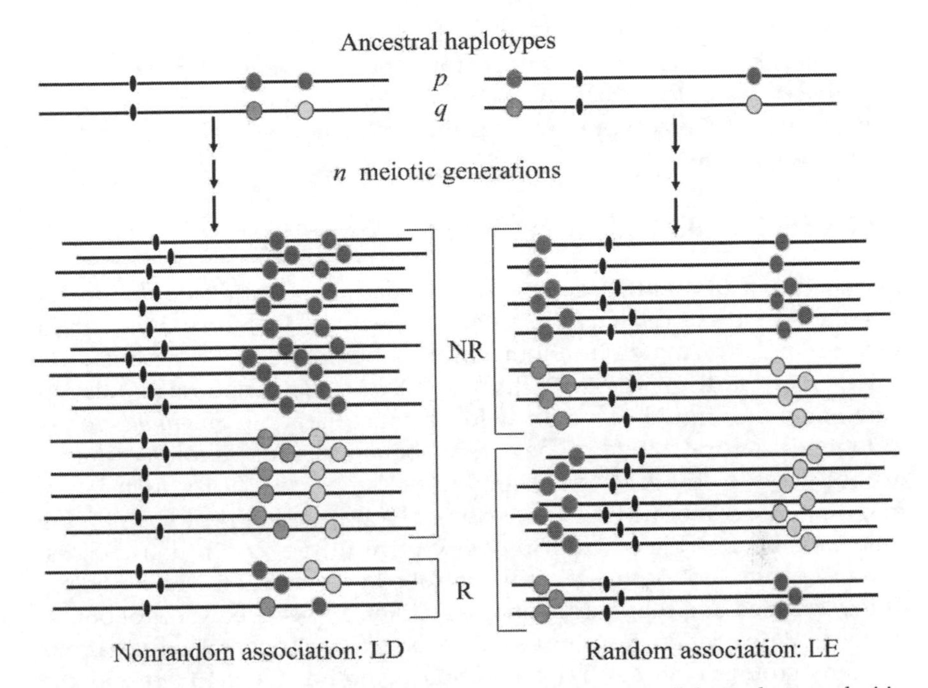

Figure 7-1 Nonrandom versus random association between two alleles each at two loci in populations of chromosomes descending from two ancestral haplotypes. Loci are symbolized by spheres and alleles by different colored spheres. On the left, the loci are closely linked. After *n* meiotic generations, the majority of the descendant chromosomes carry the same, non-recombinant (NR) allele configurations (haplotypes) as the ancestral chromosomes with haplotype frequencies *p* and *q*. The alleles are in linkage disequilibrium (LD). On the right, the loci are more distantly linked or unlinked. After *n* meiotic generations, non-recombinant (NR) and recombinant (R) haplotypes are equally distributed in the descendant chromosomes. The alleles are in linkage equilibrium (LE).

Color image of this figure appears in the color plate section at the end of the book.

are not available. DNA-based markers and the human genome sequence provide the basis for large scale association mapping experiments that aim at the molecular identification of genes underlying complex traits (Frazer et al. 2009; Tenesa and Dunlop 2009). In plant science, association mapping was introduced with the beginning of the new millennium (Gebhardt et al. 2001; Hansen et al. 2001; Thornsberry et al. 2001; Flint-Garcia et al. 2003; Ivandic et al. 2003). There are two strategies for association studies: the genome-wide approach and the candidate gene approach. In the genome-wide approach, a large number of markers covering the whole genome at regular intervals are genotyped in a population of individuals related by descent. The same individuals are evaluated for the phenotypes of interest. Appropriate statistical methods are then used to test for significant associations between genotype and phenotype (McCarthy et al. 2008).

The candidate gene approach uses physiological, biochemical, genetic and molecular information available for a trait of interest to make learned hypotheses about the function or structure of the gene(s) underlying the trait of interest. Genotyping focuses then on genomic regions harboring such candidate genes.

7.2 Potato Populations for Association Mapping

The theory of population genetics is founded on natural populations of diploid individuals related by descent. It is assumed that the individuals in such a population mate at random with each other and that generations do not overlap (Hartl and Clark 1997). Close to this ideal of a natural population may come individual plants of wild, self-incompatible *Solanum* species growing in natural habitats. The criteria for a natural population certainly do not apply to the potato genetic material most useful for association mapping of agronomic traits, consisting of tetraploid or diploid cultivars that have been selected from progeny of multiple, controlled crosses. Landraces may not originate from controlled crosses but are nevertheless highly selected and therefore not natural either. Typical pedigrees of potato varieties often contain the same founder clones and varieties used as parents in many crosses over several generations. Another typical feature is the introgression of genetic material from one or more other *Solanum* species, wild or cultivated, in order to introduce into the breeding pool novel genes such as genes for resistance to pests and pathogens (*http://www.plantbreeding.wur.nl/potatopedigree/index.html*) (Ross 1986; Love 1999; Braun et al. 2004; Braun and Wenzel 2004; van Berloo et al. 2007). The genomes of modern potato cultivars are, therefore, a patchwork of *Solanum tuberosum* and other, closely related *Solanum* species. In genomic regions, where introgressions have occurred, the frequency of recombination might be reduced (Ballvora et al. 2007). Due to vegetative propagation of potato clones, the time elapsing between meiotic generations is quite long. Typically, less than 10 meiotic generations separate contemporary potato varieties from their ancestors 100 years ago (Love 1999; Gebhardt et al. 2004), when Mendelian genetics was introduced into potato breeding (Salaman 1910–1911). Considering the history of European and North American potato breeding, whichever cultivars from a breeding pool are assembled in an association mapping population, most of them will be related, with few meiotic generations separating the individuals from each other. The peculiarities of potato breeding and propagation explain why LD in this material can extend across large genomic segments corresponding to genetic distances of several Centi-Morgan (Simko et al. 2006; D'hoop et al. 2008; Li et al. 2008). Despite the relatedness of cultivars by pedigree, the observed molecular diversity between them is high. This is illustrated in Fig. 7-2 showing a genetic

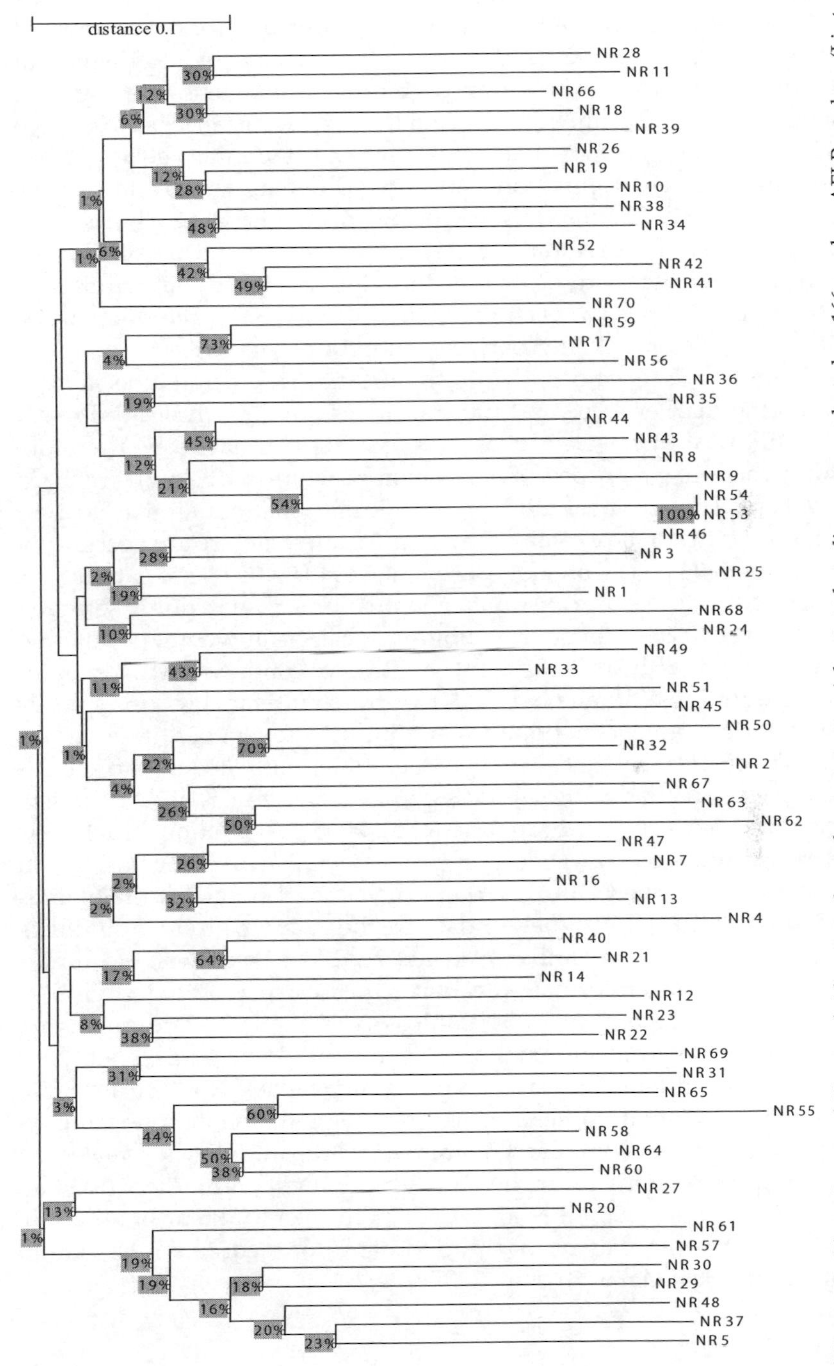

Figure 7-2 Phenetic tree of 67 tetraploid potato clones from a commercial potato breeding program based on 166 random AFLP markers (Li et al. 2005). Bootstrap values are shown at the base of the branches.

distance analysis based on 166 amplified fragment length polymorphisms (AFLP, see below) genotyped in 67 tetraploid cultivars from a commercial breeding program (Li et al. 2005). Most cultivars are highly distinct at the DNA level and separated from each other in phenetic trees by long branches of similar length. In fact, when the genetic distance between two individuals is very small, it often turns out that the identical genotype has been included in the population under two different names and the few differences observed resulted from scoring errors (Li et al. 2008).

Populations of individuals related by descent might be structured. This means that there are groups of individuals within the population, which are more similar to each other than to the rest of the population. Allele frequencies may differ between subgroups, and this can give rise to false positive associations, that is significant marker-trait associations, although the marker locus and the trait locus are not physically linked. The significance of novel marker-trait associations has to be therefore examined including the aspect of population structure (see below) (Pritchard et al. 2000b). Though it should be kept in mind, that not all marker-trait associations that can be explained by population structure are necessarily false positives. The very phenotypic character, for which association tests are performed, may be the cause of the population structure due to selection (Remington et al. 2001). In potato, subgroups can occur when a population is sampled from cultivars of different countries or continents with different breeding traditions (Braun et al. 2004), or from cultivars bred for specific markets, for example table potatoes and industrial potatoes. Populations sampled from different, but sexually compatible, cultivated potato species or subspecies such as *S. tuberosum* ssp. *tuberosum* and *S. tuberosum* ssp. *andigena* are also structured (Spooner et al. 2005). Population structure is experimentally assessed by genotyping the association mapping population with a set of unlinked DNA markers distributed on all chromosomes (Pritchard et al. 2000a). The marker data are then used to group individuals according to molecular similarity (Li et al. 2005; Pajerowska-Mukhtar et al. 2009). Pedigree information can also be used to allocate individuals to subgroups (Simko et al. 2004; Malosetti et al. 2007). Pedigrees are, however, often not available or incomplete for individual potato genotypes and cannot be verified in most cases due to unavailability of the ancestral clones. The association studies performed so far with populations sampled from varieties and contemporary, commercial breeding clones revealed no or little population structure when analyzed with DNA markers (Li et al. 2005; D'hoop et al. 2008; Li et al. 2008; Pajerowska-Mukhtar et al. 2009). Population structure has been reported based on pedigree information (Simko et al. 2004; Malosetti et al. 2007).

7.3 Phenotypic Variation

Associations between DNA variants and phenotypic traits are only detectable when the frequency of the trait alleles in the population considered is sufficient for statistic analysis. Obviously, this is not the case for rare alleles present in only one or two individuals or for nearly fixed alleles present in most individuals in the population. As a rule of thumb, association mapping is most appropriate for allele frequencies between 10% and 90% (Li et al. 2008). The power of the association test depends on the number of individuals phenotyped and genotyped (the more the better), the allele frequencies, the size of the allele effects, and last but not least the quality of the phenotypic data. A cornerstone of association mapping is, therefore, the phenotypic evaluation of the individual genotypes in the population. Potato breeding clones and varieties are naturally subject to intensive evaluation of many agronomic traits over several years. These evaluation data are documented in variety descriptions, gene banks or breeders' records and can be taken advantage of when testing DNA markers for associations with qualitative (Kasai et al. 2000; Flis et al. 2005; Song et al. 2005) and quantitative traits (Gebhardt et al. 2004; Li et al. 2005; Sattarzadeh et al. 2006; Malosetti et al. 2007; D'hoop et al. 2008). Such phenotypic data are, however, historical and originate from different sources, different environments and/or variable methodologies. Whereas this is less of a problem for highly penetrant phenotypes such as extreme resistance to Potato Virus Y, available data for truly quantitative traits can be highly unbalanced (D'hoop et al. 2008). Alternatively, phenotypic evaluation of association mapping populations under similar environmental conditions in a greenhouse (Simko et al. 2004) or in replicated field trials (Fig. 7-3) results in more balanced phenotypic data (Li et al. 2008; Pajerowska-Mukhtar et al. 2009). The difficulty of proprietary breeding clones not being available for multi-locus trials could be circumvented by including a set of varieties as common standards in field trials.

7.4 Detection and Scoring of DNA Variation

Due to the potato's heterozygosity, DNA variation is observed within individual genotypes rather than between genotypes as is the case in homozygous plants (Tenaillon et al. 2001). In tetraploid potato genotypes, DNA variants occur in four possible allele dosages: simplex (one in four), duplex (two in four), triplex (three in four) and quadruplex (all four). Point mutations and insertion-deletion polymorphisms in genomic DNA can be detected in various ways, some of which are suitable for estimating the allele dosage.

Figure 7-3 Field trial for resistance to late blight (courtesy of Dr. H.-R. Hofferbert, Böhm-Nordkartoffel Agrarproduktion GbR, Ebstorf, Germany).

Color image of this figure appears in the color plate section at the end of the book.

Restriction fragment length polymorphisms [RFLPs; (Gebhardt and Salamini 1992)] based on Southern blot hybridization (Southern 1975) are now historical. The genotyping techniques currently used in potato association studies are all based on the polymerase chain reaction [PCR; (Saiki et al. 1988)]. In the potato association studies published to date, microsatellites, single strand conformation polymorphisms (SSCPs), amplified fragment length polymorphisms (AFLPs), nucleotide binding site (NBS)-profiling, single nucleotide polymorphisms (SNPs), cleaved amplified polymorphic sequences (CAPS), sequence characterized amplified regions (SCARs) and allele specific amplification (ASAs) have been used for genotyping.

7.4.1 Microsatellite Markers

Microsatellites or simple sequence repeats (SSRs) consist of stretches of tandemly repeated di- tri- tetra- or pentanucleotide motifs that occur ubiquitously in eukaryotic genomes (Tautz and Renz 1984). The number of repeat motifs and therefore the length of the microsatellite frequently differ between alleles, causing insertion-deletion polymorphisms of few nucleotides. Allelic variation at SSR loci is detected after PCR amplification of the microsatellite region using genomic DNA as template and primers derived from unique, flanking sequences. The amplicons are size separated by high resolution electrophoresis, for example on polyacrylamide gels or on automated sequencers, and visualized by silver staining or fluorescent dyes. For well behaved SSR loci, it is possible to deduce the allele dosages from the number of alleles observed in a tetraploid individual and their relative band intensity [Fig. 7-4A; (Pajerowska-Mukhtar et al. 2009)]. Around 150 mapped SSR loci have been published (Kawchuk et al. 1996; Provan et al. 1996; Milbourne et al. 1998; Feingold et al. 2005; Ghislain et al. 2009), which represent an important resource for association genetics in potato, not only for analyzing population structure but also for detecting marker-trait associations (Simko et al. 2004). Genotyping 25 to 30 SSR loci chosen from different potato chromosomes was sufficient to analyze population substructure (Pajerowska-Mukhtar et al. 2009).

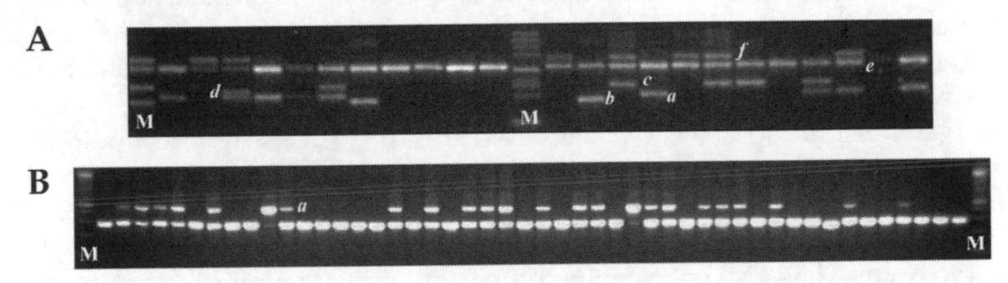

Figure 7-4 (A) Panel of tetraploid cultivars genotyped for the SSR marker *StI028* (Feingold et al. 2005). SSR alleles varying between 170 and 200 base pairs in length were separated on the Elchrom gel electrophoresis system (Elchrom Scientific AG, Cham, Switzerland). Between one and four alleles (*a* to *f*) are present in each cultivar with band intensities proportional to the allele dose. (B) Panel of tetraploid cultivars genotyped for the SCAR marker GP171. Allele *a* is negatively associated with chip quality and tuber starch content (Li et al. 2008). M: size marker.

7.4.2 Single Strand Conformational Polymorphism Markers (SSCP)

SSCPs are based on the different electrophoretic mobility of single stranded DNA's having small differences in nucleotide sequence (Orita et al. 1989a and b; Slabaugh et al. 1997; Bormann et al. 2004). Amplicons are generated from genomic DNA templates with specific primers, denatured, separated on polyacrylamide gels and silver stained. Polymorphism detection is enhanced by digesting the amplicons with four-cutter restriction enzymes prior to denaturation. SSCP detects DNA variation at specific loci with efficiency similar to that of RFLPs (Li et al. 2005; Li et al. 2008) and does not require highly sophisticated and expensive equipment. Though it does not allow the scoring of the allele dosage of an SSCP fragment.

7.4.3 Amplified Fragment Length Polymorphism (AFLP) Markers

AFLP™ (Vos et al. 1995) is a marker technology that does not require a priori sequence information and generates in a single experiment markers at multiple loci. Genomic DNA is digested with two different restriction enzymes, one having a four-base-pair, the other having a six-base-pair recognition site. The fragments are ligated with recognition site-specific adapter oligonucleotides. A subset of circa 50 to 100 restriction fragments is then selectively amplified in two consecutive PCR reactions. The amplified restriction fragments are radioactively or fluorescently labeled and separated by electrophoresis on high resolution polyacrylamide gels. Selectivity is achieved by designing PCR primers that anneal specifically to the adapter and the recognition site and carry one to three arbitrary chosen nucleotides at the 3′ end. Only those restriction fragments that have nucleotides complementary to the adapter and the recognition site plus the arbitrary nucleotide extension on both ends will be amplified. DNA fragment polymorphisms between different genotypes result from point mutations in the enzyme recognition sites or the nucleotide extensions, which either allow or prevent PCR amplification of a specific DNA fragment. Indels between recognition sites can also give rise to AFLPs. AFLP technology is best suited for quickly generating a large number of segregating DNA fragments, which can be used to analyze population structure and to detect marker trait associations (Li et al. 2005; D'hoop et al. 2008). AFLP allele dosage has also been evaluated based on measuring band intensity (D'hoop et al. 2008). The major disadvantage of AFLPs is the difficulty of re-identifying a genetic locus defined by a single AFLP fragment in one genetic background in independent genetic backgrounds. This difficulty is aggravated in association mapping populations. AFLPs are therefore best used in combination with anchor markers such as SSRs, which reliably tag the same specific loci irrespective from the genetic background.

7.4.4 Candidate Gene Markers: Nucleotide Binding Site (NBS) Profiling

NBS profiling (van der Linden et al. 2004) is a marker technique, which also does not require specific sequence information. Due to the choice of the primers used for amplification of genomic DNA fragments, polymorphisms in plant resistance genes and resistance-gene-like sequences having a nucleotide binding site (NBS) domain are selectively targeted. Fragment separation and detection is similar to AFLPs. NBS profiling is not locus specific but useful to screen genome wide for markers associated with pathogen resistance (Malosetti et al. 2007).

7.4.5 Single Nucleotide Polymorphism (SNP) Markers

The most direct and informative method for detecting DNA variation in tetraploid and diploid potato is sequence analysis. Amplicons can be obtained with the same primers from genomic DNA templates of different genotypes. The dideoxy chain-termination sequencing method used on Applied Biosystems automated DNA sequencers produces electropherograms (trace files), on which the sequence of the four deoxynucleotides G, C, T and A appear as differently colored peaks. Due to the heterozygosity of potato, SNPs are detected as alternatively colored, overlapping peaks at a given position in sequence trace files. High quality trace files allow the estimation of the allele dosage in tetraploid genotypes based on the height ratio of two overlapping nucleotide peaks. Insertion/deletion polymorphisms can also be detected by observing specific positions in the trace files, downstream of which different nucleotide sequences overlap. The SNP frequency in the potato genome is very high (Rickert et al. 2003; Simko et al. 2006). On average, one SNP occurs every 21 base pairs and one indel every 243 base pairs. Sequencing an amplicon a few hundred base pairs in length in a population allows the scoring of multiple SNPs (Pajerowska-Mukhtar et al. 2009). SNPs can be scored with a variety of techniques, for example pyrosequencing or SNuPE (single nucleotide primer extension) (Rickert et al. 2002). In the future, oligonucleotide arrays or new sequencing technologies will become available at affordable cost, which should allow the parallel scoring of thousands of SNPs in an individual. This will facilitate whole genome association studies in potato. SNPs are, therefore, the markers of choice for comprehensive population genetics and linkage disequilibrium studies. Due to the heterozygosity, SNPs in potato are unphased, which means that haplotypes cannot be deduced directly from the SNP scoring data as it is possible in homozygous individuals (Tenaillon et al. 2001). The "SATlotyper" software was developed to deduce haplotype models from unphased SNP data in polyploid species (Neigenfind et al. 2008).

7.4.6 Other PCR-based Markers

Any DNA sequence of suitable length and specific position in the potato or the closely related tomato genome can eventually be converted in an easy-to-use and cost effective PCR-based marker assay. This is useful in association studies and particularly relevant when markers associated with agronomic traits are to be used in breeding applications. Sequence information of hundreds of RFLP markers from potato, tomato and other Solanaceae is available in public databases (SOL genomics network: *http://sgn.cornell.edu/*; PoMaMo—Potato maps and more: *http://www.gabipd.org/database/maps.shtml*). Usually, amplicons are generated from genomic DNA templates using locus specific primers and size separated on agarose gels. DNA polymorphisms in the amplicons are detected directly (indels, Fig. 7-4B) or after restriction digestion. Many examples for this can be found in the literature, for example (Niewöhner et al. 1995; Oberhagemann et al. 1999; Gebhardt et al. 2004; Sliwka et al. 2007). Primers can also be designed to amplify one specific marker allele that is diagnostic for an important resistance allele (Kasai et al. 2000; Sattarzadeh et al. 2006).

7.5 Association Test Statistics

A DNA marker scored in a population of individuals related by descent is directly or indirectly associated with phenotypic variation in the same population, if the phenotypic means of contrasting marker genotype classes differ significantly, and if the marker is physically linked with the trait locus. False positive marker-trait associations can arise from unlinked markers due to population substructure (see above) or by chance alone. Various statistical models have been developed, which correct the association test for population structure and/or familial relatedness based on pedigree information or genetic distances calculated from marker data (Pritchard et al. 2000a and b; Yu et al. 2006; Malosetti et al. 2007; Stich et al. 2008; Stich and Melchinger 2009). The association studies performed so far in potato used statistical models ranging from a very simple, non-parametric Mann-Whitney-U test (Gebhardt et al. 2004) to a complex mixed model, which takes into account population structure, kinship, allele substitution and interaction effects of the marker alleles at a locus with four allele doses (Pajerowska-Mukhtar et al. 2009). At present, there is no "one and only" statistic model for association mapping in potato. Using different models to explore an experimental dataset is therefore the best way to assess the influence of the association test statistics on the number and robustness of the marker-trait associations found (Malosetti et al. 2007; D'hoop et al. 2008; Li et al. 2008; Stich and Melchinger 2009). In the author's experience , as a rule of thumb, a marker-trait association with a p-value smaller than 10^{-4} in a simple t-test or ANOVA is likely to survive the most rigorous and sophisticated statistical treatments. A marker-trait association is experimentally validated when it is reproducible in a different population. This has been found so far for the association of DNA variants at the invertase candidate locus *InvGE/InvGF* on potato chromosome IX with chip color (Li et al. 2005; Li et al. 2008). Markers that are linked to QTL for a particular trait in experimental mapping populations and are associated with the trait in association mapping populations also lend credibility (Gebhardt et al. 2004; Li et al. 2005). Finally, marker-trait associations are verified when the de novo selection of individuals based on contrasting marker alleles indeed results in significant phenotypic differences between the selected marker genotype classes.

The risk of false positive marker-trait associations occurring by chance alone increases linearly with each independent association test performed. The significance threshold for marker-trait associations should therefore be lowered when performing multiple statistical tests. In cases where no alleles with major effects are present in the population, applying the most conservative multiple-testing correction of Bonferroni (α divided by the number of tests) might eliminate all marker-trait associations, whereas without any correction a large number of marker-trait associations are obtained (D'hoop et al. 2008). Somewhere in between is the correction according to Benjamini et al. (2005) that was used by Li et al. (2008).

7.6 Association Studies in Potato

To date, marker-trait associations have been found in potato for disease resistance (Kasai et al. 2000; Gebhardt et al. 2001; Gebhardt et al. 2004; Simko et al. 2004; Flis et al. 2005; Song et al. 2005; Sattarzadeh et al. 2006; Malosetti et al. 2007; Pajerowska-Mukhtar et al. 2009) and for tuber quality traits (Li et al. 2005; D'hoop et al. 2008; Li et al. 2008). Single genes for extreme resistance to Potato Virus Y and a major QTL for resistance to the root cyst nematode *Globodera pallida* were first pinned to a single chromosomal position in experimental mapping populations. Marker alleles linked to resistance were then tested for association in populations of more distantly related cultivars with known resistance phenotype (Kasai et al. 2000; Flis et al. 2005; Song et al. 2005; Sattarzadeh et al. 2006; Witek et al. 2006). These studies represent the first examples in which highly diagnostic DNA markers suitable for breeding applications were identified.

Simko and colleagues (2004) evaluated a population of 137 North American tetraploid cultivars and advanced breeding lines for quantitative resistance to Verticillium wilt (*Verticillium dahliae*). They then took advantage of the fact that *Ve* resistance genes had been mapped and cloned in the closely related tomato (Diwan et al. 1999; Kawchuk et al. 2001), to identify the orthologous locus *StVe* on potato chromosome IX. A SSR allele closely linked to *StVe* (1.5 cM) was found associated with quantitative resistance to *Verticillium dahliae* (Simko et al. 2004). *R* genes and QTL for resistance to late blight (*Phytophthora infestans*) have been mapped in a number of diploid and tetraploid experimental mapping populations (Gebhardt and Valkonen 2001; Simko et al. 2007). Knowledge of the "know where" of *R* genes and quantitative resistance loci guided the selection of markers and candidate genes for association testing with resistance to late blight and plant maturity, which is correlated with late blight resistance. In a first experiment, a collection of 600 tetraploid potato cultivars bred between 1850 and 1990 in different countries and maintained by the IPK germplasm bank at Groß-Lüsewitz (Germany) was genotyped with five ASA, CAPS and SCAR markers linked to a known QTL for resistance to late blight and plant maturity on potato chromosome V. DNA polymorphisms were tested for association based on the phenotypic data available for 415 cultivars. Significant marker-trait associations were detected with an ASA marker derived from *R1*, a major gene for resistance to late blight (Ballvora et al. 2002), and markers flanking the *R1* locus at 0.2 cM genetic distance (Gebhardt et al. 2001; Gebhardt et al. 2004). The candidate gene approach was expanded in an association study aimed at the identification of DNA variants associated with field resistance to late blight that is not compromised by late plant maturity (maturity corrected resistance = MCR) (Pajerowska-Mukhtar et al. 2009). A population of 192 tetraploid breeding clones was phenotyped for field resistance to late blight and plant maturity and genotyped for 230 SNPs at 24 candidate loci. Nine SNPs were found significantly ($P < 0.001$) associated with maturity corrected resistance, which collectively explained 50% of the genetic variance of this trait. One major association was found at the *StAOS2* locus on potato chromosome XI, which explained 30% to 36% of the genetic variance of MCR. *StAOS2* encodes allene oxide synthase 2, a key enzyme in the biosynthesis of jasmonates, plant hormones that function in defense signaling (Liechti and Farmer 2002). This finding, and the functional characterization of natural variants of *StAOS2* (Pajerowska-Mukhtar et al. 2008) together suggest that *StAOS2* is one of those genes that are causal for natural variation of pathogen resistance in potato. In the association study of Malosetti and colleagues (2007), 204 tetraploid cultivars were genotyped by NBS profiling. Association tests with existing phenotypic data on late blight resistance revealed between two and 30 marker-trait associations, depending on the association model applied. Two NBS markers of unknown map position were consistently associated with resistance to late blight.

Most phenotypic characters of the tuber show polygenic inheritance (van Eck 2007) and genetic variation of tuber quality traits can be assumed to result from common alleles. This and the importance of tuber traits for crop quality make them highly suitable for association studies. Moreover, traits linked to carbohydrate contents such as tuber starch and sugar content can serve as models for the candidate gene approach to QTL identification in potato, as carbohydrate metabolism has been intensively studied at the physiological, biochemical and molecular levels (Isherwood 1973; Hofius and Börnke 2007). Many enzymes catalyzing the principal anabolic and catabolic reactions are known, the coding genes have been cloned and functionally characterized (Frommer and Sonnewald 1995), and in part molecularly mapped (Chen et al. 2001). QTL maps for tuber starch and sugar content are also available (Schäfer-Pregl et al. 1998; Menendez et al. 2002; Gebhardt et al. 2005). Candidate genes are therefore available in abundance, among these invertases, which convert sucrose into the reducing sugars glucose and fructose. The tuber content of reducing sugars determines the quality of deep fried products such as chips and French fries (Townsend and Hope 1960). In a first study based on 188 tetraploid cultivars, an association was found between DNA variation at the invertase locus *invGE/GF* on potato chromosome IX, which co-localized with a reducing sugar QTL (Menendez et al. 2002), and the trait "chip quality" (Li et al. 2005). Interestingly, the potato invertase gene *invGE* is orthologous to the tomato invertase gene *Lin5*, which is causal for the fruit-sugar-yield QTL Brix9-2-5 (Fridman et al. 2004), suggesting that natural variation of sugar yield in tomato fruits and sugar content of potato tubers is controlled by functional variants of orthologous invertase genes. The *invGE/GF* locus and 35 other candidate loci on 11 potato chromosomes were evaluated for natural DNA variation in an independent population of 243 tetraploid individuals, which was phenotyped in two years for chip quality before and after cold storage, tuber starch content, yield and starch yield (Li et al. 2008). Highly significant and robust marker-trait associations were identified and the association of the *invGE/GF* locus was confirmed. Most frequent were associations with chip quality and tuber starch content. Alleles increasing tuber starch content improved chip quality and vice versa. The most significant associations were observed with DNA variants in genes encoding the soluble acid invertase Pain-1 on potato chromosome III and two starch phosphorylases Stp23 and StpL on chromosome III and V, respectively. Structural and functional analysis of allelic variants of these genes is underway, which eventually will demonstrate that these genes are indeed underlying QTL for chip quality, tuber starch content and starch yield. D'hoop and colleagues (2008) genotyped a population of 221 tetraploid cultivars for 250 AFLP markers, which were tested for association with plant maturity and the tuber traits shape, flesh color, underwater weight (correlated with starch content), cooking type, darkening after cooking or baking, frying color, chipping color, blackspot bruising and enzymatic browning. The phenotypic data originated from field trials conducted over a varying number of years by five Dutch breeding companies. Depending on the association model used, 68 and 94 marker-trait associations were detected, between two and 17 per trait.

7.7 Conclusions and Outlook

Association mapping as outlined in this chapter is a new, valuable tool in potato genetics, which closes the gap between linkage mapping of quantitative and qualitative traits in experimental, mostly diploid populations and DNA marker applications in breeding programs aimed at developing commercially successful new varieties. It opens the possibility to develop PCR-based markers of general diagnostic value for parental screening and marker-assisted selection. Ultimately, these markers are derived from the genes underlying complex traits and allow the precise diagnosis of superior and inferior trait alleles. As in human genetics, whole genome association mapping with SNPs will become feasible and affordable in the future, and this will uncover many more marker-trait associations, provided that it is combined with excellent phenotypic analysis.

References

Ballvora A, Ercolano MR, Weiss J, Meksem K, Bormann CA, Oberhagemann P, Salamini F, Gebhardt C (2002) The *R1* gene for potato resistance to late blight (*Phytophthora infestans*) belongs to the leucine zipper/NBS/LRR class of plant resistance genes. Plant J 30: 361–71.

Ballvora A, Jöcker A, Viehover P, Ishihara H, Paal J, Meksem K, Bruggmann R, Schoof H, Weisshaar B, Gebhardt C (2007) Comparative sequence analysis of *Solanum* and Arabidopsis in a hot spot for pathogen resistance on potato chromosome V reveals a patchwork of conserved and rapidly evolving genome segments. BMC Genom 8: 112.

Benjamini Y, Krieger A, Yekutieli D (2005) Two Staged Linear Step Up FDR Controlling Procedure. *http: //www.math.tau.ac.il/~ybenja*

Bormann CA, Rickert AM, Ruiz RA, Paal J, Lübeck J, Strahwald J, Buhr K, Gebhardt C (2004) Tagging quantitative trait loci for maturity-corrected late blight resistance in tetraploid potato with PCR-based candidate gene markers. Mol Plant-Microbe Interact 17: 1126–38.

Braun A, Schullehner K, Wenzel G (2004) Molecular analysis of genetic variation in potato (*Solanum tuberosum* L.). II. International cultivar spectrum. Potato Res 47: 93–99.

Braun A, Wenzel G (2004) Molecular analysis of genetic variation in potato (*Solanum tuberosum* L.). I. German cultivars and advanced clones. Potato Res 47: 81–92.

Chen X, Salamini F, Gebhardt C (2001) A potato molecular-function map for carbohydrate metabolism and transport. Theor Appl Genet 102: 284–295.

D'hoop BB, Paulo MJ, Mank RA, van Eck HJ, van Eeuwijk FA (2008) Association mapping of quality traits in potato (*Solanum tuberosum* L.). Euphytica 161: 47–60.

Diwan N, Fluhr R, Eshed Y, Zamir D, Tanksley SD (1999) Mapping of *Ve* in tomato: a gene conferring resistance to the broad-spectrum pathogen, *Verticillium dahliae* race 1. Theor Appl Genet 98: 315–319.

Feingold S, Lloyd J, Norero N, Bonierbale M, Lorenzen J (2005) Mapping and characterization of new EST-derived microsatellites for potato (*Solanum tuberosum* L.). Theor Appl Genet 111: 456–466.

Flint-Garcia SA, Thornsberry JM, Buckler ES (2003) Structure of linkage disequilibrium in plants. Annu Rev Plant Biol 54: 357–374.

Flis B, Hermig J, Strzelczyk-Zyta D, Gebhardt C, Marczewski W (2005) The *Ry-f(sto)* gene from *Solanum stoloniferum* for extreme resistant to *Potato Virus Y* maps to potato chromosome XII and is diagnosed by PCR marker GP122(718) in PVY resistant potato cultivars. Mol Breed 15: 95–101.

Frazer KA, Murray SS, Schork NJ, Topol EJ (2009) Human genetic variation and its contribution to complex traits. Nat Rev Genet 10: 241–251.

Fridman E, Carrari F, Liu Y-S, Fernie AR, Zamir D (2004) Zooming in on a quantitative trait for tomato yield using interspecific introgressions. Science 305: 1786–1789.

Frommer WB, Sonnewald U (1995) Molecular analysis of carbon partitioning in solanaceous species. J Exp Bot 46: 587–607.

Gebhardt C, Salamini F (1992) Restriction fragment length polymorphism analysis of plant genomes and its application to plant breeding. Int Rev Cytol 135: 201–237.

Gebhardt C, Valkonen JP (2001) Organization of genes controlling disease resistance in the potato genome. Annu Rev Phytopathol 39: 79–102.

Gebhardt C, Schüler K, Walkemeier B, Ballvora A (2001) Association mapping in potato of a QTL for late blight resistance. In: Plant Anim Genome IX Conf, San Diego, CA, USA.

Gebhardt C, Ballvora A, Walkemeier B, Oberhagemann P, Schüler K (2004) Assessing genetic potential in germplasm collections of crop plants by marker-trait association: a case study for potatoes with quantitative variation of resistance to late blight and maturity type. Mol Breed 13: 93–102.

Gebhardt C, Menendez C, Chen X, Li L, Schäfer-Pregl R, Salamini F (2005) Genomic approaches for the improvement of tuber quality traits in potato. Acta Hort 684: 85–92.

Ghislain M, Núñez J, del Rosario Herrera M, Pignataro J, Guzman F, Bonierbale M, Spooner D (2009) Robust and highly informative microsatellite-based genetic identity kit for potato. Mol Breed 23: 377–388.

Hansen M, Kraft T, Ganestam S, Auml, Ll T, Ouml, Rn, Nilsson N-O (2001) Linkage disequilibrium mapping of the bolting gene in sea beet using AFLP markers. Genet Res 77: 61–66.

Hartl DL, Clark AG (1997) Principles of Population Genetics. Sinauer Assoc, Sunderland, MA, USA.

Hofius D, Börnke FAJ (2007) Photosynthesis, carbohydrate metabolism and source-sink relations. In: D Vreudgenhil, J Bradshaw, C Gebhardt, F Govers, DKL MacKerron, MA Taylor, HA Ross (eds) Potato Biology and Biotechnology, Advances and Perspectives. Amsterdam, Boston, Heidelberg, London, New York, Oxford, Paris, San Diego, San Francisco, Singapore, Sydney, Tokyo, pp 257–285.

Isherwood FA (1973) Starch-sugar interconversion in *Solanum tuberosum*. Phytochemistry 12: 2579–2591.

Ivandic V, Thomas WTB, Nevo E, Zhang Z, Forster BP (2003) Associations of simple sequence repeats with quantitative trait variation including biotic and abiotic stress tolerance in *Hordeum spontaneum*. Plant Breed 122: 300–304.

Kasai K, Morikawa Y, Sorri VA, Valkonen JP, Gebhardt C, Watanabe KN (2000) Development of SCAR markers to the PVY resistance gene Ryadg based on a common feature of plant disease resistance genes. Genome 43: 1–8.

Kawchuk LM, Lynch DR, Thomas J, Penner B, Sillito D, Kulcsar F (1996) Characterization of *Solanum tuberosum* simple sequence repeats and application to potato cultivar identification. Am J Potato Res 73: 325–335.

Kawchuk LM, Hachey J, Lynch DR, Kulcsar F, van Rooijen G, Waterer DR, Robertson A, Kokko E, Byers R, Howard RJ, Fischer R, Prüfer D (2001) Tomato *Ve* disease resistance genes encode cell surface-like receptors. Proc Natl Acad Sci USA 98: 6511–6515.

Kerem B-S, Rommens JM, Buchanan JA, Markiewicz D, Cox TK, Chakravarti A, Buchwald M, Tsui L-C (1989) Identification of the cystic fibrosis gene: Genetic analysis. Science 245: 1073–1080.

Li L, Strahwald J, Hofferbert HR, Lubeck J, Tacke E, Junghans H, Wunder J, Gebhardt C (2005) DNA variation at the invertase locus *invGE/GF* is associated with tuber quality traits in populations of potato breeding clones. Genetics 170: 813–21.

Li L, Paulo MJ, Strahwald J, Lubeck J, Hofferbert HR, Tacke E, Junghans H, Wunder J, Draffehn A, van Eeuwijk F, Gebhardt C (2008) Natural DNA variation at candidate loci is associated with potato chip color, tuber starch content, yield and starch yield. Theor Appl Genet 116: 1167–1181.

Liechti R, Farmer EE (2002) The jasmonate pathway. Science 296: 1649–1650.

Love SL (1999) Founding clones, major contributing ancestors, and exotic progenitors of prominent North American potato cultivars. Am J Potato Res 76: 263–272.

Malosetti M, van der Linden CG, Vosman B, van Eeuwijk FA (2007) A mixed-model approach to association mapping using pedigree information with an illustration of resistance to *Phytophthora infestans* in potato. Genetics 175: 879–889.

McCarthy MI, Abecasis GR, Cardon LR, Goldstein DB, Little J, Ioannidis JPA, Hirschhorn JN (2008) Genome-wide association studies for complex traits: consensus, uncertainty and challenges. Nat Rev Genet 9: 356–369.

Menendez CM, Ritter E, Schäfer-Pregl R, Walkemeier B, Kalde A, Salamini F, Gebhardt C (2002) Cold sweetening in diploid potato: mapping quantitative trait loci and candidate genes. Genetics 162: 1423–1434.

Milbourne D, Meyer RC, Collins AJ, Ramsay LD, Gebhardt C, Waugh R (1998) Isolation, characterisation and mapping of simple sequence repeat loci in potato. Mol Gen Genet 259: 233–245.

Neigenfind J, Gyetvai G, Basekow R, Diehl S, Achenbach U, Gebhardt C, Selbig J, Kersten B (2008) Haplotype inference from unphased SNP data in heterozygous polyploids based on SAT. BMC Genom 9: 356.

Niewöhner J, Salamini F, Gebhardt C (1995) Development of PCR assays diagnostic for RFLP marker alleles closely linked to alleles *Gro1* and *H1*, conferring resistance to the root cyst-nematode *Globodera rostochiensis* in potato. Mol Breed 1: 65–78.

Oberhagemann P, Chatot-Balandras C, Schafer-Pregl R, Wegener D, Palomino C, Salamini F, Bonnel E, Gebhardt C (1999) A genetic analysis of quantitative resistance to late blight in potato: towards marker-assisted selection. Mol Breed 5: 399–415.

Orita M, Iwahana H, Kanazawa H, Hayashi K, Sekiya T (1989a) Detection of polymorphisms of human DNA by gel electrophoresis as single-strand conformation polymorphisms. Proc Natl Acad Sci USA 86: 2766–2770.

Orita M, Suzuki Y, Sekiya T, Hayashi K (1989b) Rapid and sensitive detection of point mutations and DNA polymorphisms using the polymerase chain reaction. Genomics 5: 874–879.

Pajerowska KM, Parker JE, Gebhardt C (2005) Potato homologs of *Arabidopsis thaliana* genes functional in defense signaling—identification, genetic mapping, and molecular cloning. Mol Plant-Microbe Interact 18: 1107–19.

Pajerowska-Mukhtar KM, Mukhtar MS, Guex N, Halim VA, Rosahl S, Somssich IE, Gebhardt C (2008) Natural variation of potato *Allene Oxide Synthase 2* causes differential levels of jasmonates and pathogen resistance in Arabidopsis. Planta 228: 293–306.

Pajerowska-Mukhtar K, Stich B, Achenbach U, Ballvora A, Lubeck J, Strahwald J, Tacke E, Hofferbert H-R, Ilarionova E, Bellin D, Walkemeier B, Basekow R, Kersten B, Gebhardt C (2009) Single nucleotide polymorphisms in the *Allene Oxide Synthase 2* gene are associated with field resistance to late blight in populations of tetraploid potato cultivars. Genetics 181: 1115–1127.

Pritchard JK, Stephens M, Donnelly P (2000a) Inference of population structure using multilocus genotype data. Genetics 155: 945–959.

Pritchard JK, Stephens M, Rosenberg NA, Donnelly P (2000b) Association mapping in structured populations. Am J Hum Genet 67: 170–181.

Provan J, Powell W, Waugh R (1996) Microsatellite analysis of relationships within cultivated potato (*Solanum tuberosum*). Theor Appl Genet 92: 1078–1084.

Remington DL, Thornsberry JM, Matsuoka Y, Wilson LM, Whitt SR, Doebley J, Kresovich S, Goodman MM, Buckler ES (2001) Structure of linkage disequilibrium and phenotypic associations in the maize genome. Proc Natl Acad Sci USA 98: 11479–11484.

Rickert AM, Premstaller A, Gebhardt C, Oefner PJ (2002) Genotyping of SNPs in a polyploid genome by pyrosequencing. Biotechniques 32: 592–3, 596–8, 600 passim.

Rickert AM, Kim JH, Meyer S, Nagel A, Ballvora A, Oefner PJ, Gebhardt C (2003) First-generation SNP/InDel markers tagging loci for pathogen resistance in the potato genome. Plant Biotechnol J 1: 399–410.

Ross H (1986) Potato Breeding—Problems and Perspectives. Paul Parey, Berlin, Hamburg, Germany.

Saiki RK, Gelfand DH, Stoffel S, Scharf SJ, Higuchi R, Horn GT, Mullis KB, Erlich HA (1988) Primer-directed enzymatic amplification of DNA with a thermostable DNA polymerase. Science 239: 487–491.

Salaman RN (1910–1911) The inheritance of colour and other characters in the potato. J Genet 1: 7–46.

Sattarzadeh A, Achenbach U, Lubeck J, Strahwald J, Tacke E, Hofferbert HR, Rothsteyn T, Gebhardt C (2006) Single nucleotide polymorphism (SNP) genotyping as basis for developing a PCR-based marker highly diagnostic for potato varieties with high resistance to *Globodera pallida* pathotype Pa2/3. Mol Breed 18: 301–312.

Schäfer-Pregl R, Ritter E, Concilio L, Hesselbach J, Lovatti L, Walkemeier B, Thelen H, Salamini F, Gebhardt C (1998) Analysis of quantitative trait loci (QTLs) and quantitative trait alleles (QTAs) for potato tuber yield and starch content. Theor Appl Genet 97: 834–846.

Simko I, Costanzo S, Haynes KG, Christ BJ, Jones RW (2004) Linkage disequilibrium mapping of a *Verticillium dahliae* resistance quantitative trait locus in tetraploid potato (*Solanum tuberosum*) through a candidate gene approach. Theor Appl Genet 108: 217–224.

Simko I, Haynes KG, Jones RW (2006) Assessment of linkage disequilibrium in potato genome with single nucleotide polymorphism markers. Genetics 173: 2237–2245.

Simko I, Jansky S, Stephenson S, Spooner D (2007) Genetics of resistance to pests and disease. In: D Vreugdenhil, J Bradshaw,C Gebhardt, F Govers, DKL MacKerron, MA, Taylor, HA Ross (eds) Potato Biology and Biotechnology: Advances and Perspectives. Elsevier, Amsterdam, Boston, Heidelberg, London, New York, Oxford, Paris, San Diego, San Francisco, Singapore, Sydney, Tokyo, pp 117–155.

Slabaugh MB, Huestis GM, Leonard J, Holloway JL, Rosato C, Hongtrakul V, Martini N, Toepfer R, Voetz M, Schell J, Knapp SJ (1997) Sequence-based genetic markers for genes and gene families: single-strand conformational polymorphisms for the fatty acid synthesis genes of Cuphea. Theor Appl Genet 94: 400–408.

Sliwka J, Jakuczun H, Lebecka R, Marczewski W, Gebhardt C, Zimnoch-Guzowska E (2007) Tagging QTLs for late blight resistance and plant maturity from diploid wild relatives in a cultivated potato (*Solanum tuberosum*) background. Theor Appl Genet 115: 101–112.

Song Y-S, Leonard Hepting L, Schweizer G, Hartl L, Wenzel G, Schwarzfischer A (2005) Mapping of extreme resistance to PVY (*Ry sto*) on chromosome XII using anther-culture-derived primary dihaploid potato lines. Theor Appl Genet 111: 879–887.

Southern EM (1975) Detection of specific sequences among DNA fragments separated by gel electrophoresis. J Mol Biol 98: 503–517.

Spooner DM, McLean K, Ramsay G, Waugh R, Bryan GJ (2005) A single domestication for potato based on multilocus amplified fragment length polymorphism genotyping. Proc Natl Acad Sci USA 102: 14694–14699.

Stich B, Melchinger A (2009) Comparison of mixed-model approaches for association mapping in rapeseed, potato, sugar beet, maize, and Arabidopsis. BMC Genom 10: 94.

Stich B, Mohring J, Piepho HP, Heckenberger M, Buckler ES, Melchinger AE (2008) Comparison of mixed-model approaches for association mapping. Genetics 178: 1745–1754.

Tautz D, Renz M (1984) Simple sequences are ubiquitous repetitive components of eukaryotic genomes. Nucl Acid Res 12: 4127–4138.

Tenaillon MI, Sawkins MC, Long AD, Gaut RL, Doebley JF, Gaut BS (2001) Patterns of DNA sequence polymorphism along chromosome 1 of maize (*Zea mays* ssp. *mays* L.). Proc Natl Acad Sci USA 98: 9161–9166.

Tenesa A, Dunlop MG (2009) New insights into the aetiology of colorectal cancer from genome-wide association studies. Nat Rev Genet 10: 353–358.

Thornsberry JM, Goodman MM, Doebley J, Kresovich S, Nielsen D, Buckler ES (2001) *Dwarf8* polymorphisms associate with variation in flowering time. Nat Genet 28: 286–289.

Townsend LR, Hope GW (1960) Factors influencing the color of potato chips. Can J Plant Sci 40: 58–64.

van Berloo R, Hutten R, van Eck H, Visser R (2007) An Online Potato Pedigree Database Resource. Potato Res 50: 45–57.

van der Linden CG, Wouters DCAE, Mihalka V, Kochieva EZ, Smulders MJM, Vosman B (2004) Efficient targeting of plant disease resistance loci using NBS profiling. Theor Appl Genet 109: 384–393.

van Eck HJ (2007) Genetics of morphological and tuber traits. In: D Vreudgenhil, J Bradshaw, C Gebhardt, F Govers, DKL MacKerron, MA Taylor, HA Ross (eds) Potato Biology and Biotechnology, Advances and Perspectives. Elsevier, Amsterdam, Boston, Heidelberg, London, New York, Oxford, Paris, San Diego, San Francisco, Singapore, Sydney, Tokyo, pp 91–115.

Vos P, Hogers R, Bleeker M, Reijans M, van der Lee T, Fornes M, Frijters A, Pot J, Peleman J, Kuiper M, Zabeau M (1995) AFLP: a new technique for DNA fingerprinting. Nucl Acids Res 23: 4407–4414.

Witek K, Strzelczyk-Żyta D, Hennig J, Marczewski W (2006) A multiplex PCR approach to simultaneously genotype potato towards the resistance alleles $Ry\text{-}f_{sto}$ and *Ns*. Mol Breed 18: 273–275.

Yu J, Pressoir G, Briggs WH, Vroh Bi I, Yamasaki M, Doebley JF, McMullen MD, Gaut BS, Nielsen DM, Holland JB, Kresovich S, Buckler ES (2006) A unified mixed-model method for association mapping that accounts for multiple levels of relatedness. Nat Genet 38: 203–208.

8

Cloning of Late Blight Resistance Genes: Strategies and Progress

James M. Bradeen

ABSTRACT

In recent years, significant progress has been made in the cloning of potato disease resistance genes and the development of disease resistant transgenic potato. Genes conditioning resistance to late blight disease have been especially targeted in these efforts. Three basic approaches have been utilized, frequently in combination, in cloning late blight resistance genes: chromosome walking, candidate gene, and comparative, cross-species genomics. These approaches have been made possible by technical advances in the development of large insert libraries, an improved understanding of the molecular architecture of plant disease resistance genes and *Phythophthora infestans* effector genes, and observations that homologs of disease resistance genes frequently occupy corresponding genome locations across multiple *Solanum* species. Initially, the cloning of late blight resistance genes was slow and labor-intensive. But technical advances are making these genes more accessible and increasing the frequency with which agriculturally useful disease resistance genes are being cloned. The first late blight resistance gene to be cloned, *R1*, derives from the wild potato *S. demissum*. The recovery of this gene in 2002 represents a historically significant achievement. Unfortunately, since the effectiveness of this gene was long ago circumvented by shifting *P. infestans* population dynamics, *R1* is likely to be of limited future agricultural use. In contrast, *RB*, cloned from the wild potato *S. bulbocastanum* in 2003, is the single most promising disease resistance gene for control of potato late blight disease. Several functionally equivalent homologs of *RB* have been subsequently cloned. As of 2009, a total of 15 cloned late blight resistance genes had been reported. Importantly, all late blight resistance genes cloned to date belong to the nucleotide binding site-leucine rich repeat

University of Minnesota, Department of Plant Pathology, 495 Borlaug Hall/1991 Upper Buford Circle, St. Paul, MN 55108, USA; e-mail: *jbradeen@umn.edu* superfamily of disease resistance genes. The technical details of efforts to clone these genes nicely illustrate the various approaches that may used to isolate disease resistance genes.

Keywords: candidate genes, comparative genomics, effector biology, late blight resistance, *R* genes

8.1 Introduction

The cultivated potato (*Solanum tuberosum* L.) is host to more than 60 diseases of contemporary significance (Stevenson et al. 2001). Chief among these is late blight caused by the oomycete *Phytophthora infestans*. Late blight caused the European Potato Famine of the 1840s and today still ranks as the number one constraint of potato production worldwide, with annual losses and chemical inputs totaling billions of dollars (Kamoun 2001). Given the significance of this disease, there has been concerted effort for 100 years to develop late blight resistant potato cultivars. While *S. tuberosum* lacks significant resistance, many of the approximately 200 wild potato species are rich sources of late blight resistance (*R*) genes.

The majority of wild potato species can be crossed directly with *S. tuberosum*, although manipulation of ploidy levels or the use of 2*n* gametes is sometimes necessary (see Chapter 1). A few wild potato species have been widely used by breeders to improve the crop. For example, the late blight resistant Mexican species *S. demissum*, source of at least 11 different *R* genes, was crossed to *S. tuberosum* in early efforts to breed late blight resistant potatoes. Initially promising, large-scale deployment of potato cultivars carrying *S. demissum R* genes quickly resulted in a shift of *P. infestans* population composition to overcome resistance. Despite this, it has been estimated that 50% of modern potato cultivars carry genes from *S. demissum* (Ross 1986). Overall, however, potato breeders have made only modest use of available wild germplasm, with an estimated 15 of the approximately 200 wild potato species having been incorporated into breeding lines (Ross 1986). In part, this reflects the fact that while wild species may carry specific genes for potato improvement, crossing them with *S. tuberosum* also introduces genes conditioning unfavorable traits. Adding to this, the allogamous, complex genetic, autotetraploid nature of potato precludes the effective recovery of parental genotypes via backcrossing approaches. For a minority of wild potato species [a group of approximately 18–20 species, comprising the bulk of the taxon "superseries *Stellata*" (Hawkes 1990) and representing the tertiary genepool of cultivated potato (Harlan and de Wet 1971) (see Chapter 1)], sexual incongruity with *S. tuberosum* has severely limited breeders' access to what would otherwise be very useful *R* genes.

Research activities within the past decade have made the cloning of late blight *R* genes and the creation of late blight resistant transgenic potatoes a reality. In a recent review, we outlined technical progress in potato transformation and highlighted several potato transgenes imparting disease and pest resistance or tuber starch qualities that hold particular promise for agricultural applications (Bradeen et al. 2008). Chapter 3 of this volume provides discussion on strategies and progress towards using transgenes for potato improvement. For a discussion on the legal and societal limitations on the actual production and sale of transgenic potatoes, see Chapter 12. In the current chapter, we focus on molecular approaches used for gene isolation. Specific examples are drawn from the realm of late blight resistance, which represents a major contemporary research emphasis for potato scientists.

8.2 Potato *R* Gene Cloning Strategies

In a broad sense, efforts to clone late blight *R* genes have employed at least one, and often several, of three approaches: (1) traditional map-based cloning via large insert library chromosome walking, (2) candidate *R* gene-driven methods, and (3) comparative, cross-species genomics.

8.2.1 Large Insert Chromosome Walking Approaches

Chromosome walking, often in combination with other approaches, forms the basis for most gene cloning efforts in potato and other plant species. Chromosome walking was made possible by the development of large insert libraries, especially bacterial artificial chromosome (BAC) libraries (Shizuya et al. 1992).

A BAC library is an arrayed collection of *Escherichia coli* colonies, each carrying an individual BAC clone. A BAC clone is a circular DNA molecule conceptually akin to an exceptionally large plasmid. The BAC clone is comprised of a vector, which generally harbors an origin of replication, regions involved in plasmid stability and maintenance of copy number, the *LacZ* gene for blue/white selection, and a selectable marker (e.g., antibiotic resistance) gene. The BAC clone also includes a DNA fragment of ~ 75–300 kb derived from the target genome (e.g., potato).

Construction of a BAC library first entails extraction of high molecular weight DNA from the target organism, randomly breaking the DNA into smaller sizes using shearing, sonication, or partial restriction enzyme digestion, and size fractionation on high-resolution agarose gels. DNA in the target range for the library is recovered and cloned into an appropriate BAC vector. Multiple BAC vectors have been developed and several are commercially available. Following cloning, the BAC molecules are electroporated into an appropriate *E. coli* host. The transformed *E. coli* cells are then plated on medium containing a selectable agent (e.g., antibiotic) and individual colonies are picked either manually or by robotic means into liquid medium in library plates. Picking frequently takes advantage of blue/white selection made possible by the *LacZ* gene. Often, the library plates are comprised of 384 arrayed wells each and a typical BAC library can consist of dozens to hundreds of 384 well plates.

The concept of "genome equivalents" is used to express the degree to which a BAC library represents the genome of the target organism. Based on calculations of genome size, library insert size, and number of clones per library, an average measurement of coverage, the genome equivalent, is calculated. For example, in a library described as encompassing 10 × genome equivalents, a researcher would find each single copy marker or gene represented an average of 10 times. However, regional variation in the degree of coverage is frequently encountered.

Advantages of the BAC library over other large insert library technologies include comparative ease in extracting BAC DNA, low levels of chimerism, and general insert stability (Osoegawa and de Jong 2004). However, instability in BAC clones, especially in regions that contain clusters of related sequences, have been reported (Song et al. 2003a). This may be especially relevant to the cloning of *R* genes since the majority reside in clusters of structurally related but functionally divergent sequences (Hulbert et al. 2001). In potato and close relatives, dozens of BAC libraries have been constructed, many of which are available commercially to the broader research community. Song et al. (2000) and Dong et al. (2000) developed a BAC library from the wild potato *S. bulbocastanum* and assigned specific BAC clones to individual potato chromosomes. This resource has contributed significantly to the cytological study of potato (see Chapter 9).

As the name implies, the chromosome walking procedure entails a step-by-step reiterative approach that ultimately yields a collection of overlapping DNA fragments in the form of BAC clones that physically encompasses a chromosome region. In general, the procedure begins with using genetic approaches to identify molecular markers linked to a specific genome region. The markers associated with this region are then used to survey the BAC library. This often entails either arraying the library onto a nylon membrane and using a modified Southern hybridization approach with the genetic marker used as a probe or screening pools of BAC clone DNA using PCR-based markers. Once a BAC clone harboring the marker is identified, sequence derived from the insert ends of the clone is used to develop new probes or PCR primers. These new probes or primers are then used to again screen the library, identifying a new BAC clone that partially overlaps the first. This effort constitutes one "step" in the chromosome "walk". The procedure is repeated over and over again until a collection of partially overlapping BACs is recovered that represents the entire target region. A group of partially overlapping BACs is referred to as a contig, short for contiguous region. Variations on BAC-based chromosome walking exist and a chromosome walk may benefit from additional genetic mapping using BAC end-derived markers (Bradeen et al. 2003) or cytological approaches to orient the walk relative to genetic markers (Song et al. 2003b). Figure 8.1 provides a graphical representation of a BAC-based chromosome walking effort in potato.

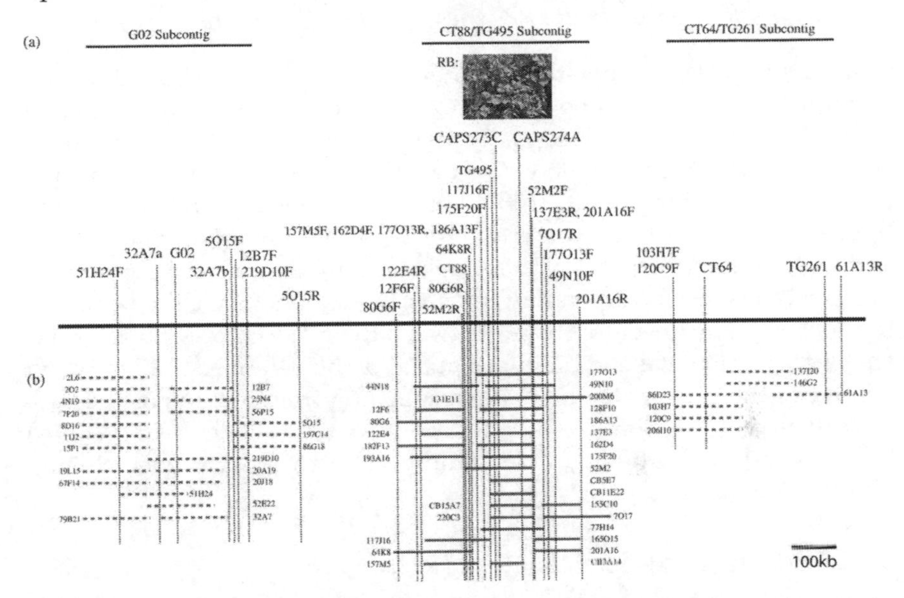

Figure 8-1 A representation of chromosome walking efforts at the *S. bulbocastanum* late blight resistance locus *RB*. (a) An integrated physical map of the *S. bulbocastanum* chromosome VIII in the *RB* vicinity. Genetically mapped markers include RFLP (CT64, CT88, TG261, TG495), RAPD (G02), and BAC-derived markers. RFLP markers were selected to initiate chromosome walking efforts; BAC markers are the result of iterative steps in the walk. Marker locations are based on combined genetic and physical data for the region. All measurements between markers reflect actual or calculated physical distances. Subcontig groupings are indicated. Late blight resistance (*RB*) is mapped to a region flanked by BAC-derived markers CAPS273C and CAPS274A. (b) BAC contigs. All BAC clones were isolated from the *S. bulbocastanum* BAC library developed by Song et al. (2000). BAC names indicate plate of origin and register location within a plate. Individual BAC clones are derived from the *RB* (resistant) homolog (*green*), the *rb* (susceptible) homolog (*blue*), or are of unknown homolog origin (*red*). BACs represented by *dashed lines* are of unknown insert size and their placement is tentative. Insert sizes have been estimated for BACs represented by *solid lines;* these BACs are drawn to scale (note 100-kb marker). [Reprinted with kind permission from Springer Science+Business Media: Bradeen et al. (2003) Concomitant reiterative BAC walking and fine genetic mapping enable physical map development for the broad-spectrum late blight resistance region, *RB*. *Mol Genet Genom* 269:603-611 (Fig. 2).].

Color image of this figure appears in the color plate section at the end of the book.

8.2.2 Candidate Gene Approaches

The term "candidate" gene is used to describe a predicted gene or gene fragment thought to condition a specific phenotype. Improved understanding of *R* gene architecture and pathogen effector biology have led to widely used approaches for the structural or functional identification of candidate *R* genes.

8.2.2.1 Identifying Candidate R Genes Based on Structure

Gene cloning efforts in potato and other plant species have been aided by an improved understanding of the molecular architecture of *R* genes and methods for isolating related sequences, independent of function, from genomic DNA. Approximately 75% of all *R* genes cloned to date belong to the nucleotide binding site (NBS)-leucine rich repeat (LRR) superfamily (Hulbert et al. 2001). While specific conserved protein motifs define the NBS and LRR domains, overall NBS sequence similarity between different R proteins (and their corresponding *R* genes) can be quite low. Understanding the conserved architecture shared between the majority of R proteins has enabled experimental and bioinformatic approaches aimed at identifying genes or gene fragments putatively originating from functional *R* genes. Two related approaches, construction of resistance gene analog (RGA) libraries and NBS profiling, hold significant promise for *R* gene cloning efforts in potato.

8.2.2.1.1 Resistance Gene Analogs (RGAs)

Leister et al. (1996) noted that the deduced NBS region encoded by the tobacco *N* gene and the NBS region encoded by the *Arabidopsis thaliana RPS2* gene showed only 50% overall sequence homology. However, two short deduced protein motifs within the NBS region of these largely disparate proteins, the kinase-1a or "P-loop" motif and the downstream GLPLAL motif, shared 100% amino acid identity. Targeting these motifs, the authors designed a series of corresponding degenerate and non-degenerate PCR primers. Applying these primers to potato genomic DNA yielded a series of fragments. Of 16 unique sequences recovered, 12 could be translated in silico to yield intact deduced protein sequences. Importantly, all 12 sequences included R protein motifs predicted by a comparison of N and RPS2 (Leister et al. 1996). PCR fragments derived in this fashion have been termed resistance gene analogs (RGAs) or resistance gene homologs (RGHs) and putatively derive from functional *R* genes or homologous sequences. The RGA approach has been widely used in dicot and monocot plant species (e.g., Collins et al. 1998, 2001; Pflieger et al. 1999; Donald et al. 2002; Zhang et al. 2002; Maleki et al. 2003; He et al. 2004; Trognitz and Trognitz 2005; Quirin 2006).

Interestingly, although RGAs were first recovered from *S. tuberosum*, development of comprehensive RGA resources for potato have progressed more slowly than for other related species. Analysis of complete genome sequence revealed that the *Arabidopsis thaliana* genome contains 149 NBS-LRR sequences (Meyers et al. 2003) and the rice genome contains 480 NBS-LRR sequences (Zhou et al. 2004). Based on these results, it is widely anticipated that the potato genome will contain several hundred NBS-LRR sequences. Leister et al. (1996) reported the recovery of 12 potato RGAs. Recently, Brugmans et al. (2008) reported the development of 34 additional RGAs from cultivated potato using NBS profiling (discussed below). However, to date, sequence information for these RGAs has not been made publicly available. These two studies comprise the entirety of cultivated potato RGAs reported to date.

At least three other, more comprehensive RGA resources have been developed for *Solanum* species. Working in tomato (*S. lycopersicum*), Pan et al. (2000) generated 75 NBS-LRR RGAs. Based on phylogenetic analysis of the deduced RGA protein sequences, 35 RGA clones were selected for subsequent mapping. Many of the clones mapped to multiple genome locations, some of which had been previously associated with phenotypic disease resistance. Importantly, phylogenetic analyses revealed that tomato RGAs, in many cases, clustered with the NBS regions from *R* genes cloned from other Solanaceous species (Pan et al. 2000). Trognitz and Trognitz (2005) used primers targeting the NBS of *R* genes in general and the *S. demissum* late blight resistance gene *R1* in particular to amplify RGAs from *S. caripense*, a South American relative of tomato and potato. In total, 45 RGA sequences putatively encoding intact proteins were reported, but the authors speculate that *S. caripense* may in fact possess 10 times this many functional *R* genes (Trognitz and Trognitz 2005). More recently, our laboratory has generated 109 RGA sequences from the disease resistant wild potato *S. bulbocastanum* (Quirin 2006). Our laboratory is now working to develop a framework for cross species comparisons of RGA sequences from *Solanum* species that includes the definition of homology groups, which approximate R gene family lineages. A broader survey of *Solanum* species is also being conducted in order to understand the rate of *R* gene diversification and birth-and-death in the Solanaceae and to identify *Solanum* species harboring unique *R* gene lineages.

8.2.2.1.2 NBS Profiling

The technique known as NBS profiling combines traditional RGA approaches, which rely on degenerate or non-degenerate PCR primers targeting conserved NBS motifs, and AFLP, a marker technology that relies upon restriction enzyme digestion, ligation of adapters, and selective PCR. The method, in essence but not in name, was first employed by Hayes and Saghai Maroof (2000) to tag *Rsv1*, a soybean gene imparting resistance to soybean mosaic virus. These authors generated DNA pools from resistant and susceptible F_2 progeny, digested them with *Eco*RI and *Mse*I restriction enzymes, ligated adapters of known sequence, and employed selective PCR to amplify a subset of the generated fragments. The resulting DNA mixtures were diluted and served as a template for a second round of amplification using a degenerate primer designed from the aligned NBS P-loop motifs of the tobacco *N* gene, the *Arabidopsis Rps2* gene, and the flax *L6* gene paired with one of the adapter primers. Because the degenerate primer had been labeled with ^{32}P, the resulting marker profiles could be detected by separation on acrylamide gels and autoradiography. The method proved useful in identifying RGA fragments linked to the *Rsv1* gene. van der Linden et al. (2004) later modified the procedure for use in identifying candidate *R* genes in potato, tomato, barley, and lettuce. These authors developed a series of NBS primers from the P-loop and kinase-2 motifs, pairing each with an adapter-specific primer. Similar to the results of Hayes and Saghai Maroof (2000), van der Linden et al. (2004) showed that 50–88% of recovered fragments encoded motifs common to R proteins. These authors are also the first to use the phrase "NBS profiling" for this molecular procedure (van der Linden et al. 2004).

Brugmans et al. (2008) applied NBS profiling to potato, taking an innovative approach to efficiently map 34 potato RGAs onto an existing ultra-high density AFLP map of potato (van Os et al. 2006). NBS profiling employed a series of degenerate primers targeting the conserved NBS P-loop, kinase-2 and GLPL motifs paired with the adapter primers reported by van der Linden et al. (2004). Primer pairs were applied first to template DNA of the mapping population parents and a subset of 29 F_1 progeny selected as being highly informative on the basis of prior genome-wide mapping results (van Os et al. 2006). Of the 90 sequenced NBS profiling fragments, 60 (66.7%) showed sequence homology to *R* genes or RGAs. These fragments sometimes bore significant similarity to standard RGA fragments generated previously for potato (Leister et al. 1996) and tomato (Pan et al. 2000). Other fragments were seemingly novel. Fragments that segregated amongst the F_1 progeny were mapped relative to existing AFLP data, allowing assignment of RGAs to specific chromosome regions. Importantly, many NBS profiling fragments mapped to genome regions shown previously to condition phenotypic resistance, thus classifying specific fragments as candidates for genes conditioning specific traits.

Next, Brugmans et al. (2008) employed cDNA as a template for NBS profiling. The use of cDNA in this experiment allowed the authors to recover fragments from only transcribed loci, an approach that should limit the number of pseudogenes represented in their RGA collection. cDNA was extracted from leaves, stems, and roots of eight F_1 progeny. Results showed that most NBS profiling fragments were monomorphic, being present in all genotypes and all tissue types. This suggests that most

R genes are transcribed in all plant tissues. However, some NBS profiling fragments showed clear intensity differences between tissues of the same genotype. Thus, some *R* genes might be transcribed at different rates in different tissues.

8.2.2.2 Identifying Candidate R Genes Based on Function

R proteins interact, directly or indirectly, with cognate proteins from pathogens referred to as effectors. While R proteins have a defined function (the detection of a pathogen and activation of defense responses), effectors are diverse in function, although they are generally implicated in enabling pathogenesis, especially by overcoming plant basal defense mechanisms. Similarly, while a majority of R proteins, regardless of the plant species from which they originate, share the basic NBS-LRR structure, effectors are structurally more diverse. This has historically slowed the cloning of pathogen effectors. However, this may be changing. Recent advances in understanding the biology and molecular basis of *P. infestans* effectors has facilitated rapid in silico and in vivo identification of effectors. This in turn has spawned new approaches for identifying and isolating candidate *R* genes based on functional recognition of *P. infestans* effectors.

In recent years, at least four oomycete effector genes have been cloned (Allen et al. 2004; Shan et al. 2004; Armstrong et al. 2005). Analysis of these genes revealed that they encode functionally divergent proteins. However, all deduced proteins share a signal peptide for secretion and a downstream "RXLR" motif, where "X" indicates any amino acid. The motif "EER" is often, but not always, present as well (Birch et al. 2008). The RXLR motif has been demonstrated to be essential for translocation of the *P. infestans* effector AVR3a inside of a host cell (Whisson et al. 2007), confirming a hypothesized functional role for this conserved motif. Survey of complete genome sequence of *P. infestans* revealed 425 genes encoding the RXLR motif (Whisson et al. 2007). These genes represent possible *P. infestans* effectors.

Research strategies aimed at first identifying effectors that are both non-redundant and essential to a pathogen's ability to infect and then identifying *R* genes with which they interact may yield long term durability (Birch et al. 2008). These approaches appear to be gaining favor among potato researchers. Vleeshouwers et al. (2008) provide a concrete example. These authors made use of candidate *P. infestans* effector genes first to screen wild *Solanum* germplasm to functionally characterize *R* genes present and then, in transient expression assays, to test candidate *R* genes generated on the basis of homology to a previously cloned gene. This approach enabled the rapid and efficient cloning of the late blight *R* genes *Rpi-sto1* and *Rpi-pta1* (Vleeshouwers et al. 2008; see also in detail below). Importantly, Vleeshouwers et al. (2008) also propose the use of candidate *P. infestans* effector libraries to catalog *R* genes based on which effectors are detected. Pyramiding *R* genes that detect different *P. infestans* effectors may enhance the likelihood of achieving durable late blight resistance. This application of knowledge of effector biology to identify, characterize, clone, and deploy late blight *R* genes represents a significant shift in research strategies and is likely to yield significant results in coming years.

8.2.3 Cross-species Comparative Genomics Approaches

Bonierbale et al. (1988) noted on the basis of common RFLP markers mapped in tomato and potato, that these two species share a significant degree of genomic structural similarity. Specifically, with few exceptions, RFLP markers mapped in the same order along corresponding chromosomes of tomato and potato. This conservation of marker order is referred to as synteny. Subsequent work refined the relationships of the tomato and potato genomes (Tanksley et al. 1992) and extended observations of synteny to other *Solanum* species (Doganlar et al. 2002) and other Solanaceae species (Prince et al. 1993; Livingstone et al. 1999). In these analyses, observed differences in marker order between species generally encompass blocks of markers, suggesting that chromosome rearrangement events (e.g., translocations, inversions, etc.) occurred during species divergence. Despite these rearrangements, however, similarities in genome structure across Solanaceous species are clearly decipherable.

8.2.3.1 Use of Reference Heterologous Sequence

The observations of Bonierbale et al. (1988) and others revealed that the order of molecular markers is conserved along tomato and potato chromosomes. This research led to the hypothesis that the order of homologous genes might also be conserved in tomato, potato, and more distantly related Solanaceous species. This hypothesis was first tested for disease resistance genes at a phenotypic level. In what might be considered a watershed paper in Solanaceae *R* gene research, Grube et al. (2000) aligned the chromosomes of potato, tomato, and pepper, correcting for chromosome rearrangements. Next, these authors culled the literature to identify genome regions in any of the three species to which disease resistance phenotypes had been mapped. Approximating genome locations conditioning resistance on the aligned maps revealed that phenotypic resistance mapped to corresponding genome regions in potato, tomato, and pepper more often that would be expected by chance (Fig. 8-2). Interestingly, this research revealed that homologous genome regions across these species were often associated with resistance to very different pathogens (e.g., a bacteria, a virus, and a nematode). However, there was some tendency for resistance to oomycete pathogens (especially

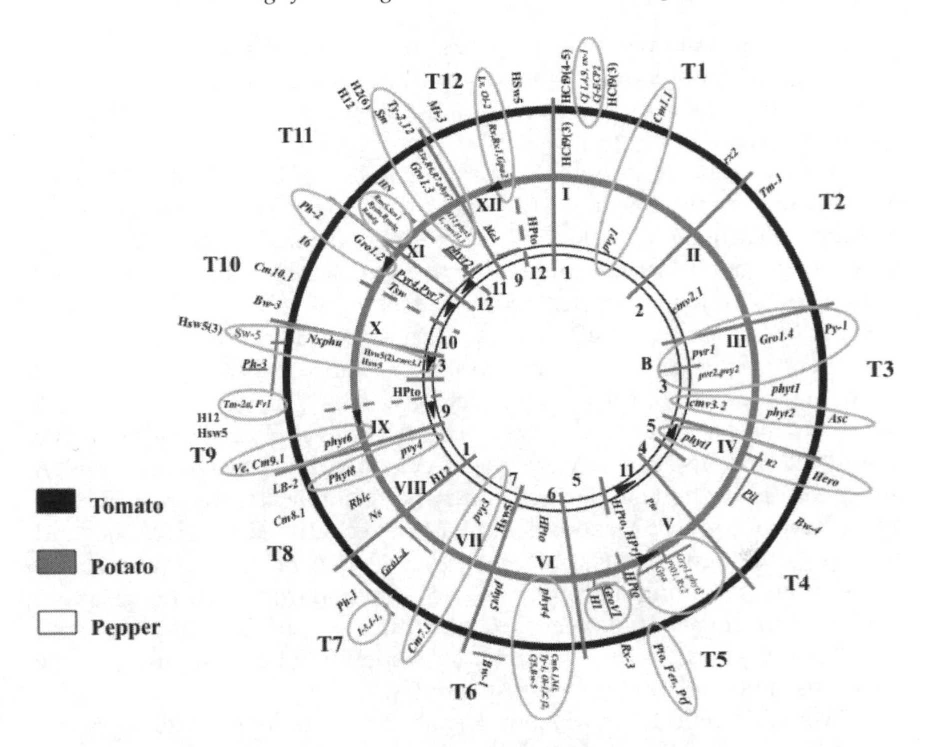

Figure 8-2 Resistance gene organization in the Solanaceae. The tomato genome is represented by the outer circle, proceeding clockwise from top to bottom of each chromosome. The potato and pepper genomes are represented by the middle and inner circles, respectively. Inversions that distinguish the three genomes are indicated by arrows within the rearranged portions of the potato and pepper maps. Solid lines indicate breaks between chromosomes, and dashed lines indicate the borders of rearranged regions. The five translocations between the pepper and tomato genomes are shown by splitting each of pepper chromosomes 3, 5, 9, 11, and 12 into two halves, each half syntenous to different tomato chromosomes. Resistance genes are designated by solid type; resistance gene homologues are designated by the letter H in outlined type followed by the gene name in subscript. Dark bars along the length of the chromosome are used to denote regions of a chromosome associated with a gene if imprecisely located; names of genes excluded from analyses due to imprecise location are underlined. Ovals are drawn around cross-generic and intrageneric *R* gene clusters ($\leq$ 15 cM) identified by this analysis. Note that genomes are not drawn to scale. This figure reproduced with permission from Grube et al. (2000).

Phytophthora spp.) to be conditioned by homologous genome regions in each of the three species, suggesting a long and significant evolutionary relationship between the Solanaceae and oomycetes (Grube et al. 2000).

The research of Grube et al. (2000) was significant in that it suggested that *R* genes of a common evolutionary origin (but divergent function) might reside in corresponding genome locations throughout Solanaceous species. However, this research was conducted at a phenotypic level, leaving some question as to whether the genes underlying phenotypic resistance mapping to homologous chromosome regions were in fact themselves homologous. Recently, Mazourek et al. (2009) demonstrated that this is indeed the case.

The *Bs2* gene of pepper (*Capsicum chacoense*) is an NBS-LRR gene conferring resistance against the bacterium *Xanthomonas campestris* pv. *vesicatoria*. The gene was cloned from pepper (Tai et al. 1999), but its map location was not fully determined due to the presence of numerous related sequences within the pepper genome. Mazourek et al. (2009) observed that the *Bs2* gene had significant sequence homology to the potato *Rx* and *Gpa2* genes, *R* gene paralogs that confer resistance to Potato Virus X and the nematode *Globodera pallida*, respectively. In fact, survey of public DNA databases revealed that, at a sequence similarity level, *Rx* and *Gpa2* are more closely related to *Bs2* than any other gene cloned to date. On the basis of DNA and deduced protein similarities and striking similarity in gene structure and genomic context, Mazourek et al. (2009) concluded that *Rx/Gpa2* and *Bs2* are orthologous genes, having descended from a distant common ancestor and retaining disease resistance function throughout speciation. Given the findings of Grube et al. (2000), these authors also hypothesized that *Rx/Gpa2* and *Bs2* should occupy corresponding homologous genome locations in potato and pepper, respectively.

To test this hypothesis, Mazourek et al. (2009) employed a multi-species mapping approach that included mapping markers linked to *Bs2* in tomato and markers linked to *Gpa2* in pepper. Although the *Capsicum* genome contains at least 22 structural rearrangements compared to that of tomato (Livingstone et al. 1999), knowledge of breakpoints of rearrangements allowed Mazourek et al. (2009) to align map locations of *Bs2* and *Gpa2* across all three species. Significantly, these authors concluded that, as hypothesized, homologs of the potato *Rx/Gpa2* genes map to the pepper *Bs2* region, confirming the orthologous relationship of these genes. Thus, the observations of Grube et al. (2000) that similar genome regions condition disease resistance across Solanaceous species, translate into homologous *R* genes residing in common genome locations across species.

The study of Mazourek et al. (2009) demonstrates the power of using knowledge of genome structure and sequence from one species to map or clone a homologous gene from a related species. This concept, known as comparative genomics, has already facilitated the cloning of several *Solanum* *R* genes (detailed below) and will likely become more important once the genome sequences of tomato and potato are released.

8.2.3.2 Isolation of R Genes by Homology

As demonstrated by the comparison of the pepper *Bs2* and the potato *Rx/Gpa2* genes, DNA sequence similarity between *R* gene homologs is largely conserved, even among distantly related Solanaceous species (Mazourek et al. 2009). This conservation of DNA sequence has led many researchers to isolate homologs of cloned *R* genes from an array of species, enabling evolutionary and functional studies. This has been especially true for late blight resistance genes with agricultural deployment potential. In most cases, these efforts have entailed PCR-based amplification from genomic DNA using primers designed from the cloned *R* genes. All late blight *R* genes cloned to date are members of the NBS-LRR superfamily. Because most NBS-LRR genes are harbored in clusters of structurally related but functionally divergent genes, Sanchez and Bradeen (2006) advocated for design of PCR primers that flank the coding regions of *R* genes. This approach worked effectively to isolate probable *R* gene orthologs from multiple genotypes within a single species (Sanchez 2005) but is expected to be less effective across species boundaries since *R* gene flanking regions may diverge quickly, resulting in loss of priming sites. Several researchers have instead used PCR primers designed directly from *R* gene coding regions to amplify corresponding fragments from related species (detailed below). This approach has been coupled with functional testing of the recovered fragments to establish orthology.

8.3 Progress Towards Cloning Late Blight *R* Genes: A Timeline

Advances in the cloning of late blight *R* genes in the past eight years have been astounding. Significant *R* gene cloning efforts are summarized in Table 8-1 and are detailed below in chronological order. Early *R* gene cloning efforts were slow, laborious, and far between. But advances in both biology and technology have led to more rapid, and more frequent, *R* gene recovery in recent years.

8.3.1 The Cloning of **R1***—2002*

R1, a late blight resistance gene originating from the Mexican wild potato, *S. demissum*, was the first late blight *R* gene cloned. *R1* is one of at least 11 *P. infestans* race-specific *R* genes introgressed into cultivated potato from *S. demissum* using traditional breeding approaches (Wastie 1991; Umaerus and Umaerus 1994). All were quickly defeated. Although it proved ultimately not to be durable, the cloning of *R1* represents a significant first in understanding the molecular basis of late blight resistance.

Table 8-1 Timeline of late blight resistance gene cloning efforts.

Year	Gene	Species of Origin	Strategy	Reference
2002	*R1*	*S. demissum* introgressed into cultivated potato	Candidate gene + chromosome walking	Ballvora et al. 2002
2003	*RB*	*S. bulbocastanum*	Candidate gene + chromosome walking	Song et al. 2003
2003	*RB* (*Rpi-blb1* allele)	*S. bulbocastanum*	Chromosome walking	Vossen et al. 2003
2005	*R3a*	*S. demissum* introgressed into cultivated potato	Comparative genomics	Huang et al. 2005
2005	*Rpi-blb2*	*S. bulbocastanum*	Candidate gene + chromosome walking	Vossen et al. 2005
2006	RB^{ver}	*S. verrucosum*	Homology to *RB*	Liu and Halterman 2006
2008	*Rpi-sto1*	*S. stoloniferum*	Effector biology + homology to *RB*	Vleeshouwers et al. 2008
2008	*Rpi-pta1*	*S. papita*	Effector biology + homology to *RB*	Vleeshouwers et al. 2008
2009	*Rpi-vnt1.1*	*S. venturii*	Chromosome walking + comparative genomics + candidate gene	Pel et al. 2009; Foster et al. 2009
2009	*Rpi-vnt1.3*	*S. venturii*	Chromosome walking + comparative genomics + candidate gene	Pel et al. 2009; Foster et al. 2009
2009	*Rpi-blb3*	*S. bulbocastanum*	Chromosome walking + comparative genomics	Lokossou et al. 2009
2009	*Rpi-abpt*	*S. bulbocastanum* (?)	Chromosome walking + comparative genomics	Lokossou et al. 2009
2009	*R2*	*S. demissum*	Homology to *Rpi-abpt*	Lokossou et al. 2009
2009	*R2-like*	*S. demissum*	Homology to *Rpi-abpt*	Lokossou et al. 2009
2009	*Rpi-bt1*	*S. bulbocastanum*	Chromosome walking + candidate gene	Oosumi et al. 2009

In cloning *R1*, Ballvora et al. (2002) combined BAC-based chromosome walking with a candidate gene approach. First, fine mapping of the *R1* region on potato chromosome V identified markers flanking the resistance gene and encompassing a genetic distance of 0.2 cM. Screening of a BAC library identified a single BAC clone. Markers developed from the end sequences of the clone were genetically mapped and used to initiate a walking effort. Ultimately, BAC clones whose end sequences co-segregated with resistance were identified. Importantly, BAC end sequencing revealed a gene copy with significant homology to the tomato *Prf* gene for resistance to *Pseudomonas syringae.* A fragment derived from this *R* gene homolog was used to identify additional BAC clones that were subsequently shown through genetic mapping to encompass the *R1* region. The *R1* gene cluster was ultimately estimated to reside in a 200 kb region on chromosome V. Complementation studies involving transforming late blight susceptible potato cultivar Desiree with an *R1*-containing BAC clone confirmed that the resistance allele resided on that clone. The BAC clone was next used to generate a sublibrary comprised of ~ 10 kb inserts. Probing of the library with the *Prf* homolog identified clones carrying intact *R1* candidates, one of which was sequenced to confirm presence of the gene. This subclone was then used to transform 'Desiree', yielding the expected late blight resistant phenotype. Next, the researchers used two BAC clones known to contain the functional *R1* locus as probes to screen a cDNA library prepared from *P. infestans*-infected potato leaves. Fourteen clones were identified, eight of which showed sequence homology to known *R* genes, especially *Prf.* The predicted coding region of one cDNA clone was identical to the *Prf* homolog present in the subclone used to successfully transform 'Desiree'.

8.3.2 The Cloning of **RB**—*2003*

Solanum bulbocastanum is a diploid potato native to central Mexico. This region of the world is also widely recognized as a genotypic center of diversity for *P. infestans* and local environmental conditions are frequently ideal for late blight epidemics. Recognizing the potential for a long co-evolutionary relationship between *S. bulbocastanum* and *P. infestans,* breeders and researchers long ago described *S. bulbocastanum* as a potential source of durable late blight resistance (Reddick 1939; Niederhauser and Mills 1953; Black and Gallegly 1957; Graham et al. 1959). However, attempts to cross *S. bulbocastanum* directly with cultivated potato have uniformly failed (Jackson and Hanneman 1999). Hermsen and Ramanna (1973) were successful in moving late blight resistance genes from *S. bulbocastanum* into *S. tuberosum* using a complicated breeding scheme and two additional bridge species, *S. acaule* and *S. phureja.* Helgeson et al. (1998) were also successful in accessing late blight resistance using *S. bulbocastanum* + *S. tuberosum* somatic hybrids. These approaches, however, have the disadvantage of introducing the entire *S. bulbocastanum* genome into cultivated potato, incorporating not just late blight resistance but potentially a slew of undesired characteristics as well.

The somatic hybrids generated by Helgeson et al. (1998) enabled subsequent genetic mapping and gene isolation efforts. Similar to *R1* cloning efforts (Ballvora et al. 2002), the cloning of resistance from *S. bulbocastanum* combined BAC-based chromosome walking with a candidate gene approach. First, resistance was mapped to *S. bulbocastanum* chromosome VIII (Naess et al. 2000). Mapping efforts employed BC_2 populations segregating for late blight resistance generated by backcrossing resistant somatic hybrids (*S. bulbocastanum* + *S. tubersoum*) to susceptible cultivated potato for two generations [((*S. bulbocastanum* + *S. tuberosum*) x *S. tuberosum*) x *S. tuberosum*]. RAPD and RFLP markers mapped resistance to a single locus, *RB*, on chromosome VIII. Markers both flanking and co-segregating with resistance were identified (Naess et al. 2000). A BAC library of *S. bulbocastanum* genotype PT29, the original source of late blight resistance in the segregating populations, was constructed (Song et al. 2000) and a BAC-based chromosome walking effort using markers closely associated with resistance was initiated (Bradeen et al. 2003). Chromosome walking entailed reiterative BAC screening, development of markers from BAC end sequences, and fine scale genetic mapping on segregating progeny. Mapping was aided by screening seedlings at the cotyledon stage for recombinants between markers flanking resistance; only those individuals with recombination events in this region were advanced to phenotypic screening for resistance. This procedure enabled the researchers to genetically narrow the late blight resistance region to approximately 55 kb. The *S. bulbocastanum* genotype used in construction of the original somatic hybrid and the BAC library is heterozygous for resistance (i.e. *RB/rb*). Unfortunately, despite identifying multiple BAC clones that spanned the genetically-defined 55 kb resistance region, Bradeen et al. (2003) were unable to identify BAC clones carrying the *RB* allele; all identified BAC clones originated from the *rb* haplotype. Subsequent work demonstrated that BAC clones carrying the functional *RB* allele were unstable, with structural rearrangements of the BAC insert resulting in destruction of the *RB* allele (Song et al. 2003a). Sequencing of a BAC clone from the *rb* haplotype that spanned the 55 kb resistance region identified a cluster of four seemingly intact NBS-LRR genes and an obvious pseudogene (Song et al. 2003b).

To recover the functional *RB* allele, Song et al. (Song et al. 2003b) employed long range PCR (LR-PCR), amplifying up to 22 kb fragments from the genome of *S. bulbocastanum.* PCR primers flanking each candidate gene were designed using BAC sequence from the *rb* haplotype. Each of the four candidate genes was amplified separately, along with predicted promoter and downstream elements. Transformed into late blight susceptible potato 'Katahdin', one of the fragments imparted broad-spectrum late blight resistance (Song et al. 2003b). *RB* was confirmed to be an NBS-LRR type *R* gene. The original LR-PCR construct has been subsequently transformed into important US cultivars and extensively field tested, demonstrating that this single gene can impart agriculturally useful levels of late blight resistance, even in the absence of fungicides (Bradeen et al. 2009; Fig. 8-3). *RB* was the first late blight resistance gene with actual deployment potential to be cloned. Despite significant progress in the cloning of additional *R* genes since that time, *RB* (including its homologs subsequently derived from other species) remains the single most promising gene for long-term genetic control of late blight disease.

Figure 8-3 The transgene *RB* imparts to potato agriculturally meaningful levels of resistance to foliar late blight under field conditions in the absence of fungicides. Photograph of a late blight nursery in Minnesota in 2005. This picture was taken approximately four weeks after inoculation of the field with *Phytophthora infestans* US8. Transgenic plants (green) remain healthy and relatively disease free compared to non-transgenic potato cultivar NorChip (rows of dead, infected plants). Higher copy numbers of the potato *RB* transgene correspond to enhanced transcript and late blight resistance levels. [Reprinted with permission from Bradeen et al. (2009) Mol Plant-Microbe Interact 22: 437–446].

Color image of this figure appears in the color plate section at the end of the book.

8.3.3 The Cloning of **Rpi-blb1***, an Allele of the* **RB** *Locus—2003*

A classic example of gene cloning by BAC-based chromosome walking is the isolation of the *Rpi-blb1* allele of the *RB* locus from *S. bulbocastanum*. Working independently from Song et al. (2003b), Vossen et al. (2003) isolated a gene conferring late blight resistance and located on chromosome VIII of *S. bulbocastanum*. These authors first generated an F_1 family segregating for late blight resistance by crossing *S. bulbocastanum* genotype 8005-8, a late blight resistant plant, with *S. bulbocastanum* genotype 8006-9, a late blight susceptible plant. Segregation ratios approximated a 1:1 ratio (resistant:susceptible), suggesting that late blight resistance in genotype 8005-8 is conferred by a single locus. Importantly, these authors demonstrated resistance against a panel of *P. infestans* isolates, supporting the conclusion of Helgeson et al. (1998) and Song et al. (2003b) that *S. bulbocastanum* is a source of broad-spectrum resistance. Next, Vossen et al. (2003) mapped resistance to the *RB* region on *S. bulbocastanum* chromosome VIII using RFLP-derived CAPS markers. Simultaneously, a BAC library was constructed from the late blight resistant parent and a multistep chromosome walking approach was initiated using RFLP markers associated with the resistance phenotype. BAC end-derived markers ultimately mapped resistance to a 0.1 cM region. Importantly, BAC end sequencing in this region revealed homology with tomato RGAs isolated by Pan et al. (2000). A sublibrary was prepared from a BAC clone spanning the *S. bulbocastanum* resistance region. One of the tested subclones imparted late blight resistance to potato cultivar Impala and tomato cultivar Moneymaker. Sequence analysis revealed that the coding region of the newly cloned gene is 100% identical to that of the previously cloned (Song et al. 2003b) *RB* gene. Based on SNPs present in the intron and flanking regions, Vossen et al. (2003) dubbed the newly cloned gene *Rpi-blb1*, describing it as an allele of the *RB* locus. The authors speculate that *RB* plays a major role in imparting late blight resistance to *S. bulbocastanum* and that functional alleles of this locus are likely to be widespread in native populations.

8.3.4 The Cloning of R3a*—2005*

An elegant example of the potential for comparative, cross-species genomics to facilitate the efficient isolation of late blight resistance genes is provided by the cloning of *R3a*, a late blight resistance gene originating from *S. demissum*. In cloning *R3a*, Huang et al. (2005) capitalized upon observations of synteny between tomato and potato on chromosome XI and structural knowledge about the complex tomato *I2* locus. In tomato, the *I2* locus confers resistance and *I2* paralogs confer partial resistance to race 2 of *Fusarium oxysporum* f. sp. *lycopersici* (Ori et al. 1997; Simons et al. 1998; Sela-Buurlage et al. 2001). *I2* homologs in the tomato genome are arrayed in large clusters, three of which reside on chromosome 11. Previously, fine mapping of the *R3a* region in potato allowed Huang et al. (2004) to conclude that *R3a* resided in an *I2*-like cluster and might be an *I2* homolog, and that potato and tomato are predominantly collinear in this region. Building upon these observations, Huang et al. (2005) designed PCR primers targeting the LRR region of tomato *I2* homologs. Based upon in silico analyses, the primers were predicted to be specific to *I2*-like sequences but to amplify all

I2 homologs and alleles. The authors developed pools of 384 BAC clones from the resistant potato genotype. Each pool represented, on average, 0.05 genome equivalents. Next, the authors screened the pools both with markers associated with *R3a* and with the PCR primers targeting *I2*-like sequences. In this analysis, co-occurrence of *R3a* and *I2*-like sequences in a common BAC pool was interpreted to indicate that one of the comprising BACs contained the *R3a* region. This analysis was used ultimately to identify BAC clones harboring the *R3a* region, distinguishing it from other *I2*-like regions (which would show amplification using the *I2*-like primers but would lack the markers linked to *R3a*). Sequence analysis of two BAC clones revealed four *I2*-like candidates for *R3a*. Transformation of late blight susceptible potato 'Desiree' with constructs from each candidate gene revealed that only one imparted late blight resistance. The authors had successfully cloned late blight resistance gene *R3a* (Huang et al. 2005). This cloning effort was greatly facilitated by knowledge of corresponding tomato genome structure and sequence.

8.3.5 The Cloning of Rpi-blb2*—2005*

Solanum bulbocastanum is the source of a second cloned late blight resistance locus, *Rpi-blb2* (Vossen et al. 2005). Phenotypic testing of genetic stocks derived from the ABPT population developed by Hermsen and Ramanna (1973) revealed that broad spectrum late blight resistance had been successfully transferred from *S. bulbocastanum* to cultivated potato. Backcrossing a resistant genotype to susceptible *S. tuberosum* yielded progeny that segregated for late blight resistance, leading the authors to conclude that a single resistance gene was responsible. Using bulked segregant analysis (BSA; Michelmore et al. 1991), the authors identified 58 AFLP markers putatively linked to resistance. Two AFLP markers that were confirmed to co-segregate with resistance were converted to CAPS markers. The CAPS markers, and by linkage, *Rpi-blb2,* were subsequently mapped to chromosome VI in a reference population. Marker E40M58e, a SCAR marker originating from an AFLP fragment, was found closely associated with *Rpi-blb2*. Capitalizing upon the design of RGA primers by Leister et al. (1996), Vossen et al. (2005) next used one PCR primer derived from SCAR marker E40M58e in combination with the NBS-specific primer, S1, designed by Leister et al. (1996). The resulting amplicon co-segregated with resistance in the ABPT-derived populations and showed sequence homology to the tomato *Mi-1* gene, which imparts resistance to root knot nematode, potato aphid, and whitefly (Milligan et al. 1998; Rossi et al. 1998; Vos et al. 1998; Nombela et al. 2003). A BAC library was constructed from a late blight resistant ABPT-derived diploid clone known to be heterozygous for *Rpi-blb2*. Multistep walking ensued. Additional markers were developed from BAC ends, allowing finer mapping of the *Rpi-blb2* region. Importantly, these efforts confirmed that the *Rpi-blb2* region harbors multiple *Mi-1* homologs. Unfortunately, however, these efforts failed in building a complete BAC contig spanning the region. In response, the authors turned to an intraspecific F_1 *S. bulbocastanum* population that also segregated for *Rpi-blb2*. Recombinants in the previously defined *Rpi-blb2* region allowed identification of markers that further delimited resistance. By screening an *S. bulbocastanum* BAC library, the authors identified two BAC clones that together represented the entire *Rpi-blb2* region. Nine *Mi-1* homologs identified on a total of five BAC clones from the region were subcloned and tested as transgenes in late blight susceptible potato. One of the subclones imparted late blight resistance to potato 'Impala' and 'Kondor'. The same subclone imparted late blight resistance to tomato 'Moneymaker'. The subclone was sequenced and *Rpi-blb2*, an NBS-LRR R gene, was identified. The *Rpi-blb2* coding region shares 89.7% homology with the tomato *Mi-1* gene (Vossen et al. 2005). As was true in the case of *R3a* (Huang et al. 2005), the cloning efforts of Vossen et al. (2005) were aided by observations of shared synteny between potato and tomato and by the availability of sequence information from tomato.

8.3.6 The Cloning of **RB^{ver}**, **Rpi-sto1**, *and* **Rpi-pta1**—*2006, 2008*

Researchers have used a variety of "allele mining" approaches to isolate functional homologs of previously cloned late blight *R* genes from an array of *Solanum* species. This has been especially true for the agriculturally promising *RB* locus. Efforts to isolate *RB* from multiple species have allowed improved understanding of the evolution of this important locus. But the efforts have a more significant, practical application. *RB* was originally cloned from *S. bulbocastanum* (Song et al. 2003b), a wild potato species that is sexually isolated from cultivated potato (Jackson and Hanneman 1999). The cloning of *RB* opens the way for transgenic transfer of the gene into commercially prominent potato varieties (Bradeen et al. 2009), although this approach is not without controversy (see Chapter 12). Identification of functionally equivalent loci in species other than *S. bulbocastanum* may allow scientists to transfer genes into *S. tuberosum* without the use of transgenic technologies or laborious bridge crosses (Hermsen and Ramanna 1973) and somatic hybridization (Helgeson et al. 1998). Three significant examples of *RB* homologs recovered from wild *Solanum* species that can be directly crossed with cultivated potato have been reported.

Liu and Halterman (2006) studied the genetic nature of broad spectrum late blight resistance in the Mexican wild potato, *S. verrucosum*. Using whole plant, greenhouse inoculations, these authors demonstrated that most accessions of *S. verrucosum* are late blight resistant, although the level of resistance varied. Next, PCR primers were designed from publicly available *RB* sequence data and used to amplify homologous sequences from each accession of *S. verrucosum*. Analysis of the eight recovered sequences led the authors to conclude that two of the tested *S. verrucosum* genotypes carry full length, and potentially functional, alleles at a locus they dubbed RB^{ver}. The alleles were up to 83.4% identical at the nucleotide level to *RB* from *S. bulbocastanum*. Interestingly, Liu and Halterman (2006) indicate that the LRR region of the RB^{ver} allele encodes one additional repeat unit relative to *RB*. Transcribed, seemingly intact paralogs of the *S. bulbocastanum* locus also encode for an additional LRR repeat unit but do not function for late blight resistance (Song et al. 2003b). Nevertheless, Liu and Halterman (2006) demonstrate through complementation analysis that the RB^{ver} construct imparted resistance to four of 36 transgenic plants. These results, however, must be interpreted cautiously as the authors failed to isolate a full length RB^{ver} allele from *S. verrucosum* since the PCR amplicon lacked a predicted 50bp from the 3′ end of the gene. Instead, the construct tested was generated as a fusion product of the RB^{ver} amplicon and the 3′ end of the functional *S. bulbocastanum RB* allele. It is especially important to note that the LRR (3′ end of an NBS-LRR gene) has been widely implicated in the determination of pathogen specificity [reviewed in (Ellis et al. 2000)].

Vleeshouwers et al. (2008) reported the cloning of late blight resistance loci, *Rpi-sto1* from *S. stoloniferum* and *Rpi-pta1* from *S. papita*. First, these authors used an in vivo assay to determine whether any of 54 putative *P. infestans* effector genes, initially identified as encoding an RXLR motif and a signaling peptide, initiated *RB*-mediated resistance responses. The assay consisted of transiently co-expressing *RB* and each putative effector gene in leaves of *Nicotiana benthamiana*. The presence of a hypersensitivity response identified two effector genes, *IpiO1* and *IpiO2*, sequence-similar members of a small gene family, as *Avr-blb1* (syn. *Avr-RB*). Simultaneously, an F_1 mapping population was generated by crossing a late blight resistant genotype of *S. stoloniferum* with a late blight susceptible potato genotype. Testing of progeny with multiple isolates of *P. infestans* confirmed that resistance from *S. stoloniferum* is broad-spectrum. Resistance segregated amongst progeny at a 1:1 (resistant:susceptible) ratio, suggesting control by a single gene, *Rpi-sto1*. Next, the authors tested parents and F_1 progeny against the 54 putative *P. infestans* effectors. Of the 54 tested, only effectors corresponding to *IpiO1* and *IpiO2* triggered a hypersensitive response in late blight resistant, but not late blight susceptible, genotypes. Given that *RB* and *Rpi-sto1* both recognize *IpiO1* and *IpiO2*, but not other tested effectors, the authors concluded that these two resistance genes might be homologous.

Building upon this hypothesis, Vleeshouwers et al. (2008) next confirmed that *Rpi-sto1* mapped to a genome location corresponding to the *RB* locus by testing for co-segregation between marker CT88 (linked to *RB* in *S. bulbocastanum*) and late blight resistance derived from *S. stoloniferum*; the marker co-segregated perfectly with resistance. PCR using primers designed from the *S. bulbocastanum RB* sequence were used to amplify corresponding fragments from *S. stoloniferum*. Functional testing, initially involving in planta co-expression of R gene candidates along with *IpiO2* and later involving challenging transgenic potato plants with *P. infestans* isolates, identified *Rpi-sto1*. A similar strategy was employed to isolate *Rpi-pta1*, an *RB* homolog from the late blight resistant *S. papita*. Also using similar approaches, Wang et al. (2008) documented that late blight resistance co-segregates with *RB* homologs in *S. polytrichon* and speculated that the gene responsible, dubbed *Rpi-plt1*, is also related to *RB*. (The cloning of *Rpi-plt1* has not yet been reported). However, Spooner et al. (2004) considers *S. stoloniferum*, *S. papita*, and *S. polytrichon* to be conspecific, suggesting that *Rpi-sto1*, *Rpi-pta1*, and *Rpi-plt1* may in fact be the same gene. Regardless of taxonomic classification, it is important to note that *S. stoloniferum* and *S. papita*, both of which harbor a late blight resistance gene functionally equivalent to *RB*, are sexually compatible with cultivated potato while *S. bulbocastanum* is not. Thus, the research of Vleeshouwers et al. (2008) provides a roadmap to potato improvement using the functional equivalence of *RB*-mediated late blight resistance without relying on transgenic approaches.

8.3.7 The Cloning of Rpi-vnt1.1 *and* Rpi-vnt1.3*—2009*

Solanum venturii is a diploid wild potato from Argentina. Two research groups recently coordinated efforts to clone the late blight resistance genes *Rpi-vnt1.1* and *Rpi-vnt1.3* from this species.

Pel et al. (2009) screened a collection of *S. venurii* accessions for late blight resistance. Two resistant genotypes were used to create two separate F_1 populations, each one segregating for late blight resistance. Segregation ratios in the two populations suggested a single gene was responsible for resistance. These genes were dubbed *Rpi-vnt1.1* and *Rpi-vnt1.3*. Testing of the segregating populations with an array of *P. infestans* isolates revealed that *Rpi-vnt1.1* and *Rpi-vnt1.3* share a common resistance spectrum. NBS profiling in combination with BSA followed by mapping on the complete populations identified two markers that co-segregated with *Rpi-vnt1.1* and *Rpi-vnt1.3*. Sequence analysis of these markers revealed significant homology to the tomato *Tm-2*^{2} gene for tomato mosaic virus resistance. This gene is located on tomato chromosome 9 (Lanfermeijer et al. 2003). Mapping of RFLP markers relative to NBS profiling markers confirmed that *Rpi-vnt1.1* and *Rpi-vnt1.3* reside in a common region on potato chromosome IX and could be *Tm-2*^{2} homologs (Pel et al. 2009). Additional PCR-based markers developed using EST data from public databases were used to screen a large number of *S. venturii* progeny for recombination events in the region. This approach allowed the authors to map *Rpi-vnt1.1* to a 4.0 cM region and *Rpi-vnt1.3* to a 3.7 cM region.

Building upon the hypothesis that *Rpi-vnt1.1* and *Rpi-vnt1.3* might be homologs of tomato *Tm-2*^{2}, Pel et al. (2009) next designed a series of PCR primers targeting *Tm-2*^{2}–like sequences. One primer pair yielded a 2.5 kb fragment from late blight resistant, but not susceptible, genotypes. By sequencing multiple clones from each genotype, amplicons could be assigned to one of five classes. Each class was 75–80% similar to tomato *Tm-2*^{2}. Additional PCR-based efforts ultimately yielded putative full length gene copies for each amplicon class. Two clones, one derived from a genotype carrying *Rpi-vnt1.1* and one derived from a genotype carrying *Rpi-vnt1.3*, contained predicted open reading frames and were considered as candidate late blight resistance genes.

Functionality was tested by transient gene expression in *N. benthamiana* followed by challenge with *P. infestans*. Results seemed to confirm the successful cloning of *Rpi-vnt1.1* and *Rpi-vnt1.3*. However, stable transformants of potato 'Desiree' told a different story. While eight of nine transformants carrying the candidate *Rpi-vnt1.1* gene were late blight resistant, none of 17 tested candidate *Rpi-vnt1.3* transformants were resistant. Thus, *Rpi-vnt1.1* but not *Rpi-vnt1.3* had been cloned. Sequence analysis suggested that the extreme 5′ end of the *Rpi-vnt1.3* gene had not been recovered. LR-PCR in combination with new primers was employed, allowing recovery of a new *Rpi-vnt1.3* fragment with an additional 42nt at the 5′ end of the coding sequence. Transient expression of this fragment in *N. benthamiana* and stable transformation of potato 'Desiree' confirmed that *Rpi-vnt1.3* had also been cloned (Pel et al. 2009).

Foster et al. (2009) simultaneously reported the cloning of *Rpi-vnt1.1* and identification of candidates for *Rpi-vnt1.2* and *Rpi-vnt1.3*. These authors generated a series of F_1 mapping populations by crossing late blight resistant *S. venturii* genotypes with late blight susceptible genotypes. Segregation in each population confirmed the presence of a single major *R* gene. BSA was applied to three of the populations and AFLP markers linked to resistance were identified. Several AFLP markers were converted to simple PCR markers. Mapping one of the PCR markers in a tomato population localized *Rpi-vnt1.1* to chromosome IX, results that were subsequently supported by SSR mapping. Additional marker development in the region ultimately localized *Rpi-vnt1* to a 6.0 cM region and revealed that *Rpi-vnt1.1* maps 0.7 cM away from *Rpi-vnt1.2* and 1.2 cM away from *Rpi-vnt1.3*.

A BAC library was constructed from an *S. venturii* genotype carrying *Rpi-vnt1.1* (Foster et al. 2009). BAC walking initiated from markers linked to *Rpi-vnt1.1* generated a BAC contig spanning the genetically-defined *Rpi-*

vnt1.1 region. BAC end sequencing revealed homology to the tomato *Tm-2*^{2} locus. In parallel with BAC walking, hybridization of a marker linked to *Rpi-vnt1.1* first to pools of BAC clones and then to individual BAC clones followed by fingerprinting of recovered clones generated a high-resolution physical map of the *Rpi-vnt1.1* region. PCR markers designed from BAC ends allowed finer genetic mapping of the region as well, ultimately localizing *Rpi-vnt1.1* to a 0.33 cM region. Sequencing of two BAC clones spanning this region revealed a single NBS-LRR locus, a candidate for *Rpi-vnt1.1*.

Hypothesizing that *Rpi-vnt1.2* and *Rpi-vnt1.3* would share homology with *Rpi-vnt1.1*, Foster et al. (2009) used PCR primers designed to amplify the complete *Rpi-vnt1.1* locus to recover candidates for *Rpi-vnt1.2* and *Rpi-vnt1.3* from appropriate genotypes. Of 29 tested 'Desiree' potato plants carrying the putative *Rpi-vnt1.1* transgene, 24 were resistant to *P. infestans*, confirming that *Rpi-vnt1.1* had been cloned. Further testing revealed that these transgenic plants were resistant to 10 of 11 tested *P. infestans* isolates. Functional testing of *Rpi-vnt1.2* and *Rpi-vnt1.3* candidates was not reported.

The cloned *Rpi-vnt1.1* gene is highly similar to tomato *Tm-2*^{2} (80.9% identity at the nucleotide level). The *Rpi-vnt1.2* candidate is identical to *Rpi-vnt1.1* except for the insertion of 42nt at the 5′ end of the gene and three SNPs, two of which result in amino acid changes. *Rpi-vnt1.3* differs from *Rpi-vnt1.2* by only a single SNP near the 5′ end of the gene (Foster et al. 2009; Pel et al. 2009).

8.3.8 The Cloning of Rpi-blb3, Rpi-abpt, R2, *and* R2-like*—2009*

Rpi-blb3, *Rpi-abpt*, *R2*, and *R2-like* have each been mapped to a common region on potato chromosome IV (Li et al. 1998; Park et al. 2005a, b, c). Lokossou et al. (2009) cloned *Rpi-blb3* from *S. bulbocastanum* using a reiterative process of BAC-based chromosome walking and fine genetic mapping using markers developed from BAC ends. Ultimately, *Rpi-blb3* was localized to a 0.1 cM region that was spanned by two partially overlapping BAC clones. Similarly, these same authors identified two partially overlapping BAC clones thought to span the *Rpi-abpt* region. The species of origin for *Rpi-abpt* is not known with certainty, but is probably *S. bulbocastanum*. The BAC clones harboring *Rpi-blb3* and *Rpi-abpt* originated from separate libraries. Sequencing of one BAC clone thought to harbor *Rpi-abpt* revealed two *R* gene-like loci with significant homology to those harbored on a previously sequenced tomato BAC clone. Lokossou et al. (2009) next designed PCR primers from regions conserved between the *Rpi-abpt* candidates and the homologous tomato sequences. Subclone libraries were constructed for BAC clones harboring *Rpi-blb3* and *Rpi-abpt* candidates. PCR screening coupled with restriction enzyme digestion allowed the authors to classify the recovered candidates into separate structural classes and representatives of each class were tested by transformation of the late blight susceptible potato 'Desiree'. Seven of eight plants harboring one class of *Rpi-blb3* candidates were resistant to two different *P. infestans* isolates and the authors concluded that *Rpi-blb3* had been cloned (Lokossou et al. 2009). However, none of the tested candidates for *Rpi-abpt* conferred resistance. Additional BAC-based chromosome walking ensued, leading to the discovery of an additional *Rpi-abpt* candidate. This candidate was tested by transient expression in *N. benthamiana* leaves. Challenge with *P. infestans* isolates revealed that *Rpi-abpt* had also been cloned.

R2, which originates from *S. demissum*, and *R2-like*, of unknown origin, are late blight resistance genes that map to the same region of potato chromosome IV as *Rpi-blb3* and *Rpi-abpt*. All four *R* genes share a common pathogen race specificity. It has been demonstrated for several pathosystems that pathogen specificity is determined largely by the LRR region of R proteins (reviewed in Ellis et al. 2000). Importantly, Lokossou et al. (2009) noted that the deduced LRR regions encoded by *Rpi-blb3* and *Rpi-abpt* were highly similar, suggesting that *R2* and *R2-like* might also share structural similarities in this region. To test this hypothesis, the authors used the PCR primers designed for the amplification of *Rpi-abpt* candidates to amplify fragments from genotypes carrying *R2* and *R2-like*. Nineteen candidates for *R2* and eight candidates for *R2-like* were identified in this manner. Phylogenetic analyses of the deduced amino acid sequences revealed that a single *R2-like* candidate and five *R2* candidates closely clustered with the functional *Rpi-blb3* and *Rpi-abpt* genes. These candidates were functionally tested by transient expression in *N. benthamiana* as described above. The single *R2-like* candidate was confirmed to be the functional *R2-like* gene. The deduced R2-like protein shares 97.3% amino acid identity with Rpi-blb3. One of the five tested *R2* candidates was confirmed to be the functional *R2* gene. R2 shares 92.6% amino acid identity with Rpi-blb3. All four cloned R genes, *Rpi-blb3*, *Rpi-abpt*, *R2*, and *R2-like*, recognize the *P. infestans* effector *PiAVR2* and are thus functionally equivalent (Lokossou et al. 2009).

8.3.9 The Cloning of **Rpi-bt1***—2009*

Recently, a fourth late blight resistance gene, *Rpi-bt1*, was cloned from *S. bulbocastanum*. Oosumi et al. (2009) studied backcross progeny originating from the *S. tuberosum* + *S. bulbocastanum* somatic hybrids generated by Helgeson et al. (1998). Starting with a BC_2 population reported by Naess et al. (2000), Oosumi et al. (2009) generated BC_3 progeny that segregated for resistance to *P. infestans*. AFLP and RFLP markers were used to map a single resistance gene to the *S. bulbocastanum* chromosome VIII to a region near to but distinct from the *RB* region. Specifically, Naess et al. (2000) reported that *RB* co-segregated with RFLP marker CT88 in a BC_2 population of 64 individuals but Oosumi et al. (2009), using 100 BC_3 individuals, reported that *Rpi-bt1* maps 1.6 cM away from CT88. Using a traditional BAC-based chromosome walking approach, *Rpi-bt1* was ultimately localized to two BAC clones. BAC sequencing revealed a cluster of nine NBS-LRR genes, including three truncated gene copies, four pseudogene copies (predicted on the basis of frameshift mutations or premature stop codons), and two seemingly intact gene copies. The two intact gene copies were initially considered as candidates for *Rpi-bt1*. However, transformation of late blight susceptible potato with each of the two candidate genes failed to produce resistant plants. RACE-PCR using primers designed from the candidate genes and mRNA from *P. infestans*-challenged *S. bulbocastanum* generated products with homology, but not exact identity, to the candidate genes. Given these results, the authors speculated that *S. bulbocastanum* genotype PT29, the source of late blight resistance, may have been heterozygous at the *Rpi-bt1* locus (i.e., *Rpi-bt1/rpi-bt1*) and that BAC clones spanning the region could have originated from the homolog carrying the susceptible (*rpi-bt1*) allele, similar to what Song et al. (2003b) encountered in cloning *RB*. In particular, the authors speculate that one of the four pseudogenes identified in their analysis could have a corresponding functional allele on the alternative homolog. To test this hypothesis, the authors designed primers targeting the predicted pseudogenes. These primers amplified a fragment from mRNA of a late blight resistant BC_3 genotype with homology to the pseudogene but lacking the frameshift mutation. Sequence analysis of the recovered fragment suggested it originated from a full length *R* gene transcript. The corresponding sequence was also recovered from genomic DNA. This fragment was used as a transgene to transform late blight susceptible potato 'Atlantic'. In greenhouse tests, eight of eight transgenic plants showed reduced late blight lesion development relative to untransformed controls; two of these plants never developed disease symptoms. The researchers had successfully cloned *Rpi-bt1* (Oosumi et al. 2009). The deduced Rpi-bt1 protein is 76–78% identical to RB and RB paralogs. Testing of the efficacy of *Rpi-bt1* against additional *P. infestans* isolates and stacking the gene with *RB* should be future research priorities.

8.4 Future Prospects

Late blight disease continues to impact potato production on a global scale. Recent advances in understanding the molecular basis of genetic resistance against *P. infestans*, the architecture of *R* genes, and effector biology have greatly sped up *R* gene mapping and cloning efforts. Especially exciting is the potential for late blight *R* genes to be assigned to functional classes on the basis of which effectors they detect, thereby enabling efficient stacking of *R* genes with complementary modes of action (Vleeshouwers et al. 2008). Additionally, studying the population dynamics of cognate effector genes in *P. infestans* populations may hold the promise of designing *R* gene deployment strategies that will favor long-term durability. The near future is likely to see the cloning of additional *R* genes and more detailed study of *R* gene function. These efforts will undoubtedly be assisted by the completion of genome sequencing efforts in potato and tomato. Because the transgenic deployment of *R* genes remains controversial in many countries, strategies to incorporate *R* genes derived from wild species into cultivated potato using non-transgenic approaches are likely to have a major impact on the control of late blight disease.

References

Allen RL, Bittner-Eddy PD, Grenville-Briggs LJ, Meitz JC, Rehmany AP, Rose LE, Beynon JL (2004) Host-parasite co-evolutionary conflict between *Arabidopsis* and downy mildew. Science 306: 1957–1960.

Armstrong MR, Whisson SC, Pritchard L, Bos JIB, Venter E, Avrova AO, Rehmany AP, Böhme U, Brooks K, Cherevach I, Hamlin N, White B, Fraser A, Lord A, Quail MA, Churcher C, Hall N, Berriman M, Huang S, Kamoun S, Beynon JL, Birch PRJ (2005) An ancestral oomycete locus contains late blight avirulence gene *Avr3a*, encoding a protein that is recognised in the host cytoplasm. Proc Natl Acad Sci USA 102: 7766–7771.

Ballvora A, Ercolano MR, Weiss J, Meksem K, Bormann CA, Oberhagemann P, Salamini F, Gebhardt C (2002) The *R1* gene for potato resistance to late blight (*Phytophthora infestans)* belongs to the leucine zipper/NBS/LRR class of plant resistance genes. Plant J 30: 361–371.

Birch PRJ, Boevink PC, Gilroy EM, Hein I, Pritchard L, Whisson SC (2008) Oomycete RXLR effectors: delivery, functional redundancy and durable disease resistance. Curr Opin Plant Biol 11: 373–379.

Black W, Gallegly ME (1957) Screening of *Solanum* species for resistance to physiologic races of *Phytophthora infestans*. Am Potato J 34: 273–281.

Bonierbale MW, Plaisted RL, Tanksley SD (1988) RFLP maps based on a common set of clones reveal modes of chromosomal evolution in potato and tomato. Genetics 120: 1095–1103.

Bradeen JM, Naess SK, Song J, Haberlach GT, Wielgus SM, Buell CR, Jiang J, Helgeson JP (2003) Concomitant reiterative BAC walking and fine genetic mapping enable physical map development for the broad-spectrum late blight resistance region, RB. Mol Gen Genet 269: 603–611.

Bradeen JM, Carputo D, Douches DS (2008) Potato. In: C Kole,TC Hall (eds) Compendium of Transgenic Crop Plants, vol 7: Transgenic Sugar, Tuber and Fiber Crops. Wiley-Blackwell, West Sussex, UK, pp 117–156.

Bradeen JM, Iorizzo M, Mollov DS, Raasch J, Colton Kramer L, Millett BP, Austin-Phillips S, Jiang J, Carputo D (2009) Higher copy numbers of the potato *RB* transgene correspond to enhanced transcript and late blight resistance levels. Molecular Plant-Microbe Interact 22: 437–446.

Brugmans B, Wouters D, van Os H, Hutten R, van der Linden G, Visser RGF, Van Eck HJ, Van der Vossen EAG (2008) Genetic mapping and transcription analyses of resistance gene loci in potato using NBS profiling. Theor Appl Genet 117: 1379–1388.

Collins NC, Webb CA, Seah S, Ellis JG, Hulbert SH, Pryor A (1998) The isolation and mapping of disease resistance gene analogs in maize. Mol Plant-Microbe Interact 11: 968–978.

Collins N, Park R, Spielmeyer W, Ellis J, Pryor AJ (2001) Resistance genes analogs in barley and their relationship to rust resistance genes. Genome 44: 375–381.

Doganlar S, Frary A, Daunay M-C, Lester RN, Tanksley SD (2002) A comparative genetic linkage map of eggplant (*Solanum melongena*) and its implications for genome evolution in the Solanaceae. Genetics 161: 1697–1711.

Donald TM, Pellerone F, Adam Blondon AF, Bouquet A, Thomas MR, Dry IB (2002) Identification of resistance gene analogs linked to a powdery mildew resistance locus in grapevine. Theor Appl Genet 104: 610–618.

Dong F, Song J, Naess SK, Helgeson JP, Gebhardt C, Jiang J (2000) Development and applications of a set of chromosome-specific cytogenetic DNA markers in potato. Theor Appl Genet 101: 1001–1007.

Ellis J, Dodds P, Pryor T (2000) Structure, function and evolution of plant disease resistance genes. Curr Opin Plant Biol 3: 278–284.

Foster SJ, Park T, Pel MA, Brigneti G, Sliwka J, Jagger L, van der vossen E, Jones JDG (2009) *Rpi-vnt1.1*, a *Tm-2²* homolog from *Solanum venturii*, confers resistance to potato late blight. Mol Plant-Microbe Interact 22: 589–600.

Graham KM, Niederhauser JS, Servin L (1959) Studies on fertility and late blight resistance in *Solanum bulbocastanum* Dun. in Mexico. Can J Bot 37: 41–49.

Grube RC, Radwanski ER, Jahn M (2000) Comparative genetics of disease resistance within the Solanaceae. Genetics 155: 873–887.

Harlan JR, de Wet JMJ (1971) Toward a rational classification of cultivated plants. Taxon 20: 509–517.

Hawkes JG (1990) The Potato: Evolution, Biodiversity and Genetic Resources. Smithsonian Institution Press, Washington DC, USA.

Hayes AJ, Saghai Maroof MA (2000) Targeted resistance gene mapping in soybean using modified AFLPs. Theor Appl Genet 100: 1279–1283.

He L, Du C, Covaleda L, Xu Z, Robinson AF, Yu JZ, Kohel RJ, Zhang H (2004) Cloning, characterization, and evolution of the NBS-LRR-encoding resistance gene analogue family in polyploid cotton (*Gossypium hirsutum* L.). Mol Plant-Microbe Interact 17: 1234–1241.

Helgeson JP, Pohlman JD, Austin S, Haberlach GT, Wielgus SM, Ronis D, Zambolim L, Tooley P, McGrath JM, James RV (1998) Somatic hybrids between *Solanum bulbocastanum* and potato: a new source of resistance to late blight. Theor Appl Genet 96: 738–742.

Hermsen JGT, Ramanna MS (1973) Double-bridge hybrids of *Solanum bulbocastanum* and cultivars of *Solanum tuberosum*. Euphytica 22: 457–466.

Huang S, Vleeshouwers VGAA, Werij JS, Hutten RCB, van Eck HJ, Visser RGF, Jacobsen E (2004) The *R3* resistance to *Phytophthora infestans* in potato is conferred by two closely linked *R* genes with distinct specificities. Mol Plant-Microbe Interact 17: 428–435.

Huang S, Van der Vossen EAG, Kuang H, Vleeshouwers VGAA, Zhang N, Borm TJA, van Eck HJ, Baker B, Jacobsen E, Visser RGF (2005) Comparative genomics enabled the isolation of the *R3a* late blight resistance gene in potato. Plant J 42: 251–261.

Hulbert SH, Webb CA, Smith SM, Sun Q (2001) Resistance gene complexes: evolution and utilization. Annu Rev Phytopathol 39: 285–312.

Jackson SA, Hanneman RE Jr (1999) Crossability between cultivated and wild tuber- and non-tuber-bearing Solanums. Euphytica 109: 51–67.

Kamoun S (2001) Nonhost resistance to Phytophthora: novel prospects for a classical problem. Curr Opin Plant Biol 4: 295–300.

Lanfermeijer F, Dijkhuis J, Sturre M, de Haan P, Hille J (2003) Cloning and characterization of the durable tomato mosaic virus resistance gene *Tm-2(2)* from *Lycopersicon esculentum*. Plant Mol Biol 52: 1037–1049.

Leister D, Ballvora A, Salamini F, Gebhardt C (1996) A PCR-based approach for isolating pathogen resistance genes from potato with potential for wide application in plants. Nat Genet 14: 421–429.

Li X, van Eck HJ, Rouppe van der Voort JNAM, Huigen DJ, Stam P, Jacobsen E (1998) Autotetraploids and genetic mapping using common AFLP markers: the *R2* allele conferring resistance to *Phytophthora infestans* mapped on potato chromosome 4. Theor Appl Genet 96: 1121–1128.

Liu Z, Halterman D (2006) Identification and characterization of *RB*-orthologous genes from the late blight resistant wild potato species *Solanum verrucosum*. Physiol Mol Plant Pathol 69: 230–239.

Livingstone KD, Lackney VK, Blauth JR, van Wijk R, Jahn MK (1999) Genome mapping in *Capsicum* and the evolution of genome structure in the Solanaceae. Genetics 152: 1183–1202.

Lokossou AA, Park T, van Arkel G, Arens M, Ruyter-Spira C, Morales J, Whisson SC, Birch PRJ, Visser RGF, Jacobsen E, van der Vossen EAG (2009) Exploiting knowledge of R/Avr genes to rapidly clone a new LZ-NBS-LRR family of late blight resistance genes from potato linkage group IV. Mol Plant-Microbe Interact 22: 630–641.

Maleki L, Faris JD, Bowden RL, Gill BS, Fellers JP (2003) Physical and genetic mapping of wheat kinase analogs and NBS-LRR resistance gene analogs. Crop Sci 43: 660–670.

Mazourek M, Cirulli E, Collier SM, Landry LG, Kang B-C, Quirin EA, Bradeen JM, Moffett P, Jahn M (2009) The fractionated orthology of *Bs2* and *Rx/Gpa2* and the comparative genomics model of disease resistance in the Solanaceae. Genetics 182: 1351–1364.

Meyers BC, Kozik A, Griego A, Kuang H, Michelmore RW (2003) Genome-wide analysis of NBS-LRR-encoding genes in *Arabidopsis*. Plant cell 15: 809–934.

Michelmore RW, Paran I, Kesseli RV (1991) Identification of markers linked to disease-resistance genes by bulked segregant analysis: a rapid method to detect markers in specific genomic regions by using segregating populations. Proc Natl Acad Sci USA 88: 9828–9832.

Milligan SB, Bodeau J, Yaghoobi J, Kaloshian I, Zabel P, Williamson VM (1998) The root knot nematode resistance gene Mi from tomato is a member of the leucine zipper, nucleotide binding, leucine-rich repeat family of plant genes. Plant Cell 10: 1307–1319.

Naess SK, Bradeen JM, Wielgus SM, Haberlach GT, McGrath JM, Helgeson JP (2000) Resistance to late blight in Solanum bulbocastanum is mapped to chromosome 8. Theor Appl Genet 101: 697–704.

Niederhauser JS, Mills WR (1953) Resistance of *Solanum* species to *Phytophthora infestans* in Mexico. Phytopathology 43: 456–457.

Nombela G, Williamson VM, Muniz M (2003) The root-knot nematode resistance gene *Mi-1.2* of tomato is responsible for resistance against the whitefly *Bemisia tabaci*. Mol Plant-Microbe Interact 16: 645–649.

Oosumi T, Rockhold DR, Maccree MM, Deahl KL, McCue KF, Belknap WR (2009) Gene *Rpi-bt1* from *Solanum bulbocastanum* confers resistance to late blight in transgenic potatoes. Am J Potato Res 86: 456–465.

Ori N, Eshed Y, Paran I, Presting G, Aviv D, Tanksley SD, Zamir D, Fluhr R (1997) The *I2C* family from the wilt disease resistance locus *I2* belongs to the nucleotide binding, leucine-rice repeat superfamily of plant resistance genes. Plant Cell 9: 521–532.

Osoegawa K, de Jong PJ (2004) BAC library construction. In: S Zhao, M Stodolsky (eds) Bacterial Artificial Chromosomes: Library Construction, Physical Mapping, and Sequencing. Humana Press, Totowa, NJ, USA, pp 1–46.

Pan Q, Liu YS, Budai Hadrian O, Sela M, Carmel Goren L, Zamir D, Fluhr R (2000) Comparative genetics of nucleotide binding site-leucine rich repeat resistance gene homologues in the genomes of two dicotyledons: tomato and Arabidopsis. Genetics 155: 309–322.

Park TH, Vleeshouwers VGAA, Huigen DJ, Van der Vossen EAG, van Eck HJ, Visser RGF (2005a) Characterization and high-resolution mapping of a late blight resistance locus similar to *R2* in potato. Theor Appl Genet 111: 591–597.

Park TH, Gros J, Sikkema A, Vleeshouwers VGAA, Muskens M, Allefs S, Jacobsen E, Visser RGF, Vossen EAGvd (2005b) The late blight resistance locus *Rpi-blb3* from *Solanum bulbocastanum* belongs to a major late blight *R* gene cluster on chromosome 4 of potato. Mol Plant-Microbe Interact 18: 722–729.

Park TH, Vleeshouwers VGAA, Hutten RCB, van Eck HJ, Van der Vossen E, Jacobsen E, Visser RGF (2005c) High-resolution mapping and analysis of the resistance locus *Rpi-abpt* against *Phytophthora infestans* in potato. Mol Breed 16: 33–43.

Pel MA, Foster SJ, Park T, Rietman H, van Arkel G, Jones JDG, Van Eck HJ, Jacobsen E, Visser RGF, Van der Vossen EAG (2009) Mapping and cloning of late blight resistance genes from *Solanum venturii* using an interspecific candidate gene approach. Mol Plant-Microbe Interact 22: 601–615.

Pflieger S, Lefebvre V, Caranta C, Blattes A, Goffinet B, Palloix A (1999) Disease resistance gene analogs as candidates for QTLs involved in pepper-pathogen interactions. Genome 42: 1100–1110.

Prince JP, Pochard E, Tanksley SD (1993) Construction of a molecular linkage map of pepper and a comparison of synteny with tomato. Genome 36: 404–417.

Quirin EA (2006) Resistance Gene Analog (RGA) Library Development for the Wild Potato Species, *Solanum bulbocastanum* and the Analysis of Non-RGA Sequences Obtained via the Polymerace Chain Reaction (PCR) Using Degenerate Primers. MS Thesis: Plant Pathology. University of Minnesota, St. Paul, MN, USA.

Reddick D (1939) Whence came Phytophthora infestans? Chron Bot 4: 410–412.

Ross H (1986) Potato breeding—problems and perspectives. J Plant Breed (Suppl) 13 Verlag Paul Parey, Berlin and Hamburg, Germany.

Rossi M, Goggin FL, Milligan SB, Kaloshian I, Ullman DE, Williamson VM (1998) The nematode resistance gene Mi of tomato confers resistance against the potato aphid. Proc Natl Acad Sci USA 95: 9750–9754.

Sanchez MJ (2005) Allelic mining for late blight resistance in wild *Solanum* species beloging to series *Bulbocastana*. MS Thesis: Plant Pathology. University of Minnesota, St. Paul, MN, USA.

Sanchez MJ, Bradeen JM (2006) Towards efficient isolation of R gene orthologs from multiple genotypes: optimization of Long Range-PCR. Mol Breed 17: 137–148.

Sela-Buurlage MB, Budai-Hadrian O, Pan Q, Carmel-Goren L, Vunsch R, Zamir D, Fluhr R (2001) Genome-wide dissection of *Fusarium* resistance in tomato reveals multiple complex loci. Mol Gen Genom 265: 1104–1111.

Shan W, Cao M, Leung D, Tyler BM (2004) The *Avr1b* locus of *Phytophthora sojae* encodes an elicitor and a regulator required for avirulence on soybean plants carrying resistance gene *Rps1b*. Mol Plant-Microbe Interact 17: 394–403.

Shizuya H, Birren B, Kim UJ, Mancino V, Slepak T, Tachiiri Y, Simon M (1992) Cloning and stable maintenance of 300-kilobase-pair fragments of human DNA in *Escherichia coli* using an F-factor-based vector. Proc Natl Acad Sci USA 39: 8794–8797.

Simons G, Groenendijk J, Wijbrandi J, Reijans M, Groenen J, Diergaarde P, Van der Lee T, Bleeker M, Onstenk J, de Both M, Haring M, Mes J, Cornelissen B, Zabeau M, Vos P (1998) Dissection of the Fusarium *I2* gene cluster in tomato reveals six homologs and one active gene copy. Plant Cell 10: 1055–1068.

Song J, Dong F, Jiang J (2000) Construction of a bacterial artificial chromosome (BAC) library for potato molecular cytogenetics research. Genome 43: 199–204.

Song J, Bradeen JM, Naess SK, Helgeson JP, Jiang J (2003a) BIBAC and TAC clones containing potato genomic DNA fragments larger than 100 kb are not stable in *Agrobacterium*. Theor Appl Genet 107: 958–964.

Song J, Bradeen JM, Naess SK, Raasch JA, Wielgus SM, Haberlach GT, Liu J, Kuang H, Austin-Phillips S, Buell CR, Helgeson JP, Jiang J (2003b) Gene *RB* from *Solanum bulbocastanum* confers broad spectrum resistance against potato late blight pathogen *Phytophthora infestans*. Proc Natl Acad Sci USA 100: 9128–9133.

Spooner DM, van den Berg RG, Rodriguez A, Bamberg J, Hijmans RJ, Lara-Cabrera SI (2004) Wild Potato (*Solanum* Section *Petota;* Solanaceae) of North and Central America. American Society of Plant Taxonomists, Ann Arbor, MI, USA.

Stevenson WR, Loria R, Franc GD, Weingartner DP (2001) Compendium of Potato Diseases. 2nd edn. APS Press, St. Paul, MN, USA.

Tai TH, Dahlbeck D, Clark ET, Gajiwala P, Pasion R, Whalen MC, Stall RE, Staskawicz BJ (1999) Expression of the *Bs2* pepper gene confers resistance to bacterial spot disease in tomato. Proc Natl Acad Sci USA 96: 14153–14158.

Tanksley SD, Ganal MW, Prince JP, Vicente MCd, Bonierbale MW, Broun P, Fulton TM, Giovannoni JJ, Grandillo S, Martin GB (1992) High density molecular linkage maps of the tomato and potato genomes. Genetics 132: 1141–1160.

Trognitz F, Trognitz B (2005) Survey of resistance gene analogs in *Solanum caripense*, a relative of potato and tomato, and update on *R* gene genealogy. Mol Genet Genom 274: 595–605.

Umaerus V, Umaerus M (1994) Inheritance of resistance to late blight. In: JE Bradshaw,GR Mackay (eds) Potato Genetics. CABI Publ, Wallingford, Oxon, UK, pp 365–401.

van der Linden CG, Wouters DCAE, Mihalka V, Kochieva EZ, Smulders MJM, Vosman B (2004) Efficient targeting of plant disease resistance loci using NBS profiling. Theor Appl Genet 109: 384–393.

van Os H, Andrzejewski S, Bakker E, Barrena I, Bryan GJ, Caromel B, Ghareeb B, Isidore E, de Jong W, van Koert P, Lefebvre V, Milbourne D, Ritter E, Rouppe van der Voort JNAM, Rousselle-Bourgeois F, van Vliet J, Waugh R, Visser RGF, Bakker J, van Eck HJ (2006) Construction of a 10,000-marker ultradense genetic recombination map of potato: providing a framework for accelerated gene isolation and a genomewide physical map. Genetics 173: 1075–1087.

Vleeshouwers VGAA, Rietman H, Krenek P, Champouret N, Young C, Oh S-K, Wang M, Bouwmeester K, Vosman B, Visser RGF, Jacobsen E, Govers F, Kamoun S, Van der Vossen EAG (2008) Effector genetics accelerates discovery and functional profiling of potato disease resistance and *Phytophthora infestans* avirulence genes. PLoS One 3: e2875.

Vos P, Simons G, Jesse T, Wijbrandi J, Heinen L, Hogers R, Frijters A, Groenendijk J, Diergaarde P, Reijans M, Fierens-Onstenk J, de Both M, Peleman J, Liharska T, Hontelez J, Zabeau M (1998) The tomato *Mi-1* gene confers resistance to both root-knot nematodes and potato aphids. Nat Biotechnol 16: 1365–1369.

Vossen Evd, Sikkema A, Hekkert BTL, Gros J, Stevens P, Muskens M, Wouters D, Pereira A, Stiekema W, Allefs S (2003) An ancient R gene from the wild potato species *Solanum bulbocastanum* confers broad-spectrum resistance to *Phytophthora infestans* in cultivated potato and tomato. Plant J 37: 867–882.

Vossen EAGvd, Gros J, Sikkema A, Muskens M, Wouters D, Wolters P, Pereira A, Allefs S (2005) The *Rpi-blb2* gene from *Solanum bulbocastanum* is an *Mi-1* gene homolog conferring broad-spectrum late blight resistance in potato. Plant J 44: 208–222.

Wang M, Allefs S, van den Berg RG, Vleeshouwers VGAA, van der Vossen EAG, Vosman B (2008) Allele mining in *Solanum:* conserved homologues of *Rpi-blb1* are identified in *Solanum stoloniferum*. Theor Appl Genet 116: 933–943.

Wastie RL (1991) Breeding for resistance. Adv Plant Pathol 7: 193–224.

Whisson SC, Boevink PC, Moleleki L, Avrova AO, Morales JG, Gilroy EM, Armstrong MR, Grouffaud S, van West P, Chapman S, Hein I, Toth IK, Pritchard L, Birch PRJ (2007) A translocation signal for delivery of oomycete effector proteins into host plant cells. Nature 450: 115–118.

Zhang LP, Khan A, Nino-Liu D, Foolad MR (2002) A molecular linkage map of tomato displaying chromosomal locations of resistance gene analogs based on a *Lycopersicon esculentum x Lycopersicon hirsutum* cross. Genome 45: 133–146.

Zhou T, Wang Y, Chen JQ, Araki H, Jing Z, Jiang K, Shen J, Tian D (2004) Genome—wide distribution of NBS genes in *japonica* rice reveals significant expansion of divergent non-TIR NBS-LRR genes. Mol Gen Genom 271: 402–415.

9

Application of Molecular Cytogenetics in Fundamental and Applied Research of Potato

Tatjana Gavrilenko

ABSTRACT

Potato is the fourth most important crop worldwide. An integrated genetic and physical map is needed to advance genomics research of this important crop species. In this chapter we describe recent results of the application of fluorescence in situ hybridization (FISH) analyses for development of a molecular cytogenetic map of potato, physical mapping in conjunction with the potato genome sequencing project, improvement of structural genomics studies, and comparative mapping of potato and related species. Like FISH, genomic in situ hybridization (GISH) is an important tool for evolutionary genome studies. GISH has been successfully applied to distinguish between subgenomes of wild allopolyploid potato species. Application of molecular cytogenetics to potato genome analysis has greatly advanced both fundamental and applied research of this important crop species.

Keywords: BAC, cytogenetics, FISH, potato

9.1 Introduction

Development of plant molecular cytogenetics was reviewed by Puertas and Naranjo (2008), Jiang and Gill (1994, 2006), and Schwarzacher (2003). The history of potato cytogenetics was reviewed in our previous paper (Gavrilenko 2007). Here we consider achievements in this field of potato research made during recent years with special attention to the application of fluorescence in situ hybridization (FISH) in potato genetics and breeding.

N.I. Vavilov Institute of Plant Industry, B. Morskaya Str. 42/44, 190000, St. Petersburg, Russia; e-mail: *tatjana9972@yandex.ru*

9.2 FISH Karyotyping and Physical Mapping of Potato Chromosomes

Potatoes encompass the tuber-bearing species of section *Petota* of the genus *Solanum* and include diploid, triploid, tetraploid, pentaploid and hexaploid species, with the basic chromosome number of twelve ($x = 12$). Within the karyotype of common potato (*Solanum tuberosum*; $2n = 4x = 48$), identification of all individual chromosomes by conventional methods (such as pachytene karyotyping or Giemsa C-banding) can be difficult (Gavrilenko 2007). Another limitation is the absence of cytogenetic stocks in potato such as monosomic and nullisomic lines or structural chromosome mutants, which allow scientists to locate genetically-mapped DNA markers or genes to specific chromosomes or chromosomal segments. Such an approach was widely used to construct cytogenetic maps in cereals (Künzel et al. 2000; Werner et al. 1992) and *Solanum lycopersicum* (tomato), a close relative of potato (Khush and Rick 1968). These limitations in potato cytogenetics have been overcome by application of fluorescent in situ hybridization (FISH).

FISH is the main technique of molecular cytogenetics. FISH methods use fluorochrome-labeled probes to detect specific DNA sequences in cytological preparations and to localize physically on chromosomes unique, low-copy-number or repetitive DNA sequences (Jiang and Gill 2006). FISH techniques allow one to identify chromosomes and chromosomal segments in plant species with small, numerous chromosomes and in species with morphologically indistinguishable chromosomes. Recently, modifications of basic FISH methods have extended their use in plant genomic research (Jiang and Gill 2006). During recent years, rapid progress has been made in potato FISH-based karyotyping and in construction of integrated chromosomal maps of potato. Application of molecular cytogenetics to potato genome analysis has greatly advanced both fundamental and applied research of this important crop species.

9.2.1 FISH-based Chromosome Identification and Integration of Chromosomal and Genetic Maps of Potato

FISH provides an alternative approach for identification of small plant chromosomes (Jiang and Gill 1994; Jiang et al. 1995). This approach is based on direct hybridization of genetic marker-anchored bacterial artificial chromosome (BAC) clones to mitotic metaphase chromosomes. FISH mapping of BAC clones anchored with previously genetically mapped restriction fragment length polymorphism (RFLP) markers helped to select a set of chromosome-specific cytogenetic DNA markers (CSCDM) (Dong et al. 2000; Song et al. 2000). Using CSCDMs, all 12 linkage groups have been associated with specific potato chromosomes (Dong et al. 2000). This study first integrated genetic linkage groups with chromosomes of potato.

BAC-FISH mapping has also been used for high-resolution mapping of potato pachytene chromosomes. Iovene et al. (2008) reported construction of a cytogenetic map of potato chromosome VI by FISH mapping of BAC clones that contained different amplified fragment length polymorphism (AFLP) markers from linkage group 6 of potato. Potato genotypes included in this FISH mapping effort were the same as those used to develop an ultradense genetic (AFLP) recombinant map of potato (van Os et al. 2006). Two approaches were used for BAC-FISH mapping—reprobing of the same preparations with different probes and mapping of multiple (up to eight different) BAC probes onto the same preparation (Fig. 9-1a, b). In total, 13 BACs were mapped on the short arm and 17 BACs on the long arm of the pachytene chromosome VI with an average interval spacing of ~ 3 cM. Nine BACs were also placed in the pericentromeric heterochromatin region (Iovene et al. 2008).

This chromosomal map helped to reveal the euchromatic or heterochromatic locations of 30 AFLP markers and to evaluate recombination frequencies along chromosome VI. While marker order along both the long and short arms was consistent between the genetic linkage and cytological maps, comparison of the maps revealed that the pericentromeric heterochromatin of potato chromosome VI experiences a suppression of genetic recombination (Iovene et al. 2008). Significant suppression of genetic recombination in the pericentromeric regions of potato has been reported previously, being described as "cold spots for recombination" and revealed as regions with a strong clustering of genetically mapped markers (van Os et al. 2006). However, linkage maps alone are less effective at revealing suppression of recombination than comparison of linkage and physical maps.

In an independent study, another six potato AFLP marker-anchored BAC clones were cytogenetically mapped on the short arm of the chromosome VI of six different diploid potato genotypes (Tang et al. 2008). BAC-FISH mapping revealed minor putative rearrangements in the short arm of chromosome VI in one diploid genotype derived from hybridization of *S. sparsipilum* and *S. tuberosum* (Fig. 9-1c; Tang et al. 2008).

Recently a cytogenetic map of all 12 chromosomes was developed for potato (*S. tuberosum*) using BAC-FISH technology (Tang et al. 2009). In this study a set of 60 AFLP anchored BAC clones (five BACs per chromosome) was located on meiotic chromosomes of a diploid genotype, yielding a standard potato pachytene karyotype.

BAC-FISH mapping has also been an important complement to the potato genome sequencing project. One hundred-fifty-eight BAC clones were assigned to the 12 potato pachytene chromosomes by BAC-FISH (Visser et al. 2009). This helped define the boundary between the euchromatin and pericentromeric heterochromatin and to verify the distal terminal regions of

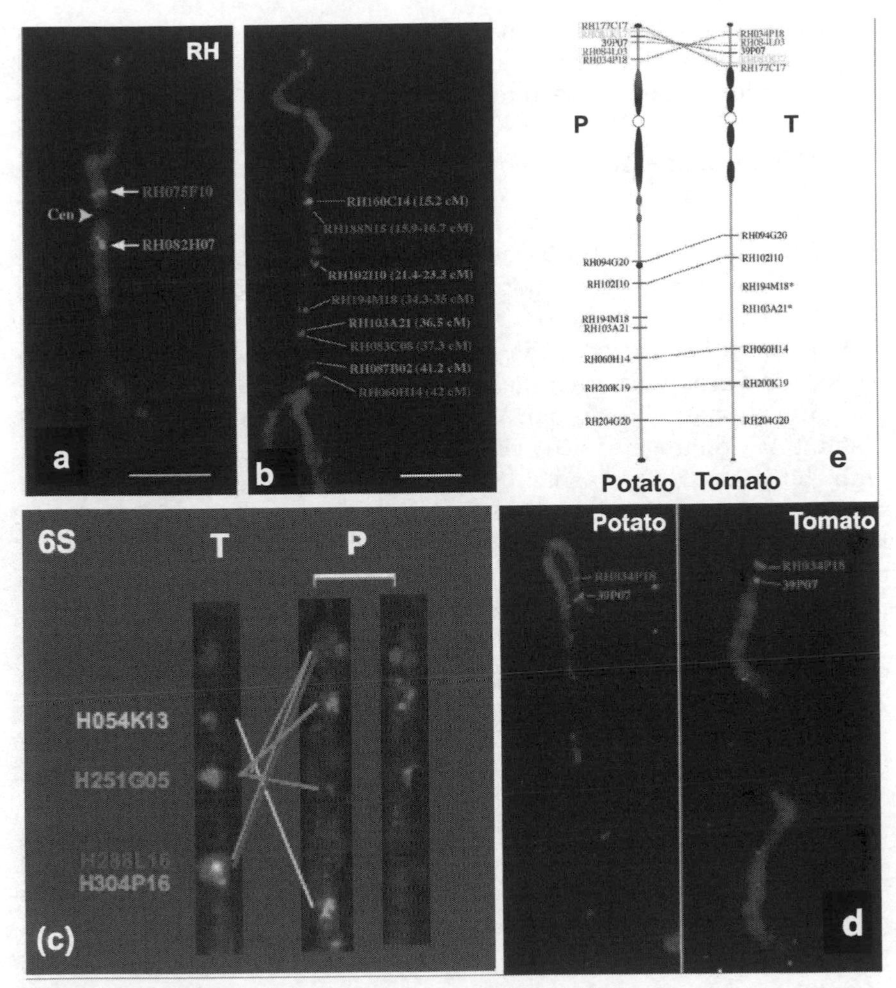

Figure 9-1 FISH mapping of AFLP marker-anchored BAC clones on potato pachytene chromosome VI.

(Panels a-b and d-e – reprinted with permission from Iovene et al. 2008; panel c—reprinted with permission from Tang et al. 2008). (a) Determination of the genetic position of the centromere of potato chromosome VI of genotype RH by FISH mapping of BACs RH075F10 and RH082H07. Scale bar = 5 µm. (b) FISH mapping of eight BACs located in the euchromatic region on the long arm of the chromosome. Scale bar = 5 µm. (c) Cross-species FISH of tomato BACs on the short arm of pachytene chromosome VI of tomato (T) and potato (P) showed inverted order between the homeologues. BAC H251G05 (green) produced a large and small focus on the potato chromosome, suggesting a breakpoint in this BAC and a putative chromosomal rearrangement (Tang et al. 2008). (d) Cross-species FISH of potato BACs on the short arm of pachytene chromosome VI of potato (P) and tomato (T). (e) The comparative chromosomal positions of potato BACs on potato (P) and tomato (T) pachytene chromosome VI. Reproduced with permission of Genetics Society of America, Genetics Editorial Office.

Color image of this figure appears in the color plate section at the end of the book.

potato chromosomes (Visser et al. 2009). BAC-FISH also helped to estimate the physical sizes of genetic gaps on chromosome V and VIII and to verify the centromere position of chromosome XII (Tang et al. 2009).

9.2.2 Comparative Mapping of Potato and Related Species using BAC-FISH

BAC-FISH mapping has also been used to study karyotype evolution and the genetic colinearity between genomes of related plant species. In comparative BAC-FISH mapping (or cross-species BAC-FISH painting) DNA probes of one species are hybridized to chromosomes of another species. Recently two independent groups have reported the comparative cytogenetic mapping of potato and tomato chromosome VI. FISH mapping was performed with 15 potato BACs on tomato pachytene chromosome 6 (Iovene et al. 2008) and with 25 tomato BACs on potato chromosomes VI (Tang et al. 2008). Both studies revealed colinearity of the BACs between tomato and potato in the long arm of chromosome VI and an inversion in the euchromatic region of the short arm (6S) (Fig. 9-1c-e). The 6S inversion was not detected by previous comparative genetic linkage mapping in potato and tomato that revealed only paracentric inversions (inversions that did not involve the centromere) in chromosomes V, IX, X, XI and XII (Bonierbale et al. 1988; Gebhardt et al. 1991; Tanksley et al. 1992). Construction of the cytogenetic maps for the remaining chromosomes of potato and their comparative cytogenetic mapping with tomato may provide new evidence of karyotype evolution of two of the most important crops in the Solanaceae.

Besides chromosomal rearrangements, comparative analysis of BAC end sequences revealed differences between potato and tomato at the genomic level in ribosomal DNA (rDNA) content, in telomere-related sequences, and in a number of unclassified sequences, suggesting the importance of repetitive sequences divergence in evolution of these species (Zhu et al. 2008).

9.2.3 FISH Mapping of Chromosomal Domains and Various Repetitive Sequences

The FISH technique has been successfully applied for physical mapping on potato chromosomes of specific chromosomal domains, such as centromere, heterochromatin, and nucleolar organizer regions and different types of repeats.

FISH mapping of (AFLP) marker-anchored BAC clones revealed the genetic position of the centromeres on potato chromosomes VI (Fig. 9-1a; Iovene et al. 2008) and XII (Tang et al. 2009). This approach also helped to establish the distribution of heterochromatic regions along chromosome

VI: the genetic position of the euchromatin–heterochromatin boundaries, the pericentromeric heterochromatin and the interstitial heterochromatic knob on the long arm of this chromosome (Iovene et al. 2008; Visser et al. 2009).

The FISH technique has also been used to assess the chromosomal position and number of ribosomal DNA tandem repeats in the potato karyotype. 5S rDNA and 45S rDNA loci were positioned as distinct loci on potato chromosomes I and II by simultaneous hybridization of rDNA probes and CSCDMs, each labeled with a different fluorochrome (Dong et al. 2000). Thus, both repeats (45S and 5S rDNA) can also be used for identification of individual chromosomes in potato. Similar numbers and locations of 5S rDNA and 45S rDNA loci have been detected in other tested diploid potato species; in the analyzed polyploid species the number of these loci corresponds to species ploidy level (Fig. 9-2c-e).

FISH has also revealed the number and chromosomal distribution of other tandem repeats on potato chromosomes. For example, intergenic spacer (IGS)-related repeats have been located by FISH to a pericentromeric heterochromatic region on a single chromosome of *S. tuberosum* (Stupar et al. 2002). Another type of tandem repeat, the interstitial telomeric repeat (ITR), has been found mainly in centromeric and pericentromeric regions of potato chromosomes (Tek and Jiang 2004). Tang et al. (2008) suggested that these telomere sequences possibly resulted from inversion events. Both repeats (IGS and ITR) were mapped by FISH to a different number of chromosomes in several potato species and were highly diverged in structure and copy number among the tested species (Stupar et al. 2002; Tek and Jiang 2004). Also using FISH mapping, BACs containing telomere-related repeats have been localized in centromeric and pericentromeric regions of several chromosomes of *S. tuberosum* (Zhu et al. 2008).

Besides ITRs and IGSs, repeats that are widely distributed among potato species, species-specific repeats have also been found in *Solanum* using Southern blot analysis. FISH has been subsequently used to visualize their chromosomal locations in specific species. The tandemly repeated element Sobo from the diploid potato species *S. bulbocastanum* was not detected in closely related Mexican potatoes or in other, more distantly related *Solanum* species (Tek et al. 2005). At the same time, intraspecific variation was detected in the presence/absence of this repeat among the tested accessions of *S. bulbocastanum*. Sequence analysis revealed that the Sobo repeat includes regions which display similarity to the long terminal repeat (LTR) of a retrotransposon (Tek et al. 2005). Sobo was mapped as a single locus in the pericentromeric region of chromosome VII using simultaneous BAC-FISH mapping of CSCDMs and a BAC clone containing the Sobo repeat.

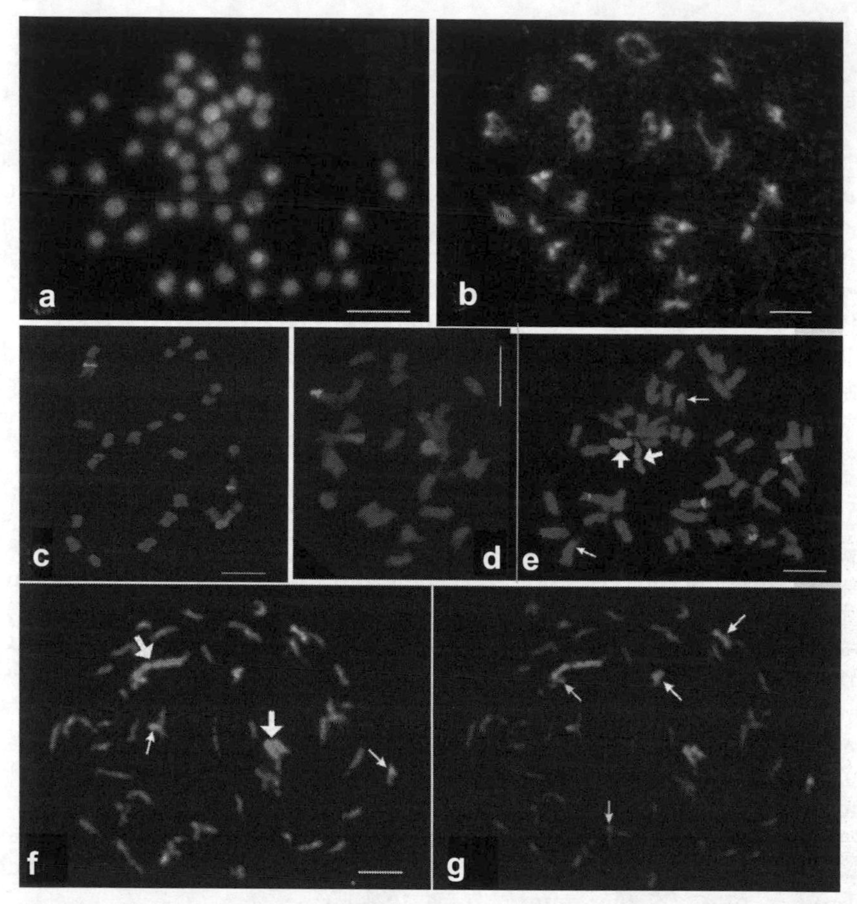

Figure 9-2 GISH analysis of series *Longipedicellata* tetraploid species *Solanum stoloniferum* ($2n = 4x = 48$) reprinted with permission from Pendinen et al. 2008. (a) Somatic chromosomes of *S. stoloniferum* probed with labeled DNA from its putative diploid ($2n = 2x = 24$) progenitor species - *S. verrucosum* (red) and *S. jamesii* (green). (b) GISH analysis of chromosomal pairing in diakinesis of *S. stoloniferum.* Pairing between *A* genome chromosomes (red, detected by labeled DNA of *S. verrucosum*) and *B* genome chromosomes (green, detected by labeled DNA of *S. jamesii*) were not observed.

(c-e) FISH mapping of 45S rDNA (red) and 5S rDNA (green) in (e) tetraploid *S. stoloniferum* and its putative diploid progenitor species (c) *S. verrucosum* and (d) *S. jamesii.* In *S. stoloniferum*, 45S rDNA hybridization sites were observed (large and small arrows). (f) Somatic chromosomes of *S. stoloniferum* probed with labeled genomic DNA from *S. verrucosum* (red) and *S. andreanum* (green). A large fragment (big arrows) and a small fragment (small arrows) from two pairs of *S. stoloniferum* chromosomes showed bright green color. Color differentiation was not observed on the rest of the chromosomes. (g) The same metaphase cell as in (f) was hybridized with 45S ribosomal RNA gene probe (red). The two large white arrows and two small yellow arrows FISH sites are not located on the chromosomes with color differentiation in GISH analysis. Scale bars = 5 μm. © 2008 NRC Canada or its licensors. Reproduced with permission.

Color image of this figure appears in the color plate section at the end of the book.

9.2.4 Physical Mapping of Genes

BAC-FISH has also been used for mapping of genes on specific chromosomes of potato. Thus, the major late blight resistance gene of wild diploid species *S. bulbocastanum* was physically mapped to a single region on chromosome VIII, a region subsequently dubbed *RB* ("Resistance from *S. bulbocastanum*") (Dong et al. 2000; Song et al. 2003). Then, physical mapping and contig construction for the *RB* region were performed using a BAC walking method with high-resolution genetic mapping (Bradeen et al. 2003). The *RB* gene was subsequently cloned and transformed into susceptible potato varieties (Song et al. 2003).

Given the significance of potato late blight disease, from a practical point of view, this is the most important example of the integration of BAC-FISH mapping together and other potato genomic resources and approaches.

9.3 Studying Natural and Artificial Polyploids by Genomic In Situ Hybridization

A special type of FISH—genomic in situ hybridization (GISH) is an excellent technique to differentiate genomes of parental species in natural and artificial polyploids. Now this method substitutes for conventional meiotic pairing analysis of polyploids.

GISH is based on the use of total genomic DNA of one of the parental species as a labeled probe and non-labeled DNA of the second parent as a competitor. The ability to discriminate chromatin of different genomes depends on the amount and divergence of repeated sequences. The standard GISH protocol allows one to distinguish genomes sharing 80–85% or less sequence homology (Schwarzacher et al. 1989). Cross hybridization in GISH experiments is suppressed by an excess of unlabeled blocking DNA. This standard method helps to discriminate genomes of species that are not very closely related by homology such as potato and *S. etuberosum* or *S. brevidens* (from *Solanum* section *Etuberosum*) (Gavrilenko 2007; Gavrilenko et al. 2002, 2003) or potato and *S. nigrum* (from *Solanum* section *Morella*) (Horsman et al. 2001). As a guideline, GISH can easily discriminate between parental chromatin of potato and distantly related species belonging to another section, however, distinguishing chromatin of species that are more closely related, such as species within section *Petota*, is more difficult.

Multicolor GISH might be helpful in discriminating genomes of species that have closer affinities to each other. This modification uses mixtures of differently labeled genomic DNA of both parental species and simultaneous hybridization of both probes with chromosomes of allopolyploids. The extent of cross hybridization between genomic DNA of parental species can be decreased by adjusting the stringency conditions in the GISH experiments. Recently, Iovene et al. (2008) reported successful GISH painting of somatic hybrids between species within section *Petota*—*S. tuberosum* and *S. bulbocastanum*, using multicolor GISH with some modification (i.e. adjusting the time of post-hybridization rinses).

Besides GISH, FISH was successfully used to differentiate closely related parental genomes in polyploids using as probes dispersed repetitive DNA sequences which are specific to different component genomes of allopolyploid (Zhang et al. 2004).

9.3.1 Studying Polyploidy in Potato Species—Traditional and Molecular Cytogenetics

GISH is a powerful technique to study allopolyploidy and to analyze genome affinity between polyploid species and their diploid progenitors. In this section we review polyploidy cytogenetics in potato and highlight achievements that were made in the study of polyploidy using GISH and FISH.

9.3.1.1 Polyploid Complexes in Relation to Species Distribution and Genetic Diversity

Petota is a complex section comprised of about 200 tuber-bearing potato species. Potato species are distributed widely in 16 countries and grow under extremely diverse climatic conditions from sea level to 4,500 m and from the southwestern USA, throughout the tropical highlands of Mexico, Central America and the Andes, to Argentina, Chile and Uruguay (Hijmans and Spooner 2001).

Potato species exist as a polyploid series ranging from diploid to hexaploid (Rybin 1929, 1933) with the same basic chromosome number of $x = 12$ for all species. Of 191 (184 wild and 7 cultivated) potato species with known chromosome numbers, 126 (66%) are exclusively diploids ($2n = 2x = 24$) (Table 9-1). Diploids occupy the greatest geographic area within the section *Petota* as well as the northern and southern extremes of the distribution of potato species (Hijmans et al. 2007). Forty-four potato species (23%) are exclusively polyploids [4.2% are triploid ($2n = 3x = 36$), 12.6% are tetraploid ($2n = 4x = 48$), 1% are pentaploid ($2n = 5x = 60$) and 5.2% are hexaploid ($2n = 6x = 72$)] (Table 9-1). Multiple cytotypes exist in 21 species (11%), mostly at the diploid and triploid levels; five species have three different cytotypes (Hijmans et al. 2007).

Polyploidization has played an important role in the distribution of plant species and in their adaptive evolution (Grant 1971; Soltis et al. 2004). Analysis of the ecological and geographic distribution of wild potato species with different ploidy levels showed that polyploids occupy a smaller geographic area than diploid species, but polyploids more frequently occur at ecological extremes (Hijmans et al. 2007). Thus, higher-level polyploids of wild potato species consistently occur in colder areas than diploids. Polyploids also occur in areas that are wetter than those occupied by diploids, although triploids tend to occur in warmer and drier areas than diploids of wild potato species (Hijmans et al. 2007).

It is well known that environmental factors stimulate the production of unreduced gametes, resulting in polyploidization (Grant 1971; McHale 1983; Ramsey and Schemske 1998; Soltis et al. 2004). For potato species, a high frequency of unreduced gametes has been reported: 1.9-36.3% for unreduced pollen, and 4.9–22.6% for unreduced eggs (Werner and Peloquin 1991). These facts suggest an important role of sexual polyploidization in the formation of polyploid complexes in potatoes (den Nijs and Peloquin 1977).

Millions of years of adaptation to various ecological and geographic areas have created significant genetic diversity among the wild potato species, including genomic divergence, creation of polyploid complexes, and auto- and allopolyploid formation. The genetic mechanisms involved in diversification of diploid potato species are largely unresolved. Matsubayashi (1991) hypothesized that genomic variants of diploid potato species differ from each other by cryptic structural differences. Dvořák (1983) proposed that rapid evolution of non-coding sequences plays an important role in genome differentiation of diploid potato species. Recent FISH studies indicate that genome differentiation of potato species both at inter- and intraspecific levels might be due to divergence in nucleotide sequences and amplification of different classes of repetitive DNA (Stupar et al. 2002; Tek and Jiang 2004; Tek et al. 2005).

At larger taxonomic scales, gross chromosomal rearrangements including inversions, translocations and transpositions have occurred during evolutionary divergence of potato species from other, non-potato *Solanum* species. Genome rearrangements have been confirmed between potato and *Solanum* subgenus *Pachystemonum* (sections *Lycopersicum, Etuberosum, Juglandifolium*) and *Solanum* subgenus *Leptostemonum* through comparative mapping studies (Tanksley et al. 1992; Perez et al. 1999; Donganlar et al. 2002; Pertuze et al. 2002) and by BAC-FISH mapping (Iovene et al. 2008; Tang et al. 2008). Despite gross structural rearrangements, a high level of genetic colinearity between these various *Solanum* species is noted. It appears that differences in repetitive sequence (rather than gene) content and composition may have a higher importance in divergence of potato and related *Solanum* species (Zhu et al. 2008). Future research combining molecular cytogenetics with other genomic resources will benefit our understanding of genetic mechanisms involved in diversification of diploid potato species and in genome evolution within potato species as well as between potato and related *Solanum* species.

Table 9-1 Chromosome numbers in wild and **cultivated** potato species (Hijmans et al. 2007, with modifications).

Chromosome number	No. (%)	Species
Species with exclusive chromosome number		
$2n = 2x = 24$	126 (66%)	*S. achacachense, S. acroglossum, S. acroscopicum, S. alandiae, S. albornozii, S. amayanum, S. ambosinum, S. anamatophilum, S. ancophilum, S. ancoripae, S. x arachuayum, S. ariduphilum, S .arnezii, S. augustii, S. avilesii, S. ayacuchense, S. aymaraesense, S. berthaultii, S. bill-hookeri, S. x blanco-galdosii, S. boliviense, S. x bruecheri, S. buesii, S. bukasovii, S. burkartii, S. cajamarquense, S. calacalinum, S. cantense, S. chacoense, S. chancayense, S. chilliasense, S. chillonanum, S. chiquidenum, S. chomatophilum, S. clarum, S. coelestipetalum, S. contumazaense, S. x doddsii, S. dolichocremastrum, S. ehrenbergii, S. gandarillasii, S. gracilifrons, S. guzmanguense, S. hastiforme, S. hintonii, S. huancabambense, S. huancavelicae, S. huarochiriense, S. humectophilum, S. hypacrarthrum, S. incahuasinum, S. incamayoense, S. incasicum, S. infundibuliforme, S. ingifolium, S. irosinum, S. jalcae, S. jamesii, S. kurtzianum, S. laxissimum, S. lesteri, S. lignicaule, S. limbaniense, S. x litusinum, S. longiusculus, S. lopez-camarenae, S. marinasense, S. megistacrolobum, S. x michoacanum, S. minutifoliolum, S. mochiquense, S. morelliforme, S. neocardenasii, S. neorossii, S. neovavilovii, S. okadae, S. olmosense, S. orophilum, S. ortegae, S. pampasense, S. paucissectum, S. peloquinianum, S. pillahuatense, S. pinnatisectum, S. piurae, S. puchupuchense, S. raphanifolium, S. raquialatum, S. regularifolium, S. rhomboideilanceolatum, S. x ruiz-lealii, S. salasianum, S. x sambucinum, S. sanctae-rosae, S. sandemanii, S. santolallae, S. sarasarae, S. sawyeri, S. saxatilis, S. scabrifolium, S. schenckii, S. x setulosistylum, S. simplicissimum, S. soestii, S. spegazzinii, S. stenophyllidium, S. tacnaense, S. tapojense, S. tarapatanum, S. tarijense, S. tarnii, S. taulisense, S. trifidum, S. trinitense, S. urubambae, S. velardei, S. venturii, S. vernei, S. vidaurrei, S. violaceimarmoratum, S. virgultorum, S. wittmackii, S. yamobambense,* ***S. ajanhurii, S. phureja, S. stenotomum***

Exclusively polyploids with single cytotype		
$2n = 3x = 36$	8 (4.2%)	*S. burtonii, S. calvescens, S. flavoviridens, S. x indunii, S. x neoweberbaueri, S. x viirsoii,* ***S. chaucha, S. juzepczukii***
$2n = 4x = 48$	24 (12.6%)	*S. acaule, S. agrimonifolium, S. bombycinum, S. colombianum, S. flahaultii, S. garcia-barrigae, S. hjertingii, S. hoopesii, S. lobbianum, S. longiconicum, S. neovalenzuelae, S. nubicola, S. orocense, S. otites, S. oxycarpum, S. pamplonense, S. paucijugum, S. solisii, S. stoloniferum, S. subpanduratum, S. x sucrense, S. tuquerrense, S. ugentii,* ***S. tuberosum***
$2n = 5x = 60$	2 (1%)	*S. x edinense,* ***S. curtilobum***
$2n = 6x = 72$	10 (5.2%)	*S. albicans, S. guerreroense, S. hougasii, S. iopetalum, S. jaenense, S. moscopanum, S. nemorosum, S. neovargasii, S. sucubunense, S. tundalomense*
Subtotal	44 (23%)	
Species with multiple cytotypes		
$2x + 3x$	12	*S. bulbocastanum, S. candolleanum, S. cardiophyllum, S. commersonii, S. immite, S. maglia, S. medians, S. microdontum, S. multiinterruptum, S. x rechei, S. sogarandinum, S. yungasense*
$2x + 4x$	5	*S. andreanum, S. brevicaule, S. leptophyes, S. polyadenium, S. sparsipilum*
$3x + 6x$	1	*S. vallis-mexici*
$5x + 6x$	1	*S. demissum*
$2x + 3x + 4x$	1	*S. verrucosum*
$2x + 4x + 6x$	1	*S. oplocense*
Subtotal	21 (11%)	
Total	191	

9.3.1.2 Traditional Concepts of Types and Origin of Polyploid Potato Species

Polyploidy is an important mechanism in plant speciation (Stebbins 1950; Grant 1971). There are two major types of polyploids: (1) autopolyploids (e.g., *AAAA*, in which *A* represents a copy of the entire genome complement received from species parent *A*) received their homologous chromosomal sets from the same species; they may show irregular meiosis, with a high frequency of multivalents at metaphase I, sterility, or a very low level of fertility. Autopolyploids may also undergo bivalent pairing, but there is no restriction of which two homologs pair. (2) Allopolyploids (e.g., *AABB*, in which *A* and *B* represent a copy of the entire genome complement received from species parent *A* and *B*, respectively) received their homoeologous chromosomal sets from different species and show regular meiosis with bivalent pairing and an extremely low frequency of multivalents. Some of the polyploids are classified as segmental allopolyploids (e.g., $A_1A_1A_2A_2$) representing species with chromosome homology somewhere in between that of allopolyploids and autopolyploids. The component genomes of segmental allopolyploids are somewhat differentiated, but pairing of chromosomes from different component genomes (e.g., A_1A_2) is still possible.

In formation of natural autopolyploids, somatic doubling has been less common than sexual polyploidization through production of unreduced gametes (Soltis et al. 2004). Self-pollination or intraspecific crosses through fertilization of unreduced gametes leads to formation of autopolyploids. Additionally, the ability to form unreduced gametes provides opportunity for hybridization between species with divergent genomes that possess different ploidy levels and/or different endosperm balance numbers (EBNs). Traditionally, identification of the type of polyploidy and development of genome concepts have been based on the analysis of meiosis of species and interspecific hybrids. According to early genome concepts, autopolyploids are less common in potatoes than strict or segmental allopolyploids (Table 9-2). Multiple cytotypes (as *AAA*, *AAAA* or higher levels) of diploid potato species (*AA*) may represent strict autopolyploids (Gavrilenko 2007).

Polyploid species of the series *Tuberosa* and *Acaulia* have been classified as segmental allopolyploids (Matsubayashi 1991). Segmental allopolyploidy can also be proposed for the 4*x* species *S. tuquerrense* of the *Piurana* series (Gavrilenko 2007). Diploid progenitors (A^aA^a, A^tA^t, A^sA^s) of segmental polyploids have been proposed but not positively identified by Matsubayashi (1991). Wild polyploid species of the series *Longipedicellata*, *Demissa*, *Conicibaccata*, and *Piurana* have been considered by several researchers to be strict allopolyploids based on their regular bivalent pairing (Hawkes 1958, 1990; Marks 1955, 1965; Irikura 1976; Ramanna and Hermsen 1979; Lopez and Hawkes 1991; Matsubayashi 1991).

Table 9-2 Genomic relationships of potato species based on different genomic concepts.

Series (Hawkes 1990)	Ploidy	Genome formulae according to the concept of: Classical cytogenetics		*Waxy, NIA* sequences	GISH
		Irikura 1976	Matsubayashi 1991	Spooner et al. 2008; Rodríguez and Spooner 2009	Pendinen et al. 2008
Cuneoalata, Megistacroloba, Commersoniana	2*x*		AA	AA	
Yungasensa	2*x*	AA	AA	AA	
Ingaefolia, Olmosiana	2*x*		A^iA^i, A^oA^{od}		
Morelliformia	2*x*		A^mA^m	BB	
Polyadenia	2*x*		$A^{po}A^{po}$	BB	
Bulbocastana	2*x*		A^bA^b	BB	
Pinnatisecta	2*x*	BB	$A^{pi}A^{pi'}$	BB	B^xB^x
Acaulia	4*x*	AAB^aB^a	AAA^aA^a		
	6*x*		$AAA^aA^aXX^b$		
Tuberosa	2*x*	AA	AA, AA^a	AA	AA
	3*x*		AAA^t		
	4*x*	AAAA	AAA^tA^t, AAA^sA^s	AAAA	
	5*x*		$AAAAA^aA^t$		
Conicibaccata	2*x*		$A^{c1}A^{c1}$, $A^{c2}A^{c2}$	AA	
	4*x*		$A^{c1}A^{c1}C'C'$	AAPP	
Longipedicellata	4*x*	AAB^sB^s	AABB	AABB	AAB^xB^x
Piurana	2*x*		A^pA^p	PP	PP
	4*x*		A^pA^pPP	PPPP	
Demissa	6*x*	$AAB^sB^sB^dB^d$	$AADDD'D'$	AABBPP for *Iopetala* Group	

In general, there are frequent contradictions in the determination of the origin of the natural potato allopolyploids (Gavrilenko 2007). Here we consider two of the most distinctive classical hypotheses developed by Irikura (1976) and Matsubayashi (1991) (Table 9-2). According to Irikura (1976), allopolyploid species share one common component genome *A* but differ from one another by the genomic variants of the second component genome *B*. The 2*x* species *S. verrucosum* was proposed as a donor of the *A* component genome in allopolyploids and the 2*x* species *S. cardiophyllum* was proposed as a possible donor of the *B* component genome of Mexican tetraploids. However, no experimental evidence establishing the *B* genome progenitor was provided by Irikura (1976). Matsubayashi (1991) recognized five genomes (*A, B, C, D, P*) in potato species. No existing diploid species with the *B, C, D* and *P* genomes nor putative progenitors of these distinct genomes have been identified. According to Matsubayashi (1991), all diploid potato species comprise one major genomic group *A* that combines very similar genomic variants designated as genome formula *A* with superscripts corresponding to each taxonomical series (Table 9-2). Allopolyploid species share one common component genome *A* and differed from each other by their second merged component genome *B, C, D* or *P* (Matsubayashi 1991) (Table 9-2). Diploid species *S. verrucosum* was suggested as the contributor of the *A* component genome to Central American allopolyploids (Bains 1951; Marks 1955, 1965; Matsubayashi 1991).

9.3.1.3 Genome Evolution of Potato Polyploids and Their Relationships with Putative Diploid Progenitors—Genome Research

9.3.1.3.1 Molecular Phylogenetics

The evolutionary history of potato polyploid species has been studied using DNA markers and organellar and nuclear phylogenies. Close relationships between the diploid species *S. verrucosum* ($2n = 2x = 24$, *AA*) and polyploid species of the series *Longipedicellata*, *Demissa* and *Acaulia* have been supported by plastid DNA phylogenies (Spooner and Sytsma 1992) and AFLP results (Kardolus 1998).

DNA sequence analyses of the single copy nuclear Granule-Bound Starch Synthase I (*GBSSI*) or *Waxy* gene (Spooner et al. 2008) and the nitrate reductase (*NIA*) gene (Rodríguez and Spooner 2009) revealed the type of polyploidy and the putative diploid progenitors of polyploid potato species. The allopolyploid nature was supported for species of the series *Longipedicellata*, *Conicibaccata*, and the informal *Iopetala* group consisting of *S. hougasii*, *S. iopetalum* and *S. schenckii* (Table 9-2). *GBSSI* and *NIA* sequence analyses (Spooner et al. 2008; Rodríguez and Spooner 2009) support a close phylogenetic relationship of polyploids within series *Conicibaccata* and

Piurana and the *Iopetala* Group, with the diploid species in series *Piurana* possessing the *P* genome (A^P genome of Matsubayashi 1991; Table 9-2).

Based on *GBSSI* and NIA data, the diploid species *S. verrucosum* (or a species closely related to *S. verrucosum*) could be the *A* genome contributor to tetraploid species of the series *Longipedicellata* (*AABB*) and hexaploids of the series *Demissa*. Clade-specific (not species-specific) progenitors have been detected for the *B* genome donor of *Longipedicellata* species which encompass 13 remaining North and Central American diploid species of series *Pinnatisecta, Polyadenia* and *Bulbocastana* (Clade 1+2 in Spooner et al. 2008; Rodríguez and Spooner 2009, correspondingly). Based on a high degree of sequence conservation at the *RB* (syn. *Rpi-blb1*) locus, Wang et al. (2008) postulate that *S. bulbocastanum* may be one of the progenitors of the 4x *S. stoloniferum* (series *Longipedicellata*).

9.3.1.3.2 GISH and FISH of Potato Polyploid Species

***Polyploid origin of the tetraploid species of the series* Longipedicellata:** In the study of natural polyploids of potato, GISH analysis was first applied to investigate genome origin of Mexican tetraploid species of the series *Longipedicellata* (*S. stoloniferum* and *S. hjertingii*) (Pendinen et al. 2008). Diploid putative progenitors for *S. stoloniferum* and *S. hjertingii* were selected based on prior hypotheses of classical genome analyses (Irikura 1976; Marks 1965; Matsubayashi 1991) and the nuclear DNA sequence-based phylogenies of Spooner et al. (2008) and Rodríguez and Spooner (2009). Specifically, *S. verrucosum* was considered as the donor of the *A* genome and diploid Mexican species from the series *Pinnatisecta* (*S. jamesii, S. cardiophyllum, S. ehrenbergii*) were considered as donors of the *B* genome (Table 9-2). Multicolor GISH was used to discriminate genomes of putative diploid ancestors of Mexican tetraploid species using a mixture of differently labeled genomic DNA of both parental diploid species and simultaneous hybridization of both probes to chromosomes of tetraploids. The differential painting of chromosomes of two component genomes of all five tested accessions of the tetraploid species *S. stoloniferum* and *S. hjertingii* indicates the phylogenetic distances between ancestral genomes (*A* and *B*) (Fig. 9-2, a-b). Thus, GISH results showed that species of the series *Longipedicellata* originated through merging two divergent ancestral genomes (*A* and *B*) and supported an *AABB* genome constitution predicted by other authors (Table 9-2). Symbol *B* (rather than $A^{pi}A^{pi}$, as used by Matsubayashi 1991) has been subsequently adopted to denote the genomes of Mexican diploid species of the series *Pinnatisecta* ($2n = 2x = 24$, *BB*), reflecting their homology to the second subgenome *B* of the allotetraploid Mexican species *S. stoloniferum* and *S. hjertingii* (Pendinen et al. 2008; Table 9-2). GISH provides evidence of the strict allopolyploid nature of Mexican tetraploid species of the series

Longipedicellata because bivalent pairing of *S. stoloniferum* and *S. hjertingii* is restricted to the pairing within ancestral parental homologues (Pendinen et al. 2008; Fig. 9-2b).

In addition to GISH analyses, the species listed above were also analyzed by FISH using 5S and 45S rDNA probes. All analyzed diploid species have one terminally located 45S rDNA locus on one chromosome pair and one 5S rDNA locus on another chromosome pair. Allotetraploid species have two chromosome pairs with one 45S rDNA locus and two other pairs of chromosomes with one 5S rDNA locus (Fig. 9-2, c-e). FISH analysis revealed different sizes of the 45S rDNA regions in allotetraploid species, indicating that the 45S rDNA regions of subgenomes *A* and *B* of *S. stoloniferum* changed during coevolution of *A* and *B* component genomes (Pendinen et al. 2008). GISH and FISH results provide important information about the nature and origin of wild polyploid potato species, as well as about their relationships with putative diploid progenitors.

In another series of multicolor GISH experiments, the DNA from *S. verrucosum* was labeled in red and the DNA from diploid South American species *S. andreanum* and *S. piurae* in green and both probes were hybridized with chromosomes of Mexican tetraploids (Pendinen et al. 2008). Most of the *S. stoloniferum* and *S. hjertingii* chromosomes were not preferentially labeled in either red or green in GISH (Fig. 9-2f). This result indicates that the genomes of *S. andreanum* and *S. piurae* have diverged from both the *A* genome of *S. verrucosum* and the *B* genome of diploid Mexican species (*S. cardiophyllum*, *S. ehrenbergii*, and *S. jamesii*). So GISH results do not contradict the genome formula *PP* (not equivalent to A^pA^p of Matsubayashi 1991) proposed by Spooner et al. (2008) for diploid species of the series *Piurana*. Only two pairs of chromosomes in *S. stoloniferum* (as well as *S. hjertingii*) had segments labeled in green due to the hybridization to the *S. andreanum* probe (Fig. 9-2f). FISH analysis (Fig. 9-2g) showed that these two pairs of chromosomes were not associated with the 45S rDNA (Pendinen et al. 2008). Thus, GISH results indicate that the genomes of diploid South-American species of the series *Piurana* [*S. piurae* and *S. andreanum* ($2n = 2x = 24$, *PP*)] and the genomes of tetraploid Mexican species [*S. stoloniferum* and *S. hjertingii* ($2n = 4x = 48$, *AABB*)] have homologous segments only on two chromosome pairs of allotetraploids (Pendinen et al. 2008). Partial homology of the two chromosome pairs of the *B* genome of Mexican tetraploid species with the *PP* genomes of *S. andreanum* (as well as *S. piurae*) indicates that the *B* genome itself may be of hybrid origin (B^x—Table 9-2) and that possible hybridization and introgression occurred between *S. andreanum*, *S. piurae* (or their progenitors or other species closely related to them) and the *B*-genome donor species.

9.4 Practical Applications of Molecular Cytogenetics

Despite the wide genetic diversity that exists in wild potatoes and closely related species of the genus *Solanum*, only 10% of potato species have been explored for use in breeding programs (Budin and Gavrilenko 1994). Many of the wild species that are highly resistant to pathogens and pests are also reproductively isolated from cultivated tetraploid potato and hence difficult to include in classical breeding programs. Therefore most genes for resistance to pathogens and pests present in modern potato varieties have been introgressed from closely related potato species. Utilization of closely related germplasm in practical breeding has resulted in the narrow genetic basis of modern potato varieties (Ross 1986). To broaden the genetic base of the common potato gene pool and to combine different resistance genes introgressed from wild *Solanum* species, various methods have been used including ploidy manipulations and bridge crosses, embryo rescue, hormone treatments, reciprocal crosses and protoplast fusion (Jansky 2006). However, this is only the first step towards successful introgression of alien genetic material into potato. The potato breeder must, over several subsequent generations, select for genotypes carrying the desired genes or alleles from the wild donor while simultaneously restoring or achieving all desirable phenotypic attributes of cultivated potato.

Success of introgression of alien genetic material into the cultivated potato genome can be monitored using cytogenetical studies of species and interspecific hybrids and their derivatives. Conventional cytogenetic analysis of potato is hampered by the relatively small size of potato chromosomes, their slight differences in morphology and similar karyotypes across potato species. FISH and GISH have greatly advanced researchers' abilities to monitor alien chromatin.

Significant progress in introgressive hybridization and breeding may be achieved when we have knowledge about genome evolution of natural wild polyploids and about genome composition of artificial polyploids. Several examples illustrate this statement.

Knowledge about the ancestors of wild polyploid species can provide excellent support for searching for desirable genes in numerous diverse germplasm collections and predict the success of introgressive hybridization. Thus, GISH and nuclear phylogeny studies undoubtedly indicate that one of the diploid progenitors of Mexican polyploids is a *B* genome diploid species of series *Pinnatisecta* or *Bulbocastana* (Clade 1 in Spooner et al. 2008). The diploids of these series are important sources of useful genes for resistance to pathogens and pests (Hawkes 1994) and there is a high probability that their resistance genes could be also found in Mexican allotetraploids. Experimental confirmation of this hypothesis was provided by Wang et al. (2008) who showed the presence of functional homologues of the *RB* (syn. *Rpi-blb1*) gene of *S. bulbocastanum* (2*x*, *BB*) in *S. stoloniferum* (4*x*, *AABB*). Significantly, allopolyploids can be much more easily crossed directly with cultivated potato (Adiwilaga and Brown 1991) than can 1 EBN *B* genome diploids [from which genes can be accessed for potato improvement only through the use of complex, multispecies bridge crosses (Hermsen 1994; Jansky 2006) or somatic hybridization (Ward et al. 1994; Thieme et al. 1997, 2008; Helgeson et al. 1998)]. Therefore, knowledge about ancestors of natural polyploids can help to better plan strategies for potato improvement.

Application of GISH and FISH techniques to genome analysis of artificial potato polyploids was reviewed in our previous paper (Gavrilenko 2007). GISH has been successfully used for identification of alien introgression in interspecific somatic hybrids of potato and their sexual progenies (Gavrilenko 2007). GISH combined with FISH using CSCDMs as probes helped to identify alien chromosomes in addition and substitution lines in a potato background (Dong et al. 2001, 2005; Tek et al. 2004). In situ hybridization techniques helped identify not only alien chromatin but also mapped a late blight resistance gene of *S. bulbocastanum* to potato chromosome VIII (Dong et al. 2000).

FISH is very useful for programs directed at positional cloning agronomically important genes, providing information about the physical distances between markers along a chromosome and about their positional orientation relative to chromosomal landmarks.

The results of comparative BAC-FISH mapping across different taxa may have practical implications in predicting the success of introgressing genes from divergent wild species into cultivated potato.

In summary, use of molecular cytogenetics together with other genomics techniques will be of great help in better understanding genome evolution of species within the section *Petota* and will help in designing more efficient breeding programs for potato.

Acknowledgments

I thank Prof. Jiming Jiang for helpful suggestions and providing Figs. 9-1a, b, d, e; the Genetics Society of America, Genetics Editorial Office for permission to publish Figs 9-1a, b, c, d, e and National Research Council Canada Research Press for permission to publish Fig. 9-2.

References

Adiwilaga KD, Brown CR (1991) Use of 2n pollen-producing triploid hybrids to introduce tetraploid Mexican wild species germ plasm into cultivated tetraploid potato gene pool. Theor Appl Genet 81: 645–652.

Bains GS (1951) Cytogenetical studies in the genus *Solanum*, sect. *Tuberarium*. MS Thesis, Univ of Cambridge, UK.

Bonierbale MW, Plaisted RL, Tanksley SD (1988) RFLP maps based on a common set of clones reveals modes of chromosomal evolution in potato and tomato. Genetics 120: 1095–1103.

Bradeen JM, Naess SK, Song J, Haberlach GT, Wielgus SM, Buell CR, Jiang J, Helgeson JP (2003) Concomitant reiterative BAC walking and fine genetic mapping enable physical map development for the broad-spectrum late blight resistance region, *RB*. Mol Genet Genom 269: 603–611.

Budin K, Gavrilenko T (1994) Genetic basis of remote hybridization in potato. Russ J Genet 30: 1225–1233.

den Nijs TPM, Peloquin SJ (1977) 2n gametes in potato species and their function in sexual polyploidization. Euphytica 26: 585–600.

Dong D, McGrath JM, Helgeson JP, Jiang J (2001) The genetic identity of alien chromosomes in potato breeding lines revealed by sequential GISH and FISH analyses using chromosome-specific cytogenetic DNA markers. Genome 44: 729–734.

Dong F, Song J, Naess SK, Helgeson JP, Gebhardt C, Jiang J (2000) Development and applications of a set of chromosome-specific cytogentics DNA markers in potato.Theor Appl Genet 101: 1001–1007.

Dong F, Tek AL, Frasca ABL, McGrath JM, Wielgus SM, Helgeson JP, J. Jiang J (2005) Development and characterization of potato—*Solanum brevidens* chromosomal addition/ substitution lines. Cytogenet Genome Res 109: 368–372.

Donganlar S, Frary A, Daunay MC, Lester RN, Tanksley SD (2002) A comparative genetic linkage map of eggplant (*Solanum melongena*) and its implications for genome evolution in the Solanaceae. Genetics 161: 1697–1711.

Dvořák J (1983) Evidence for genetic suppression of heterogenetic chromosome pairing in polyploidy species of *Solanum*, sect. *Petota*. Can J Genet Cytol 25: 530–539.

Gavrilenko T (2007) Potato cytogenetics. In: D Vreugdenhil, J Bradshaw, C Gebhardt, F Govers, DKL MacKerron, MA Taylor, H Ross (eds) Potato Biology and Biotechnology: Advances and Perspectives. Elsevier Science, Amsterdam, Netherlands, pp 203–216.

Gavrilenko T, Larkka J, Pehu E, Rokka V-M (2002) Identification of mitotic chromosomes of tuberous and non-tuberous *Solanum* species (*Solanum tuberosum* and *Solanum brevidens*) by GISH in their interspecific hybrids. Genome 45: 442–449.

Gavrilenko T, Thieme R, Heimbach U, Thieme T (2003) Fertile somatic hybrids of *Solanum etuberosum* (+) dihaploid *Solanum tuberosum* and their backcrossing progenies: relationships of genome dosage with tuber development and resistance to potato virus Y. Euphytica 131: 323–332.

Gebhardt C, Ritter E, Barone A, Debener T, Walkemeier B, Schachtschabel U, Kaufmann H, Thompson RD, Bonierbale, MW, Ganal MW, Tanksley SD, Salamini F (1991) RFLP maps of potato and their alignment with the homoeologous tomato genome. Theor Appl Genet 83: 49–57.

Grant V (1971) Plant Speciation. Columbia Univ Press, New York, USA.

Hawkes JG (1958) Potatoes: Taxonomy, cytology and crossability. In: H Kappert, W Rudorf (eds) Handbuch Pflanzenzüchtung, vol III. Paul Parey, Hamburg, Germany, pp 1–43.

Hawkes JG (1990) The Potato: Evolution, Biodiversity and Genetic Resources. Smithsonian Institution Press, Washington, DC, USA.

Hawkes JG (1994) Origins of cultivated potatoes and species relationships. In: JE Bradshaw, GR Mackay (eds) Potato Genetics. CABI Publ, Wallingford, Oxon, UK, pp 3–42.

Helgeson JP, Pohlman JD, Austin S, Haberlach GT, Wielgus SM, Ronis D, Zambolim L, Tooley P, McGrath JM, James RV, Stevenson WR (1998) Somatic hybrids between *Solanum bulbocastanum* and potato: a new source of resistance to late blight. Theor Appl Genet 96: 738–742.

Hermsen JGT (1994) Introgression of genes from wild species, including molecular and cellular approaches. In: JE Bradshaw, GR Mackay (eds) Potato Genetics. CABI Publ, Wallingford, Oxon, UK, pp 515–538.
Hijmans RJ, Spooner DM (2001) Geographic distribution of wild potato species. Am J Bot 88: 2101–2112.
Hijmans RJ, Gavrilenko T, Stephenson S, Bamberg J, Salas A, Spooner DM (2007) Geographical and environmental range expansion through polyploidy in wild potatoes (*Solanum* section *Petota*). Global Ecol Biogeogr 16: 485–495.
Horsman K, Gavrilenko T, Bergervoet JEM, Huigen D-J, Wong Joe A, Jacobsen E (2001) Alteration of the genomic composition of *Solanum nigrum* (+) potato backcross derivatives by somatic hybridization. Plant Breed 120: 201–207.
Iovene M, Wielgus SM, Simon PW, Buell CR, Jiang J (2008) Chromatin structure and physical mapping of chromosome 6 of potato and comparative analyses with tomato. Genetics 180: 1307–1317.
Irikura Y (1976) Cytogenetic studies on the haploid plants of tuber-bearing *Solanum* species. II. Cytogenetic investigations on haploid plants and interspecific hybrids by utilizing haploidy. Res Bull Hokkaido Natl Agri Exp Sta 115: 1–80.
Jiang J, Gill BS (1994) Nonisotopic in situ hybridization and plant genome mapping: the first 10 years. Genome 37: 717–725.
Jiang J, Gill BS (2006) Current status and the future of fluorescence in situ hybridization (FISH) in plant genome research. Genome 49: 1057–68.
Jiang J, Gill BS, Wang GL, Ronald PC, Ward DC (1995) Metaphase and interphase fluorescence in situ hybridization mapping of the rice genome with bacterial artificial chromosomes. Proc Natl Acad Sci USA 92: 4487–4491.
Jansky S (2006) Overcoming hybridization barriers in potato. Plant Breed 125: 1–12.
Kardolus JP (1998) A biosystematic analysis of *Solanum acaule.* PhD Thesis, Wageningen Agricultural Univ, Wageningen, The Netherlands.
Khush GS, Rick CM (1968) Cytogenetic analysis of the tomato genome by means of induced deficiencies. Chromosoma 23: 452–484.
Künzel G, Korzun L, Meister A (2000) Cytologically integrated physical restriction fragment length polymorphism maps for the barley genome based on translocation breakpoints. Genetics 154: 397–412.
Lopez LE, Hawkes JG (1991) Cytology and genome constitution of wild tuber-bearing *Solanum* species in the series *Conicibaccata.* In: JG Hawkes, RN Lester,M Nee, R Estrada (eds) Solanaceae III: Taxonomy, Chemistry, Evolution. Royal Botanic Gardens, Kew, UK, pp 327–346.
Marks GE (1955) Cytogenetic studies in tuberous *Solanum* species. I. Genomic differentiation in the group Demissa. J Genet 53: 262–269.
Marks GE (1965) Cytogenetic studies in tuberous *Solanum* species: III. Species relationships in some South and Central American species. New Phytol 64: 293–306.
Matsubayashi M (1991) Phylogenetic relationships in the potato and its related species. In: T Tsuchiya, PK Gupta (eds) Chromosome Engineering in Plants: Genetics, Breeding, Evolution, Part B. Elsevier, Amsterdam, The Netherlands, pp 93–118.
McHale NA (1983) Environmental induction of high frequency 2*n* pollen formation in diploid *Solanum.* Can J Genet Cytol 25: 609–615.
Pendinen G, Gavrilenko T, Jiang J, Spooner DM (2008) Allopolyploid speciation of the Mexican tetraploid potato species *Solanum stoloniferum* and S. *hjertingii* revealed by genomic *in situ* hybridization. Genome 51: 714–720.
Perez F, Menendez A, Dehal P, Quiros, CF (1999) Genomic structural differentiation in *Solanum:* comparative mapping of the A- and E-genomes. Theor Appl Genet 98: 1183–1193.
Pertuze RA, Ji Y, Chetelat RT (2002) Comparative linkage map of the *Solanum lycopersicoides* and *S. sitiens* genomes and their differentiation from tomato. Genome 45: 1003–1012.
Puertas MJ, Naranjo T (eds) (2008) Reviews in Plant Cytogenetics. Karger Medical and Scientific Publ, Basel, Switzerland.

Ramanna MS, Hermsen JGT (1979) Genome relationships in tuberbearing Solanums. In JG Hawkes, RN Lester,AD Skelding (eds) The Biology and Taxonomy of the *Solanaceae*, sr 7. Academic Press, London, UK, pp 647–654.

Ramsey J, Schemske DW (1998) Pathways, mechanisms and rates of polyploid formation in flowering plants. Annu Rev Ecol Syst 29: 467–501.

Rodríguez F, Spooner DM (2009) Nitrate reductase phylogeny of potato (*Solanum* sect. *Petota)* genomes with emphasis on the origins of the polyploid species. Syst Bot 34: 207–219.

Ross H (1986) Potato breeding—problems and perspectives. J Plant Breed 13(Suppl): 1–132.

Rybin VA (1929) Karyological investigation on some wild growing and indigenous cultivated potatoes of America. Bull Appl Bot Genet Breed 20: 655–720 (in Russian).

Rybin VA (1933) Cytological investigation of the South American cultivated and wild potatoes, and its significance for plant breeding. Bull Appl Bot Genet Breed Seria II, 3–100 (in Russian).

Schwarzacher T (2003) DNA, chromosomes, and *in situ* hybridization. Genome 46: 953–962.

Schwarzacher T, Leitch AR, Bennett MD, Heslop-Harrison JS (1989) *In situ* localization of parental genomes in a wide hybrid. Ann Bot 64: 315–324.

Soltis DE, Soltis PS, Tate JA (2004) Advances in the study of polyploidy since plant speciation. New Phytol 161: 173–191.

Song J, Dong F, Jiang J (2000) Construction of a bacterial artificial chromosome (BAC) library for potato molecular cytogenetics research. Genome 43: 199–204.

Song J, Bradeen JM, Naess SK, Raasch JA, Wielgus SM, Haberlach GT, Liu J, Kuang H, Austin-Phillips S, Buell CR, Helgeson JP, Jiang J (2003) Gene *RB* cloned from *Solanum bulbocastanum* confers broad spectrum resistance to potato late blight. Proc Natl Acad Sci USA 100: 9128–9133.

Spooner DM, Sytsma, KJ (1992) Reexamination of series relationships of Mexican and Central American wild potatoes (*Solanum* sect. *Petota*): evidence from chloroplast DNA restriction site variation. Syst Bot 17: 432–448.

Spooner DM, Castillo R (1997) Reexamination of series relationships of South American wild potatoes Solanaceae: *Solanum* sect. *Petota*: Evidence from chloroplast DNA restriction site variation. Amer J Bot 84: 671–685.

Spooner DM, Rodriguez A, Polgar Z, Ballard HEJr, Jansky SH (2008) Genomic origins of potato polyploids: *GBSSI* gene sequencing data. Plant Genome (Suppl Crop Sci) 48(S1): 27–36.

Stebbins GL (1950) Variation and Evolution in Plants. Columbia Univ Press, New York, USA.

Stupar RM, Song J, Tek AL, Cheng ZK, Dong F, Jiang J (2002) Highly condensed potato pericentromeric heterochromatin contains rDNA-related tandem repeats. Genetics 162: 1435–144.

Tang X, Szinay D, Lang C, Ramanna MS, van der Vossen EAG, Datema E, Lankhorst RK, de Boer J, Peters SA, Bachem C, Stiekema W, Visser RGF, de Jong H, Bai Y (2008) Cross-species bacterial artificial chromosome-fluorescence *in situ* hybridization painting of the tomato and potato chromosome 6 reveals undescribed chromosomal rearrangements. Genetics 180: 1319–1328.

Tang X, de Boer J, van Eck HJ, Bachem C, Visser RGF, de Jong H (2009) Assignment of genetic linkage maps to diploid *Solanum tuberosum* pachytene chromosomes by BAC-FISH technology. Chrom Res 17: 899–915.

Tanksley SD, Ganal MW, Prince JP, de Vicente MC, Bonierbale MW, Broun P, Fulton TM, Giovannoni JJ, Grandillo S, Martin GB, Messeguer R, Miller JC, Miller L, Paterson AH, Pineda O, Röder MS, Wing RA, Wu W, Young ND (1992) High density molecular maps of the tomato and potato genomes. Genetics 132: 1141–1160.

Tek AL, Jiang J (2004) The centromeric regions of potato chromosomes contain megabase-sized tandem arrays of telomere-similar sequence. Chromosoma 113: 77–83.

Tek AL, Stevenson WR, Helgeson JP, Jiang J (2004) Transfer of tuber soft rot and early blight resistance from *Solanum brevidens* into cultivated potato. Theor Appl Genet 109: 249–254.

Tek AL, Song J, Macas J, Jiang J (2005) Sobo, a recently amplified satellite repeat of potato, and its implications for the origin of tandemly repeated sequences. Genetics 170: 1231–1238.

Thieme R, Darsow U, Gavrilenko T, Dorokhov D, Tiemann H (1997) Production of somatic hybrids between *S. tuberosum* L. and late blight resistant Mexican wild potato species. Euphytica 97: 189–200.

Thieme R, Rakosy-Tican E, Gavrilenko T, Antonova O, Schubert J, Nachtigall M, Heimbach U, Thieme T (2008) Novel somatic hybrids (*Solanum tuberosum* L. + *Solanum tarnii*) and their BC_1 progenies express extreme resistance to potato virus Y and late blight. Theor Appl Genet 116: 691–700.

van Os HS, Andrzejewski E, Bakker I, Barrena GJ, Bryan et al (2006) Construction of a 10,000–marker ultradense genetic recombination map of potato: providing a framework for accelerated gene isolation and a genomewide physical map. Genetics 173: 1075–1087.

Visser RGF, Bachem CWB, de Boer JM et al (2009) Sequencing the potato genome: outline and first results to come from the elucidation of the sequence of the world's third most important food crop. Am J Potato Res 17: 899–915.

Wang M, Allefs S, van den Berg RG, Vleeshouwers VGAA, van der Vossen EAG, B. Vosman B (2008) Allele mining in *Solanum:* conserved homologues of *Rpi-blb1* are identified in *Solanum stoloniferum.* Theor Appl Genet 116: 933–943.

Ward AC, Phelpstead JSt-J, Gleadle AE, Blackhall NW, Cooper-Bland S, Kumar A, Powell W, Power JB, Davey MR (1994) Interspecific somatic hybrids between dihaploid *Solanum tuberosum* L. and the wild species, *S. pinnatisectum* Dun. J Exp Bot 45: 1433–1440.

Werner JE, Peloquin SJ (1991) Occurrence and mechanisms of 2*n*-egg formation in 2x potato. Genome 34: 975–982.

Werner JE, Douches DS, Freyre R (1992) Use of half-tetrad analysis to discriminate between two types of 2n egg formation in a potato haploid. Genome 35: 741–745.

Zhang P, Li W, Friebe B, Gill BS (2004) Simultaneous painting of three genomes in hexaploid wheat by BAC-FISH. Genome 47: 979–987.

Zhu W, Ouyang S, Iovene M, O'Brien K, Vuong H, Jiang J, Buell CR (2008) Analysis of 90 Mb of the potato genome reveals conservation of gene structures and order with tomato but divergence in repetitive sequence composition. BMC Genom 9: 286.

10

Functional Genomics: Transcriptomics

Xiu-Qing Li

ABSTRACT

The transcriptome acts as the primary regulator of cellular function. Various potato transcriptomics resources and technologies have been generated, including expressed sequence tags (ESTs), serial analysis of gene expression (SAGE), cDNA microarrays, oligo microarrays (e.g., Potato Oligo Chip Initiative: POCI), cDNA-AFLP, high-resolution DNA melting curve analysis, NanoString nCounter, and second generation sequencing. An in-depth comparative discussion of these technologies would facilitate their appropriate use. This chapter summarizes the available transcriptomic resources and the utility of transcriptomics approaches, describes their appropriate applications, advantages, and disadvantages, and discusses their role in potato breeding.

Keywords: expressed sequence tag, gene ontogeny, microarray, next generation sequencing, RNA

10.1 Introduction

Transcriptomics is the study of the transcriptome, which is defined as the complete set of transcripts or a specific subset of transcripts (e.g., messenger RNA molecules) produced from the genome in one cell or a population of cells of a given organism at any one time. Recent studies of plant development and environmental stress responses have converged on the roles of RNA and its metabolism as primary regulators of gene action (Hollick 2008). Most translational regulation factors such as proteases and kinases are also products of transcripts. Importantly, the RNA-dependent

Potato Research Centre, Agriculture and Agri-Food Canada, 850 Lincoln Road, P.O. Box 20280, Fredericton, New Brunswick, E3B 4Z7, Canada; e-mail: *Xiu-Qing.Li@AGR.GC.CA* epigenome plays an essential role in both development and evolution of higher organisms. The epigenome forms a complex network, which cannot be studied only at the classical single gene level. For example, there are both positive and negative regulatory interactions among substantial numbers of transcription factors but no single transcription factor is both necessary and sufficient to drive the human cell differentiation process (Center 2009). Transcriptomics approaches are particularly suitable for analysis of the cell epigenome as a whole at the transcriptional level.

In addition to the transcripts from alleles that are responsible for certain traits, other RNA molecules might also contribute to plant phenotypic stability and adaptation to environment. Paralogous gene expression may serve as a backup rescue measure for dealing with gene mutations (Kafri et al. 2005). It has been shown, at least in the human genome, that extensive RNA processing and widespread RNA-directed rewriting of DNA results in significant RNA-directed evolution (Herbert 2004). In some organisms, it has been found that tiny RNAs with a typical length of 18 nt and preferentially associated with G+C-rich promoters of highly expressed genes and with sites of RNA polymerase II binding play a role as transcription initiation RNAs (tiRNAs) (Taft et al. 2009). Consistent with this observation, it has been clearly demonstrated that both enzyme-coding transcripts and many non-coding transcripts may serve functional roles in the cell.

Functional genomics resources and their coupled public databases for plant genomics can be used to aid the discovery of gene function (Rensink and Buell 2005; Chan et al. 2007). A number of in silico tools have been developed for reconstruction of cellular (metabolic, biochemical and signal transduction) pathways based on plant gene expression datasets (van Baarlen et al. 2008).

Since most of the cultivars of *S. tuberosum* (potato) are autotetraploid, the transcriptome and transcriptional regulation are expected to be highly complex. Accordingly, study of the carbohydrate metabolism network in potato, which involves a large number of genes and is the basis for potato productivity and product quality, needs a global approach (Li 2008). Transcriptomics approaches are not only powerful but also necessary for understanding many aspects of the cell biology responsible for the performance and quality of this tetraploid crop. Considerable progress has been made in generating functional potato genomics resources, understanding the nature and regulation of the potato transcriptome, and applying potato transcriptomics resources for crop improvement (Rensink and Buell 2005; Bryan and Hein 2008; Li et al. 2008). In this chapter, the available transcriptomics resources are summarized and the necessity of transcriptomics approaches, regulation of the transcriptome, emerging technologies, strategies for application, and research prospects are discussed.

10.2 Transcriptomics Technologies and Resources Generated

Several different technologies that have been or are likely to be used to study the potato transcriptome, including sequence tagging methodologies, various types of microarrays, PCR-based technologies, and various emerging sequencing and gene expression nanotechnologies. Additionally, numerous online resources have emerged, supporting transcriptomics research efforts in potato.

10.2.1 Expressed Sequence Tags

In expressed sequence tag (EST)-based approaches, mRNA is isolated from the tissue of interest, converted to cDNA and cloned. Random clones are then partially sequenced from either the 5′ end or 3′ end by single-pass sequencing, yielding an EST. Various potato organs or tissues of different cultivars have been used to generate EST data by sequencing mRNA-derived cDNA clones (Crookshanks et al. 2001; Li 2001; Ronning et al. 2003; Flinn et al. 2005; Germain et al. 2005; Li et al. 2007) (Table 10-1). Most ESTs have been generated from tetraploid *S. tuberosum* cultivars—Shepody (Flinn et al. 2005; Li et al. 2007), Bintje (Ronning et al. 2003), Kennebec (Ronning et al. 2003), Kuras (Crookshanks et al. 2001), and Solara (Biemelt et al. ESTs released to GenBank, *www.ncbi.nlm.nih.gov*). ESTs have also been generated from diploid potatoes, such as *Solanum chacoense* (Germain et al. 2005) and a cultivated diploid potato clone, 11379-03 (Li 2001; Li et al. 2003). EST analysis is a powerful and high-throughput approach. It generates not only gene expression information but also facilitates in silico prediction of what enzymes/proteins the corresponding gene encodes. Relative abundance of a given EST contig in a tissue sample is a predictor of the strength of gene transcription. Thus, under proper experimental conditions, EST data can be interpreted quantitatively.

A total of 237,479 *S. tuberosum* ESTs existed in GenBank as of July 19, 2010, up from 227,290 in August of 2007. The latest assembly of potato expressed sequences into UniGenes by GenBank used the sequences available on November 25, 2009 and encompasses 1,360 mRNAs, 371 high-throughput cDNA (HTC, similar to ESTs, but often containing more information), 32,750 ESTs of 3′ reads, 140,814 ESTs of 5′ reads, 37,757 ESTs of others/unknowns, and 213,052 total sequences arrayed in clusters. This analysis indicated that the number of 5′ read ESTs was about fourfold that of 3′ read ESTs and that more than 99% of the potato expressed sequences in GenBank were ESTs. Publically available datasets of more than 1,000 ESTs are detailed in Table 10-1.

Table 10-1 Publicly available *Solanum tuberosum* EST data in GenBank (>1000 ESTs)[a].

GenBank library ID	Plant part(s) or library name	Number of sequences
Lib.4639	Stolon	10404
Lib.6546	Leaves and petioles	10448
Lib.7011	Mature tuber	6075
Lib.8810	*P. infestans*-challenged leaf	3131
Lib.8811	Sprouting eyes from tubers	9230
Lib.8931	One month post-harvest dormant tuber	4756
Lib.9606	Axillary buds of stem explants	4035
Lib.9854	*P. infestans*-challenged leaf, compatible reaction	5060
Lib.9856	In vitro-grown microtubers,	1391
Lib.9977	Roots	10172
Lib.10454	*P. infestans*-challenged leaf, incompatible reaction	2303
Lib.10502	Mixed tissues	10768
Lib.10893	Microarray cDNA clones mixed tissues	15048
Lib.15047	Callus cDNA library, normalized and full-length	16412
Lib.15048	Abiotic stress cDNA library	20758
Lib.15369	Swollen stolon	7547
Lib.15441	Stolon	6565
Lib.15450	In vitro root	6496
Lib.15569	Suspension culture	6299
Lib.15582	After-cooking darkening A	6897
Lib.15627	Tuber skin	4515
Lib.16489	After-cooking darkening C	3719
Lib.16519	Common scab-challenged tubers	5386
Lib.16520	Developing tubers	7701
Lib.16525	Cold sweetening B	5881
Lib.16526	Mixed leaf	6182
Lib.16527	Mixed floral	6211
Lib.16634	Low molecular weight	1077
Lib.17158	Late blight-challenged tubers	6228
Lib.18436	Cold sweetening C	5016
Lib.20426	Tubers after ca. 3 months of storage	1128
Lib.20427	Tubers after ca. 3 months of storage, sprouting	1661
Lib.20428	Tuber, dormant buds	1419
Lib.20430	Tuber, dormant buds	2028
Lib.22154	Adult leaves	3405
Lib.24231	Eight week minituber	2528

[a]Based on the last assembly of known genes and dbEST for *Solanum tuberosum* UniGenes (Build #34) by NCBI (Nov 25, 2009).

[Note that, in the above described UniGene assembly, both ESTs and fully sequenced mRNAs were employed. Therefore, a NCBI UniGene entry is a set of transcript sequences that appear to come from the same transcription locus (see NCBI; *www.ncbi.nlm.nih.gov/*). However, in contrast to this definition, in potato EST and microarray literature, the term "unigene" has been widely used for assembled EST contigs, without locus information. A more specific term such as "transcript consensus sequence" (TCS) might be more appropriate for these contigs when the assembly was based on only expressed sequences without locus information. However, in this chapter, to facilitate the discussion and citation of the literature, the term "unigene" as defined in the original potato EST literature is still used.]

Researchers from The Institute for Genomic Research (TIGR, now J. Craig Venter Institute; *www.jcvi.org*), Rockville, MD, USA, and the Canadian Potato Genome Project (CPGP; *www.cpgp.ca*) have made the largest contribution of ESTs to GenBank to date. Ronning et al. (2003) reported a total of 61,940 ESTs generated from "Bintje" and "Kennebec" aerial tissues, below ground tissues, and tissues challenged with the late blight pathogen, *Phytophthora infestans*. Approximately 80% of the potato ESTs in GenBank were significantly similar to tomato sequences (E value < –10) (Li et al. 2008). The remaining 20% of ESTs appear to be unique to potato, possibly from new genes that have evolved or common genes that have significantly diverged since the separation between the ancestors of tomato and potato.

It was the French fry potato cultivar Shepody that was used in EST sequencing by the Canadian Potato Genome Project. "Shepody", released by the AAFC Potato Research Centre in 1980, has an earlier maturation date than "Russet Burbank", another important French fry cultivar, and is widely grown in many countries. EST assembly of the 15 libraries of "Shepody" yielded 28,600 unigenes. Non-normalized EST data from "Shepody" were generated from stolons, swollen stolons, tuber skins, harvested tubers after 1-month in cold storage, harvested tubers after 3-months in cold stored, common scab-challenged tubers, roots produced *in vitro*, four-day old suspension cell cultures (Flinn et al. 2005), and reconditioned tubers (Li et al. 2007). ESTs were also generated from full length cDNA libraries and from low molecular weight cDNAs (100 bp–500 bp) from mixed tubers, stolons, tuber skins, and mature tubers of "Shepody" (Flinn et al. 2005). Normalization reduces redundancy in the libraries and increases gene discovery efficiency. Normalized EST data (ESTs from sequencing normalized cDNA libraries) include those from developing tubers, harvested tubers after 3-months in cold storage, late-blight-challenged tubers, and leaves and floral tissues of mixed developmental stages. The normalized EST dataset from cDNA library CSWB from cold stored tubers (4°C for 3 months) is very rich in unique sequences (unigenes), with 3,486 unigenes identified from 4,911 high quality ESTs (Li et al. 2007).

MicroRNAs (miRNAs) are single-stranded non-coding RNA molecules of 21–23 nucleotides in length which regulate gene expression through RNA interference (RNAi) machinery. It has been shown that land plants produce evolutionarily diverse miRNAs (Axtell and Bowman 2008), limiting researchers' ability to identify miRNAs in cross taxa comparisons. The low molecular weight EST dataset from the potato cultivar Shepody might be a good source for identifying miRNAs in potato (Flinn et al. 2005).

The number of unigenes resulting from a given set of transcript sequences can vary, depending on the computer program settings used during sequence assembly and analysis. For example, GenBank reported assembly of 211,067 potato transcript sequences (mainly ESTs) into 18,784 unigenes on January 22, 2009 (*www.ncbi.nlm.nih.gov*). However, the previous release (June 5, 2006) from the TIGR Plant Transcript Assemblies (*plantta.jcvi.org*) assembled 221,081 potato transcripts (of which 219,485 entries were ESTs) into 81,072 assemblies. The TIGR transcript assembly criteria included a 50 bp minimum match, 95% minimum identity in the overlap region and 20 bp maximum unmatched overhangs. These stringent criteria for combining individual transcript sequences into a single unigene likely account for the observed discrepancy between TIGR and GenBank results for what is largely the same sequence dataset. An assembly of TIGR and CPGP potato EST combined over 189,000 ESTs into 44,940 unigenes, of which 11,095 were found by both the TIGR and CPGP efforts, 18,452 were unique to TIGR, and 15,393 were unique to CPGP (Li et al. 2008). This rich resource of ESTs can facilitate gene expression analysis, allelic polymorphism detection, and gene discovery.

Assembled EST sequences from other *Solanum* species include 249,138 ESTs from tomato (*Solanum lycopersicum*), 8,336 ESTs from *Solanum pennellii*, 7,997 ESTs from *Solanum habrochaites*, and 6,570 from *Solanum chacoense*. The numbers of transcript assemblies are listed in Table 10-2.

Table 10-2 Expressed sequence tags available from various *Solanum* species[a].

Species	Date	Transcript assembly components			Transcript assemblies		
		Number of ESTs	Others	Total	Assemblies	Singletons	Total
S. chacoense	06-01-2007	6,570	1,235	7,805	516	5,419	5,935
S. habrochaites	09-28-2006	7,997	32	8,029	1,199	3,106	4,305
S. lycopersicum	06-01-2007	249,138	4,703	253,841	21,523	32,268	53,791
S. pennellii	09-28-2006	8,336	21	8,357	853	2,919	3,772
S. tuberosum	06-05-2006	219,485	1,596	221,081	26,280	54,792	81,072

[a]Data source: TIGR Plant Transcript Assemblies (*plantta.jcvi.org*, September 26, 2009).

10.2.2 Microarrays

Microarray technology allows high throughput transcriptomics analysis utilizing DNA:DNA or DNA:RNA hybridization. Multiple types of microarrays exist, but all share the common feature of a high-density array of DNA spots representing different transcripts from a given organism. Collectively referred to as probes, these DNA spots are typically arrayed onto a solid surface such as a glass microscope slide at a density of several thousand spots per slide. The slide with the immobilized spots is referred to as the microarray. RNA, DNA, or cDNA preparations from the sample of interest (collectively referred to as the target) are fluorescently labeled and hybridized to the microarray. Hybridization between target and probe is detected via fluorescence and data may be interpreted qualitatively (i.e., presence or absence of hybridization) or quantitatively (i.e., amount of fluorescence at each probe spot). For potato transcriptome analysis, cDNA microarrays and oligo-based microarrays have been developed. Additionally, oligo-based tomato microarrays have been successfully used to study the potato transcriptome.

10.2.2.1 cDNA Microarrays

cDNA microarrays are constructed using probes consisting of cDNA library clone inserts. Transcriptome analysis in potato using cDNA microarrays has been employed in the study of tuber development (Kloosterman et al. 2005; Van Dijk et al. 2009), light-regulated transcription (Rutitzky et al. 2009), autopolyploidization (Stupar et al. 2007), and response to cold, heat, salt (Kim et al. 2003; Rensink et al. 2005; Oufir et al. 2008), and drought (Schafleitner et al. 2007). During cold exposure (up to 50% cells injured), transcriptome analysis revealed a differential response between the wild *Solanum phureja* and cultivated *S. tuberosum* plants (Oufir et al. 2008). Gene expression profiling of potato seedling responses to cold (4°C), heat (35°C), and salt (100 mM NaCl) stresses identified a total of 3,314 genes that had significant up- or down-regulation in response to at least one stress condition. The transcriptome suggests the existence of general response pathways as well as more stress-specific pathways (Rensink and Buell 2005; Rensink et al. 2005).

cDNA microarrays have also been employed to study the potato transcriptome in response to *P. infestans* (Restrepo et al. 2005; Tian et al. 2006; Wang et al. 2008), revealing a role for metabolic and signaling (e.g., salicylic acid, jasmonic acid, and ethylene) pathways (Wang et al. 2008). A plastidic carbonic anhydrase gene was found to have a very different expression pattern in compatible vs. incompatible interactions in potato plants during infection and, interestingly, silencing this gene increases *Nicotiana benthamiana* susceptibility to *P. infestans* (Restrepo et al. 2005).

When potato plants were infected by the tuber necrotic strain of Potato virus Y (PVY^{NTN}), significant changes in expression of several stress-related genes including those for heat shock proteins, catalase 1, beta-1,3-glucanase, wound inducing gene, and genes involved in photosynthesis were detected using cDNA microarrays (Pompe-Novak et al. 2005). The transcriptome after PVY^{NTN} infection was different in susceptible and resistant cultivars. Up-regulation of genes involved in brassinosteroid, polyamine and secondary metabolite biosynthesis, and pathogenesis-related proteins was detected in the virus resistant plants (Baebler et al. 2009).

Although cDNA-based microarrays in potato played an important early role in studying the potato transcriptome, the TIGR potato cDNA microarrays that were once available for the potato research community are no longer produced. Instead, new microarray and non-microarray technologies are favored by researchers.

10.2.2.2 Oligo-Based Microarrays

On oligo-based microarrays, individual transcripts are represented not by complete cDNA fragments but instead by one or a series of short oligomer fragments. A comparison of cDNA microarrays and oligo-based arrays has revealed that cDNA microarrays are more sensitive (i.e., can detect genes expressed at lower levels) but are less specific than oligo-based microarrays due to cross hybridization of genes of similar sequence (Zhu et al. 2005; Kreil et al. 2006). Oligo microarrays, if the oligo probes are designed specifically from unique gene regions, should largely overcome the issue of mixed signals from homologous/paralogous sequences, allowing analysis at the unigene level. Various factors other than transcript abundance, including DNA fragment length, secondary and higher level structure, GC content, location of oligos on RNA molecules, and probe-target melting temperature, can affect the hybridization signal intensity for oligo-based microarrays (Hughes et al. 2001). Increasing oligo probe length results in improved signal detection. Hybridization signals to 60-mer oligos were 10-fold higher than those to 25-mer probes, but this increased signal is accompanied by a 3.6-fold loss in specificity (Relógio et al. 2002).

A 44,000 unigene microarray was constructed by the international Potato Oligo Chip Initiative (POCI). The POCI microarray was based on the contribution of ESTs from several international potato research laboratories. A total of 246,182 ESTs from 37 cDNA libraries were collected and assembled into 45,754 high quality consensus sequences. In order to maximize the number of genes represented on the microarray, the assembly used both published data and data from various unpublished sources including a cDNA library (CSWC) prepared from cold-stored, reconditioned tubers (Li et al. unpublished). The ESTs collectively represent transcripts from a variety of different tissues, including stems, roots, leaves and tubers. From the resulting unigenes, 60-mer oligos were designed and printed using Agilent SurePrint technology (Agilent Technologies, Inc., CA, USA). The DNA sequences of unigenes on the microarray are publically available at *pgrc-35.ipk-gatersleben.de/pls/htmldb_pgrc/f?p=194:2: 6725779794573348::NO*. Each of the oligos on the microarray was designed to be specific to one unigene in the DNA database. However, oligo design was limited by the extent of potato gene sequencing at the time of the assembly process; it is expected that future sequencing will reveal small design flaws such as oligos that represent more than one gene. Li et al. (2008) speculated that in cases in which the oligo is situated in a region of the gene known to be polymorphic among alleles, the microarray may have the added advantage of detecting allele-specific expression. The POCI microarray was found to be robust in identifying differentially expressed genes in potato under different conditions (Ducreux et al. 2008; Kloosterman et al. 2008; Mane et al. 2008). Transcriptome detection using the POCI oligo microarray has led to identification of candidate genes and confirmation of known genes that are involved in different developmental stages (Kloosterman et al. 2008), flavor and texture traits (Ducreux et al. 2008), and drought tolerance (Mane et al. 2008) in potato. Agilent Technologies Inc. prints the POCI microarray commercially upon request.

10.2.2.3 Tomato Heterologous Microarray

In addition to the POCI potato oligo-based microarray, an Affymetrix GeneChip (Affymetrix Inc., CA, USA) was designed specifically to monitor gene expression in tomato (*Solanum lycopersicum*). Tomato is a closest crop plant relative of potato, suggesting the potential to use the tomato array to study the potato transcriptome. The tomato GeneChip array consists of over 10,000 tomato oligo probe sets designed to interrogate over 9,200 tomato transcripts (*www.affymetrix.com/products_services/arrays/specific/tomato.affx*). Bagnaresi et al. (2008) used the tomato Affymetrix GeneChip to identify potato genes involved in cold response. Several cold-responsive genes such as amylase and invertase were identified. Subsequently, quantitative reverse transcription polymerase chain reaction (qRT-PCR) analysis was used to document expression patterns of candidate genes over a 26-day period (Bagnaresi et al. 2008). Importantly, this work illustrates the potential to use heterologous resources for the study of the potato transcriptome.

10.2.3 PCR-Based Methods

10.2.3.1 Serial Analysis of Gene Expression

Serial analysis of gene expression (SAGE) and the related LongSAGE entail generating specific, randomly concatenated tags of ca. 14–19 nucleotides from each cDNA (Nielsen et al. 2005, 2006). Sequencing of cloned tag concatamers results both in sequences of gene specific tags and information about tag frequency, which correlates with the abundance of each mRNA transcript in the cell (Velculescu et al. 1995). Of 58,322 SAGE tags of 19 nucleotides in length generated from mature field-grown potato tubers (cv. Kuras), 22,233 different tags were identified and 695 genes were represented by 10 or more SAGE tags, indicating that these genes were very active in the tubers (Nielsen et al. 2005). Lipoxygenases and various protease inhibitors were among the dominant transcripts detected, which is in agreement with EST redundancy analysis of these genes (Crookshanks et al. 2001). The deconvolution of SAGE tags is highly bioinformatic dependent and requires good EST or genome sequence resources.

10.2.3.2 cDNA-AFLP Analysis and Differential Display RT-PCR

In addition to sequencing-based approaches, polymerase chain reaction (PCR)-based approaches are also useful technologies in potato transcriptome analysis. One PCR-based approach is the cDNA-AFLP, the amplified fragment length polymorphism (AFLP) using cDNA as template (Bachem et al. 1996). cDNA-AFLP analysis has been used to study gene expression during primary cell wall synthesis (Oomen et al. 2003). In addition, this approach might enable comparative study of cDNA-AFLP and mapped genomic AFLP markers, allowing for elucidation of genome regions involved in specific cellular functions or phenotypes. cDNA-AFLP also enables expression QTL mapping, although its utility in this capacity may be limited once the complete potato genome sequence is decoded. A disadvantage of the cDNA-AFLP approach is the labor required for isolating and sequencing specific AFLP fragments from polyacrylamide gels.

Differential display RT-PCR entails the use of a limited number of short arbitrary primers in combination with anchored oligo-dT primers during reverse transcription (RT) PCR. The method has been successfully used to characterize gene expression involved in the regulation of potato tuberization by day length and gibberellins (Amador et al. 2001). Three cDNA clones encoding the enzyme GA 20-oxidase, two cDNA clones encoding 3b-hydroxylase, and one cDNA clone encoding GA 2-oxidase were identified. Changes in the levels of expression of GA 20-oxidase and 3b-hydroxylase were found to affect tuber induction and tuber yield.

10.2.4 Emerging Technologies

10.2.4.1 Next Generation Sequencing and Other Nanotechnologies

Various novel technologies in DNA sequencing have been developed in recent years. Of particular importance is a series of approaches for sequencing by synthesis. These next-generation sequencing technologies have considerably reduced the cost and dramatically accelerated the speed of sequencing. These technologies open exciting new opportunities to study the potato transcriptome. However, technologies continue to evolve rapidly and research community standards for the application of specific technologies have not yet emerged.

At the time at which this chapter was written, 454 pyrosequencing (Roche Applied Science, Basel, Switzerland) and Solexa (Illumina, Inc., CA, USA) sequencing technologies were leading contenders for furthering the study of potato. Both are examples of sequencing by synthesis. The 454 pyrosequencing platform (Ronaghi 2001) has already had a huge and rapid impact on genome sequencing efforts (Rothberg and Leamon 2008). One of the most important advantages of 454 pyrosequencing relative to other next generation sequencing technologies is its accurate reads of relatively long sequences (400–500 bp in length), theoretically enabling de novo assembly of genome sequences and more accurate transcript annotation. However, Solexa technology yields far more sequence data and has been more widely used in genome sequencing than 454 pyrosequencing technology, despite read lengths that are, at this time, much shorter (about 75 bp). In potato, it is expected that 454/Solexa sequencing technologies will facilitate whole genome sequencing and yield transcriptome sets of multiple genotypes in the near future, greatly facilitating comparative genomic and transcriptomic studies in potato.

Another nanotechnology in gene expression analysis is the nanoString nCounter gene expression system (nanoString Technologies, WA, USA), which captures and counts individual mRNA transcripts (Geiss et al. 2008). A test performed on 509 human genes demonstrated that the nCounter system is more sensitive than microarrays and similar in sensitivity to real-time PCR (Geiss et al. 2008). This nCounter technology allows a higher throughput than real-time RT-PCR and avoids the use of enzymatic amplification. Luo et al. (2010) used NanoString nCounter technology to study the effects of methods of leaf tissue sample handling and processing on quantitative gene expression for 23 potato genes. Compared to excised leaf disks that were immediately frozen in liquid nitrogen, excised leaf disks incubated for 1 hour at field temperature showed differences in gene expression. This study demonstrates the potential of using the NanoString nCounter technology to sensitively quantify gene expression in potato.

10.2.4.2 High-Resolution DNA Melting Curve (HRM) Analysis

A high-resolution DNA melting (HRM) curve approach toward transcript SNP analysis was developed using the LightScanner (Hi-Res Melting™ system with Idaho's LC Green; Idaho Technology, Inc., UT, USA) (Yuan et al. 2009). In this method, PCR of a target locus is first performed in the presence of LCGreen dye. Next, the PCR plates are transferred to the LightScanner for analysis using software, which provides automatic calling of alleles based on DNA melting temperature difference. Genes encoding carbohydrate and amino-acid metabolism enzymes including soluble acid invertase, invertase inhibitor, ADP-glucose pyrophosphorylase, sucrose synthase, starch phosphorylase, pyruvate decarboxylase, and glutamate decarboxylase were analyzed in potato using this approach. The HRM assay profiled the differential expression of alleles between different tissues or organs, between different storage stages of tubers from the same variety, and between different varieties with the same treatment. The resulting RT-PCR amplicons were directly sequenced to assist the interpretation of HRM data. The cDNA HRM curves correlated well with the nucleotide polymorphisms of the cDNA sequences and the transcript abundance of alleles. When two alleles differing at one nucleotide position were both expressed, the HRM curves of the transcript sample varied in proportion to the allele transcripts. The HRM technology proved to be more sensitive in distinguishing variation in transcription amongst related alleles than direct PCR product sequencing.

10.2.5 Online Resources

Various online resources are available to support transcriptome analysis of potato, including a USA National Science Foundation-funded database of potato disease resistance genes, the Canadian Potato Genome Project database, the Solanaceae Genomics Network, and the Solanaceae Gene Expression Database (Table 10-3). In addition to gene expression data, some of the online databases also present related resources such as germplasm collection data, genome sequences, genetic maps, and bioinformatics software. The expression data (raw and normalized data and related images) generated using the TIGR potato cDNA microarrays are available at the Solanaceae Genomics Resource in several formats (*www.jcvi.org/potato/sol_ma_microarrays.shtml*).

Table 10-3 Online resources related to potato transcriptomics research.

Site host	Description/resources/function	url
The Canadian Potato Genome Project	ESTs, potato mutant lines, microarrays	*www.cpgp.ca*
German Resource Center for Genome Research	PoMaMo: potato genetic and physical maps, SNPs, biotools	*gabi.rzpd.de/index.shtml*
Solanum Genomics Network	Potato maps and markers, BAC/EST sequences	*www.sgn.cornell.edu*
J. Craig Venter Institute (formerly The Institute for Genomic Research, TIGR)	Gene indices, SSR database, disease resistance gene database (SOLAR), microarrays	*www.jcvi.org (formerly www.tigr.org)*
Iowa State University	PlantGDB: EST assemblies, genomes, tools	*www.plantgdb.org*
The NSF Potato Genome Project	Microarrays, gene indices, SSR database, SOLAR database, genomic sequences	*www.potatogenome.org/nsf5*
Wageningen University	High density genetic map	*potatodbase.dpw.wau.nl*
The Potato Genome Sequencing Consortium	Sequencing of the complete potato genome	*www.potatogenome.net*
National Center for Biotechnology Information (NCBI)	Clusters of gene transcripts; information on protein similarities, gene expression, cDNA clone reagents, and genomic location	*www.ncbi.nlm.nih.gov/sites/entrez?db=unigene*
The UK Crop Plant Bioinformatics Network	Information related to comparative mapping and genome research	*ukcrop.net/*
IPK Gatersleben	POCI microarray unigene sequences	*pgrc-35.ipk-gatersleben.de/ pls/htmldb_pgrc/f?p=194:2:6725779794573348::NO*
The Solanaceae Genomics Resource at Michigan State University	Potato genome sequence and annotation; comparative genomics resources for the Solanaceae.	*solanaceae.plantbiology.msu.edu*
SolCAP	Gene indexes, sequences, markers, SNPs, maps	*solcap.msu.edu/resources.shtml*
DFCI Potato Gene Index	Transcript assemblies, estimation of the number of genes; downloadable potato unigene sequences	*http://compbio.dfci.harvard.edu/tgi/cgi-bin/tgi/gimain.pl?gudb=potato*

10.3 Regulation of the Transcriptome

10.3.1 Effects of Ploidy on Gene Transcription

Stupar et al. (2007) compared the transcriptomes of a synthetic autopolyploid series in potato (*S. phureja*) that included one monoploid (1*x*) clone, two diploid (2*x*) clones, and one tetraploid (4*x*) clone. Ploidy levels were confirmed through microscopic observation of the chromosomes at metaphase and flow cytometric analyses of leaflet tissues. Three independent biological replicates, each of 6-8 plants per ploidy genotype, were grown sequentially in a growth chamber. Root tip samples were harvested seven days after planting; leaflets were harvested 20 days after planting. The leaflets of the appropriate growth stage were collected and pooled for each genotype. The TIGR cDNA microarray containing 15,264 cDNA clones was used in this gene expression analysis. Results demonstrated that cell size and organ thickness were positively correlated with ploidy level. Importantly, gene expression in leaflets and root tips was more affected by a change from 1*x* to 2*x* than from 2*x* to 4*x*. None of the ploidy-upregulated genes ($1x < 2x < 4x$) displayed matching patterns across tissue types. Only a putative gibberellin receptor gene exhibited ploidy-downregulation ($1x > 2x > 4x$), with patterns matching in both leaflets and root tips. The results suggest that there are few genes with expression linearly correlated with ploidy or with expression dramatically altered by changes in ploidy (Stupar et al. 2007).

10.3.2 Gene Ontogeny Categories of Genes Expressed at Different Ploidies and under Different Environments

Plants maintain sensitive control over gene expression in response to the environment. In EST data originating from plant organs exposed to different environmental conditions, redundancy is usually very different between treatments, revealing a large number of unique unigenes. Gene ontology (GO) assignment is widely used in functional genomics analysis of ESTs and other gene sequence databases (Pirooznia et al. 2008).

In the cDNA microarray analysis of ploidy effect discussed above, genes with modified transcription levels across ploidies mostly fell into the same GO categories in both leaflets and roots. The exception to this was a clear upregulation of nucleic acid binding genes and structural molecular activity genes in leaflets relative to roots (Stupar et al. 2007).

In our study of postharvest potato tubers, temperature treatment (21°C for 3 days) after cold storage (4°C) caused considerable change in the unigenes expressed; 2,539 unigenes were detected specifically in the EST data (CSWB) from cold stored tubers and 1,827 specifically in the EST data (CSWC) from reconditioned tubers. The Biological Process GO assignments of these EST-based specific unigenes were very similar between the two datasets with an observed correlation of R = 0.999 (Fig. 10-1). It therefore

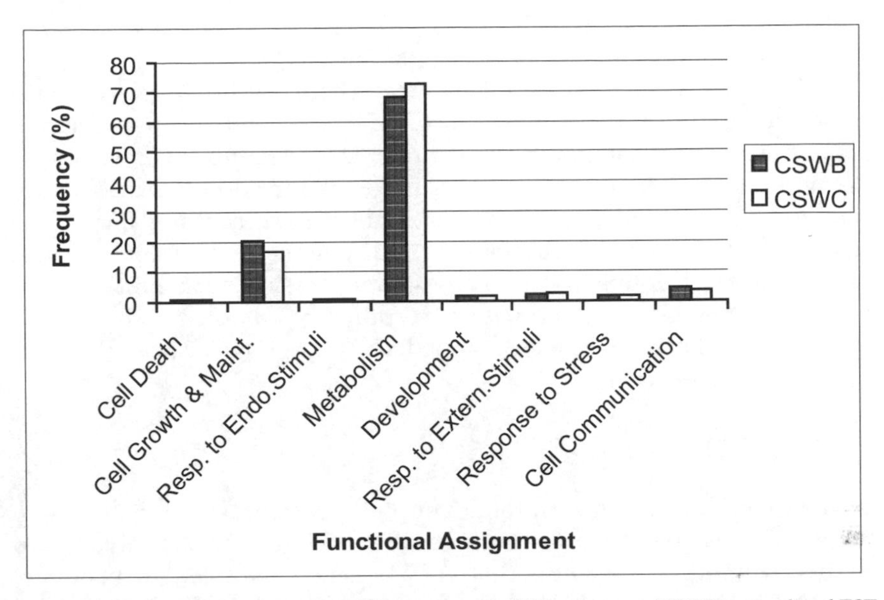

Figure 10-1 Functional assignment of library-specific EST unigenes. CSWB: normalized EST data from cDNA library generated from cold stored tubers (4°C for 3 months); CSWC: non-normalized EST data from cDNA library generated from reconditioned tubers (warmed at 21°C for 3 days after 4°C storage for 3 months). CSWB-specific unigenes: 2,539; CSWC-specific unigenes: 1,827. Functional assignments are according to the GO Biological Process. Note that the GO group percentages are very similar for the two libraries (Li et al. unpublished).

appears that gene expression activity is altered in response to varying environmental changes, such as mild temperature changes, mainly within the same GO categories of genes. It would be interesting to study whether this pattern holds for potatoes exposed to harsh environmental stresses or when the developmental stages are very different.

10.4 Application of Transcriptomics Resources

Potato transcriptomics experiments conducted to date have contributed to biological understanding and improvement of potato. This section highlights recent examples but should not be construed as a comprehensive listing of the utility of transcriptomics research.

10.4.1 Transcriptome QTL Mapping

cDNA-AFLP analysis has assisted transcriptome QTL mapping (or expression QTL, eQTL, mapping) (Brugmans et al. 2002; Ritter et al. 2008). Ritter et al. (2008) constructed an integrated 800 cM linkage map of the potato transcriptome using this technique. Most of the mapped cDNA-AFLP markers are anchored to bins of a highly saturated reference map in potato, and some co-localize with known QTL.

EST sequences are very useful in the search for candidate genes and for primer design for QTL mapping. Most of the recent genetic mapping and QTL mapping efforts in potato made use of EST sequences. For examples, Tian et al. (2008) selected 65 candidate genes for late blight resistance from an EST database and mapped 26 of them in three populations. The candidate genes were selected based on differential expression between resistant and susceptible potato plants or suppression during interactions with *P. infestans*. Genes of a putative receptor-like kinase, a lipoxygenase, and a protein phosphatase were mapped to QTL regions controlling potato resistance to late blight.

10.4.2 Gene Discovery

In most large-scale transcriptome annotations, tentative functions are assigned based on BLAST search results. EST sequence analysis also predicts whether the corresponding cDNA clone is of full length. In potato, both EST and microarray data, coupled with BLAST-based annotations, have found many applications including determination of major genes and pathways controlling traits of interest and assisting gene cloning efforts.

Microarray-based transcriptome analysis has helped identify major genes and pathways likely responsible for traits, thereby providing candidates for further verification through other approaches. A microarray study of gene expression in landraces of potato led to the hypothesis that lower accumulation of reactive oxygen species, greater mitochondrial activity, and active chloroplast defenses contributed to lower stress loads in the drought tolerant "Sullu" than in less tolerant "Negra Ojosa" (Vasquez-Robinet et al. 2008). This study suggests a role of specific mitochondrial, chloroplast, and nuclear genes in drought tolerance. The identification of other genes likely involved in important agronomic or quality traits in potato, such as tuber texture (Ducreux et al. 2008), cold sweetening (Bagnaresi et al. 2008), and resistance to late blight (Restrepo et al. 2005), clearly demonstrated the power of the transcriptome approach in gene discovery.

Gene cloning in potato has been greatly facilitated through the use of potato ESTs. One good example is the cloning of the potato glycosterol rhamnosyl transferase gene (McCue et al. 2007). Steriodalalkaloid galactosyltransferase is a key enzyme for the synthesis of potato steroidal glycoalkaloids (e.g., chaconine and solanine, two potentially harmful metabolites). ESTs of this enzyme were identified and used to isolate corresponding cDNA sequences by PCR. Antisense inhibition of this gene helped to elucidate the terminal step in formation of the potato glycoalkaloid triose side chains (McCue et al. 2007). A second illustration of the use of EST data to facilitate gene cloning comes from the realm of potato disease resistance. Type II potato protease inhibitors are involved in plant defense. All known genes of this type of protease inhibitors were thought to be tissue-specific or developmental stage specific. However, using EST data, a full length protease inhibitor homolog was generated and sequenced and found to be constitutively expressed across different developmental stages and tissue types (Zhang et al. 2008). Similarly, Solanaceae EST databases were employed by Pajerowska et al. (2005) to identify homologs of disease defense signaling genes of potato. Sixteen defense signaling homologs in potato were positioned on molecular maps, and five of them mapped to locations associated with quantitative resistance loci (QRL) against *P. infestans* and the bacterium *Erwinia carotovora,* causal agent of blackleg and tuber soft rot (Pajerowska et al. 2005).

Transcriptome data can also assist mass spectrometry protein analysis. A protein database has been established through translation of potato ESTs and was used successfully together with mass spectrometry to identify a C-terminal, vacuole targeting, signal sequence (ANKASY-COO) on patatin propeptides (Welinder and Jorgensen 2009).

10.4.3 Functional Genomics-Assisted Breeding

There are two main groups of molecular markers: genomic polymorphism markers and expression markers. The link between genomic polymorphic markers (those genetically linked to a locus) and desired traits is often population dependent because of chromosomal recombination. Environment can also impact gene expression. Thus, even when a genomic polymorphic marker accurately predicts the presence of a specific allele, it may not accurately predict manifestation of a phenotype if the allele is not expressed due to environmental factors. In contrast to genome-based markers, despite the complexity of expression patterns observed in a tetraploid species, transcriptome-based markers may help researchers target specific alleles responsible for phenotypes with improved accuracy and direct utility to plant improvement. A transcript SNP analysis using the LightScanner system reported by Yuan et al. (2009) suggests that the method can sensitively detect actively transcribed alleles, therefore allowing development of functional allele activity (FAA) markers. In some cases, it might be of interest to convert RNA-based FAA markers to Functional Allele DNA (FAD) markers, as long as the FAA marker and the FAD marker target the same functional allele. Although an allele should be functional in multiple different genotypes, occasional mutation affecting either FAA or FAD markers may reduce the prediction power of the molecular marker.

Nevertheless, FAA and FAD markers are expected to be less population dependant than genomic polymorphism DNA markers.

10.5 Concluding Remarks and Prospects

A considerable number of functional genomic resources including EST datasets, cDNA microarrays and oligo-based microarrays have been developed for the study of potato. High-throughput functional genomic approaches permit monitoring in each experiment of the expression levels of a huge number of genes—up to virtually all of the genes of the genome under ideal conditions. Importantly, EST redundancy in non-normalized, or to a degree, normalized EST datasets can provide information about the abundance of gene transcripts. Based on results of an EST study, temperature-induced genes in potato are mostly from GO categories harboring previously identified temperature-induced genes. For microarray-based studies, oligo microarrays are more specific in unigene analysis than cDNA microarrays.

Next generation sequencing and nanotechnologies will further facilitate transcriptome characterization. One of the challenges facing potato genomicists is access to informatic approaches to mine the enormous amount of data resulting from transcriptome studies. In addition to characterizing genes encoding proteins, characterization of non-coding RNA, transcriptome regulation, and application of integrative -omics approaches (proteomics, metabolomics, etc.) are needed. The nature and properties of different transcriptomics data (e.g, EST, microarray and qRT-PCR data) can be different, complicating direct comparison of different types of transcriptome data. Further research is required to characterize and expand upon the utility of different types of transcriptomics data.

Transcriptomics approaches may contribute to the understanding of the basic biology of the potato genome. Sequence-based expression maps of the genome and functional transcription unit distribution have not yet been established for potato. The technologies for this type of research have been developed in some organisms (Sémon and Duret 2004). Microarray-based gene expression profiling has led to identifying response pathways, usually incorporating a large number of genes. Similar research is needed to characterize the transcriptome related to traits influencing potato productivity and quality and to the use of transgenic or association genetic approaches to determine the function of the transcriptomics-identified genes and pathways. Although up and down regulation in gene expression has been studied using different ploidy levels of the potato, the molecular biological basis of superior performance of tetraploid potato compared to diploid potato still remain enigmatic. Little is known at this stage about how the potato transcriptome is regulated or the role of transcription factors and elongation factors. It has been demonstrated that some plant genes and repetitive DNA sequences in the genome are modulated in copy number during development or in response to environmental stimuli (Li 2009). However it is unknown, how much this type of variation influences the potato transcriptome. The function of many ESTs is still unknown and many predicted functions are based only on sequence similarity to genes of known function. Expression QTL maps integrated with genomic QTL maps might be helpful for understanding the phenotypic difference between potato clones. The utility of potato transcriptome resources can be enhanced by increased understanding of the nature and regulation of the transcriptome and by characterization and verification of transcriptomics-suggested candidate genes and pathways.

References

Amador V, Bou J, Martinez-Garcia J, Monte E, Rodriguez-Falcon M, Russo E, Prat S (2001) Regulation of potato tuberization by daylength and gibberellins. Intl J Dev Biol 45: S37–S38.

Axtell MJ, Bowman JL (2008) Evolution of plant microRNAs and their targets. Trends Plant Sci 13: 343–349.

Bachem CWB, Van Der Hoeven RS, De Bruijn SM, Vreugdenhil D, Zabeau M, Visser RGF (1996) Visualization of differential gene expression using a novel method of RNA fingerprinting based on AFLP: Analysis of gene expression during potato tuber development. Plant J 9: 745–753.

Baebler S, Krecic-Stres H, Rotter A, Kogovsek P, Cankar K, Kok EJ, Gruden K, Kovac M, Zel J, Pompe-Novak M, Ravnikar M (2009) PVY^{NTN} elicits a diverse gene expression response in different potato genotypes in the first 12 h after inoculation. Mol Plant Pathol 10: 263–275.

Bagnaresi P, Moschella A, Beretta O, Vitulli F, Ranalli P, Perata P (2008) Heterologous microarray experiments allow the identification of the early events associated with potato tuber cold sweetening. BMC Genom 9: 176–176.

Brugmans B, Del Carmen AF, Bachem CWB, Van Os H, Van Eck HJ, Visser RGF (2002) A novel method for the construction of genome wide transcriptome maps. Plant J 31: 211–222.

Bryan GJ, Hein I (2008) Genomic resources and tools for gene function analysis in potato. Int J Plant Genom 2008: e 216513.

Center TFCatROS (2009) The transcriptional network that controls growth arrest and differentiation in a human myeloid leukemia cell line. Nat Genet 41: 553–562.

Chan AP, Rabinowicz PD, Quackenbush J, Buell CR, Town CD (2007) Plant database resources at The Institute for Genomic Research. Meth Mol Biol 406: 113–136.

Crookshanks M, Emmersen J, Welinder KG, Nielsen KL (2001) The potato tuber transcriptome: Analysis of 6077 expressed sequence tags. FEBS Lett 506: 123–126.

Ducreux LJM, Morris WL, Prosser IM, Morris JA, Beale MH, Wright F, Shepherd T, Bryan GJ, Hedley PE, Taylor MA (2008) Expression profiling of potato germplasm differentiated in quality traits leads to the identification of candidate flavour and texture genes. J Exp Bot 59: 4219–4231.

Flinn B, Rothwell C, Griffiths R, Lague M, DeKoeyer D, Sardana R, Audy P, Goyer C, Li XQ, Wang-Pruski G, Regan S (2005) Potato expressed sequence tag generation and analysis using standard and unique cDNA libraries. Plant Mol Biol 59: 407–433.

Geiss GK, Bumgarner RE, Birditt B, Dahl T, Dowidar N, Dunaway DL, Fell HP, Ferree S, George RD, Grogan T, James JJ, Maysuria M, Mitton JD, Oliveri P, Osborn JL, Peng T, Ratcliffe

AL, Webster PJ, Davidson EH, Hood L (2008) Direct multiplexed measurement of gene expression with color-coded probe pairs. Nat Biotechnol 26: 317–325.

Germain H, Rudd S, Zotti C, Caron S, O'Brien M, Chantha SC, Lagace M, Major F, Matton DP (2005) A 6374 unigene set corresponding to low abundance transcripts expressed following fertilization in *Solanum chacoense* Bitt, and characterization of 30 receptor-like kinases. Plant Mol Biol 59: 515–532.

Herbert A (2004) The four Rs of RNA-directed evolution. Nat Genet 36: 19–25.

Hollick JB (2008) Sensing the epigenome. Trends Plant Sci 13: 398–404.

Hughes TR, Mao M, Jones AR, Burchard J, Marton MJ, Shannon KW, Lefkowitz SM, Ziman M, Schelter JM, Meyer MR, Kobayashi S, Davis C, Dai H, He YD, Stephaniants SB, Cavet G, Walker WL, West A, Coffey E, Shoemaker DD, Stoughton R, Blanchard AP, Friend SH, Linsley PS (2001) Expression profiling using microarrays fabricated by an ink-jet oligonucleotide synthesizer. Nat Biotechnol 19: 342–347.

Kafri R, Bar-Even A, Pilpel Y (2005) Transcription control reprogramming in genetic backup circuits. Nat Genet 37: 295–299.

Kim DY, Lee HE, Yi KW, Han SE, Kwon HB, Go SJ, Byun MO (2003) Expression pattern of potato (*Solanum tuberosum*) genes under cold stress by using cDNA microarray. Kor J Genet 25: 345–352.

Kloosterman B, Vorst O, Hall RD, Visser RGF, Bachem CW (2005) Tuber on a chip: Differential gene expression during potato tuber development. Plant Biotechnol J 3: 505–519.

Kloosterman B, De Koeyer D, Griffiths R, Flinn B, Steuernagel B, Scholz U, Sonnewald S, Sonnewald U, Bryan GJ, Prat S, Bánfalvi Z, Hammond JP, Geigenberger P, Nielsen KL, Visser RGF, Bachem CWB (2008) Genes driving potato tuber initiation and growth: Identification based on transcriptional changes using the POCI array. Funct Integr Genom 8: 329–340.

Kreil DP, Russell RR, Russell S (2006) Microarray Oligonucleotide Probes. Meth Enzymol 410: 73–98.

Li XQ (2001) Potato EST sequencing and analysis: Which genes involved in carbohydrate metabolism are most active in immature potatoes? Am J Potato Res 78: 467 (Abstr).

Li XQ (2008) Molecular characterization and biotechnological improvement of the processing quality of potatoes. Can J Plant Sci 88: 639–648.

Li XQ (2009) Developmental and environmental variation in genomes. Heredity 102: 323–329.

Li XQ, Haroon M, Bonierbale MW (2003) EST-based cloning and gene expression analysis of potato proteases. Acta Hort 619: 59–62.

Li X-Q, Griffiths R, Lague M, DeKoeyer D, Rothwell C, Haroon M, Stevens B, Xu C, Gustafson V, Bonierbale M, Regan S, Flinn B (2007) EST sequencing and analysis from cold-stored and reconditioned potato tubers. Acta Hort 745: 491–493.

Li XQ, Griffiths R, De Koeyer DL, Rothwell C, Gustafson V, Regan S, Flinn B (2008) Functional genomic resources for potato. Can J Plant Sci 88: 573–581.

Luo S, Tai H, Zebarth B, Li XQ, Millard P, De Koeyer D, Xiong XY: (2010) Sample collection protocol effects on quantification of gene expression in potato leaf tissue. Plant Mol Biol Rep. DOI 10.1007/s11105-010-0239-4.

Mane SP, Robinet CV, Ulanov A, Schafleitner R, Tincopa L, Gaudin A, Nomberto G, Alvarado C, Solis C, Bolivar LA, Blas R, Ortega O, Solis J, Panta A, Rivera C, Samolski I, Carbajulca DH, Bonierbale M, Pati A, Heath LS, Bohnert HJ, Grene R (2008) Molecular and physiological adaptation to prolonged drought stress in the leaves of two Andean potato genotypes. Funct Plant Biol 35: 669–688.

McCue KF, Allen PV, Shepherd LVT, Blake A, Malendia Maccree M, Rockhold DR, Novy RG, Stewart D, Davies HV, Belknap WR (2007) Potato glycosterol rhamnosyltransferase, the terminal step in triose side-chain biosynthesis. Phytochemistry 68: 327–334.

Nielsen KL, Grønkjær K, Welinder KG, Emmersen J (2005) Global transcript profiling of potato tuber using LongSAGE. Plant Biotechnol J 3: 175–185.

Nielsen KL, Hogh AL, Emmersen J (2006) DeepSAGE—Digital transcriptomics with high sensitivity, simple experimental protocol and multiplexing of samples. Nucl Acids Res 34: e133.

Oomen RJFJ, Bergervoet MJEM, Bachem CWB, Visser RGF, Vincken JP (2003) Exploring the use of cDNA-AFLP with leaf protoplasts as a tool to study primary cell wall biosynthesis in potato. Plant Physiol Biochem 41: 965–971.

Oufir M, Legay S, Nicot N, Van Moer K, Hoffmann L, Renaut J, Hausman JF, Evers D (2008) Gene expression in potato during cold exposure: Changes in carbohydrate and polyamine metabolisms. Plant Sci 175: 839–852.

Pajerowska KM, Parker JE, Gebhardt C (2005) Potato homologs of Arabidopsis thaliana genes functional in defense signaling—Identification, genetic mapping, and molecular cloning. Mol Plant-Microbe Interact 18: 1107–1119.

Pirooznia M, Habib T, Perkins EJ, Deng Y (2008) GOfetcher: A database with complex searching facility for gene ontology. Bioinformatics 24: 2561–2563.

Pompe-Novak M, Gruden K, Baebler S, Krecic-Stres H, Kovac M, Jongsma M, Ravnikar M (2005) Potato virus Y induced changes in the gene expression of potato (*Solanum tuberosum* L.). Physiol Mol Plant Pathol 67: 237–247.

Relógio A, Schwager C, Richter A, Ansorge W, Valcárcel J (2002) Optimization of oligonucleotide-based DNA microarrays. Nucl Acids Res 30: e51.

Rensink WA, Buell CR (2005) Microarray expression profiling resources for plant genomics. Trends Plant Sci 10: 603–609.

Rensink WA, Iobst S, Hart A, Stegalkina S, Liu J, Buell CR (2005) Gene expression profiling of potato responses to cold, heat, and salt stress. Funct Integr Genom 5: 201–207.

Restrepo S, Myers KL, Del Pozo O, Martin GB, Hart AL, Buell CR, Fry WE, Smart CD (2005) Gene profiling of a compatible interaction between *Phytophthora infestans* and *Solanum tuberosum* suggests a role for carbonic anhydrase. Mol Plant-Microbe Interact 18: 913–922.

Ritter E, Ruiz De Galarreta JI, Van Eck HJ, Sanchez I (2008) Construction of a potato transcriptome map based on the cDNA-AFLP technique. Theor Appl Genet 116: 1003–1013.

Ronaghi M (2001) Pyrosequencing sheds light on DNA sequencing. Genome Res 11: 3–11.

Ronning CM, Stegalkina SS, Ascenzi RA, Bougri O, Hart AL, Utterbach TR, Vanaken SE, Riedmuller SB, White JA, Cho J, Pertea GM, Lee Y, Karamycheva S, Sultana R, Tsai J, Quackenbush J, Griffiths HM, Restrepo S, Smart CD, Fry WE, Van Der Hoeven R, Tanksley S, Zhang P, Jin H, Yamamoto ML, Baker BJ, Buell CR (2003) Comparative analyses of potato expressed sequence tag libraries. Plant Physiol 131: 419–429.

Rothberg JM, Leamon JH (2008) The development and impact of 454 sequencing. Nat Biotechnol 26: 1117–1124.

Rutitzky M, Ghiglione HO, Curá JA, Casal JJ, Yanovsky MJ (2009) Comparative genomic analysis of light-regulated transcripts in the Solanaceae. BMC Genom 10: 60.

Schafleitner R, Gutierrez Rosales RO, Gaudin A, Alvarado Aliaga CA, Martinez GN, Tincopa Marca LR, Bolivar LA, Delgado FM, Simon R, Bonierbale M (2007) Capturing candidate drought tolerance traits in two native Andean potato clones by transcription profiling of field grown plants under water stress. Plant Physiol Biochem 45: 673–690.

Sémon M, Duret L (2004) Evidence that functional transcription units cover at least half of the human genome. Trends Genet 20: 229–232.

Stupar RM, Bhaskar PB, Yandell BS, Rensink WA, Hart AL, Ouyang S, Veilleux RE, Busse JS, Erhardt RJ, Buell CR, Jiang J (2007) Phenotypic and transcriptomic changes associated with potato autopolyploidization. Genetics 176: 2055–2067.

Taft RJ, Glazov EA, Cloonan N, Simons C, Stephen S, Faulkner GJ, Lassmann T, Forrest ARR, Grimmond SM, Schroder K, Irvine K, Arakawa T, Nakamura M, Kubosaki A, Hayashida K, Kawazu C, Murata M, Nishiyori H, Fukuda S, Kawai J, Daub CO, Hume DA, Suzuki H, Orlando V, Carninci P, Hayashizaki Y, Mattick JS (2009) Tiny RNAs associated with transcription start sites in animals. Nat Genet 41: 572–578.

Tian ZD, Liu J, Wang BL, Xie CH (2006) Screening and expression analysis of *Phytophthora infestans* induced genes in potato leaves with horizontal resistance. Plant Cell Rep 25: 1094–1103.

Tian ZD, Liu J, Portal L, Bonierbale M, Xie CH (2008) Mapping of candidate genes associated with late blight resistance in potato and comparison of their location with known quantitative trait loci. Can J Plant Sci 88: 599–610.

van Baarlen P, van Esse HP, Siezen RJ, Thomma BPHJ (2008) Challenges in plant cellular pathway reconstruction based on gene expression profiling. Trends Plant Sci 13: 44–50.

Van Dijk JP, Cankar K, Scheffer SJ, Beenen HG, Shepherd LVT, Stewart D, Davies HV, Wilkockson SJ, Leifert C, Gruden K, Kok EJ (2009) Transcriptome analysis of potato tubers-effects of different agricultural practices. J Agri Food Chem 57: 1612–1623.

Vasquez-Robinet C, Mane SP, Ulanov AV, Watkinson JI, Stromberg VK, De Koeyer D, Schafleitner R, Willmot DB, Bonierbale M, Bohnert HJ, Grene R (2008) Physiological and molecular adaptations to drought in Andean potato genotypes. J Exp Bot 59: 2109–2123.

Velculescu VE, Zhang L, Vogelstein B, Kinzler KW (1995) Serial analysis of gene expression. Science 270: 484–487.

Wang BL, Liu J, Tian ZD, Song BT, Xie CH (2008) cDNA microarray analysis of metabolism-related genes in inoculated potato leaves expressing moderate quantitative resistance to Phytophthora infestans. J Hort Sci Biotechnol 83: 419–426.

Welinder KG, Jorgensen M (2009) Covalent structures of potato tuber lipases (patatins) and implications for vacuolar import. J Biol Chem 284: 9764–9769.

Yuan J, Haroon M, Lightfoot D, Pelletier Y, Liu Q, Li XQ (2009) A high-resolution melting approach for analyzing allelic expression dynamics. Curr Iss Mol Biol 11 (Suppl 1) 11–9.

Zhang T, Fortin M, Donnelly D, Li XQ (2008) Constitutive expression of a new proteinase inhibitor II family gene in potato. J Huazhong Agri Univ 27: 10–16.

Zhu B, Ping G, Shinohara Y, Zhang Y, Baba Y (2005) Comparison of gene expression measurements from cDNA and 60-mer oligonucleotide microarrays. Genomics 85: 657–665.

Proteomics and Metabolomics

Adrian D. Hegeman

ABSTRACT

A significant body of literature details the application of metabolomics and proteomics methodologies to examine questions concerning potato tuber quality traits and potato biology especially development and stress responses. These topics have not been previously reviewed and also require significant methodological background to be meaningful. This chapter begins by providing the necessary proteomics and metabolomics background and by describing the state of the art in these two disciplines. The following literature review will focus first on the current state of potato-specific methods in proteomics and metabolomics and the specific challenges associated with potato analysis. Tuber quality traits, development and stress responses will be reviewed subsequently. The final section examines the growing body of work to provide tools for evaluating the safety of new potato cultivars especially the use of "substantial equivalence" as a criterion for assessing risk derived from genetic modifications.

Keywords: Mass spectrometry, metabolic profiling, metabolism, metabolomics, proteomics, substantial equivalence

11.1 Introduction

Dramatic improvements in nucleic acid sequencing and information technologies over the past decades have made it routine to generate genome-scale datasets enabling the descriptive evaluation of biological systems at a molecular level. Genome sequencing provides an unprecedented molecular blueprint for an organism, yielding evolutionarily-linked information about their potential metabolic and physiological behavior. The desire to extend

Departments of Horticultural Science and Plant Biology, Microbial and Plant Genomics Institute, University of Minnesota–Twin Cites, 305 Alderman Hall, 1970 Folwell Avenue, Saint Paul, MN 55108, USA; e-mail: *hegem007@umn.edu* this analysis beyond the "blueprint" to high-throughput analysis of the molecular changes underlying macroscopic behavior has spawned parallel descriptive approaches that extend along the canonical flow of biological information (DNA → RNAs → proteins → metabolites) by extension from genomics. High-throughput analytical disciplines have emerged in the order of this information flow including: genomics, transcriptomics, proteomics and metabolomics, which together constitute what is referred to as the "omics cascade" (Dettmer et al. 2007). The omics cascade provides the methodological core of systems biology—the goal of which is to make large numbers of unbiased molecular observations to construct descriptive models of biological systems that can be used to link molecular observations with macroscopic or behavioral properties of a system (Kell 2006).

While systems biology is still a rapidly developing, emerging discipline it has great potential for providing useful strategies for associating quality traits with measurable or selectable metabolic or genetic markers (Glinski and Weckwerth 2006). In this chapter we will first provide background information and an assessment of the state-of-the-art in proteomics and metabolomics methodology. As no prior reviews have specifically detailed any aspect of *Solanum tuberosum* proteomics or metabolomics subsequent sections will attempt to provide a baseline synopsis of these studies to date. The review will first detail generalist studies of potato that fall under the classification of a systems biology approach (metabolomics and proteomics). Studies that utilize comparable methodologies, but that profile known molecular targets because of their association with specific quality traits, will be covered subsequently (metabolic and protein profiling). The handful of manuscripts detailing specific abiotic or biotic stress-related experimentation will be described next, followed by a brief analysis of metabolite diversity across various potato cultivars. Finally, the role of proteomics and metabolomics for assessing food safety and other potential impacts of genetically modified potatoes will be examined.

11.2 Methodology

Genomics uses DNA sequence information to understand gene structure and function. Transcriptomics uses various methods of RNA measurement including microarrays and qRT-PCR to provide information about the regulation and kinetics of gene expression (see Chapter 10). Proteomics uses multiple biochemical methods, but especially mass spectrometry, to determine protein identity, locality, quantity, modification status, turnover and structure to understand protein regulation and function. And finally, metabolomics uses a wide array of analytical approaches to measure metabolite identity, locality, quantity, and flux through different metabolic pathways (Hollywood et al. 2006). It is important to convey that from an analytical chemistry perspective each step down the omics cascade presents new challenges, many of which have not yet been adequately addressed, especially in proteomics and metabolomics. For example, while genomics focuses on a single polymeric analyte (DNA), with virtually no dynamic range (most genes have approximately the same copy number) and no spatial variation (nucleus of most cells), metabolomics deals with huge chemical diversity over a wide range of concentrations and spatial subcellular and organismal distributions. Omics terms have been propagated (to perhaps beyond the point of absurdity) as a way to refer to the analysis of broad classes of chemical entities (e.g., "phosphoproteomics", "lipidomics" and "ionomics", to name a few). These omics subclasses often distinguish themselves from proteomics or metabolomics because of interest in specific biological questions and/or because of the presence of unique analytical challenges (Glinski and Weckwerth 2006).

11.2.1 Proteomics

The term "proteome" was first introduced at a conference on 2-dimensional gel electrophoresis (2DE) in Siena, Italy in 1994 as an efficient way to refer to the complete set of potential proteins resulting from a specific genome (Patterson and Aebersold 2003). That year also saw the coalescence of several disciplines spurred by recent technological achievements to enable rapid identification of proteins excised from 2DE gels (Fig. 11-1). Proteomics has always had 2DE at its core. First developed in the mid 1970s, technically demanding 2DE was capable of resolving and visualizing hundreds of protein spots from complex mixtures of proteins but was limited by the lack of rapid sensitive protein identification tools. With the development of the soft ionization techniques electrospray ionization (ESI) by John Fenn and matrix assisted laser desorption ionization (MALDI) by Koichi Tanaka (who shared the 2002 chemistry Nobel prize for these accomplishments) and Franz Hillenkamp and Michael Karas (who did not but arguably could have), it became possible to analyze intact proteins and peptides using mass spectrometry (Tanaka et al. 1987; Karas and Hillenkamp 1988; Fenn et al. 1989). This advance, coupled with the ever increasing availability of predicted amino acid sequence data from gene sequencing efforts, led to the near simultaneous development of multiple MS-based statistical protein identification algorithms in 1993 (Henzel et al. 1993; James et al. 1993; Mann et al. 1993; Pappin et al. 1993; Yates et al. 1993). Having the capacity to identify easily dozens of proteins from 2DE gels, the new field of proteomics was poised to take off.

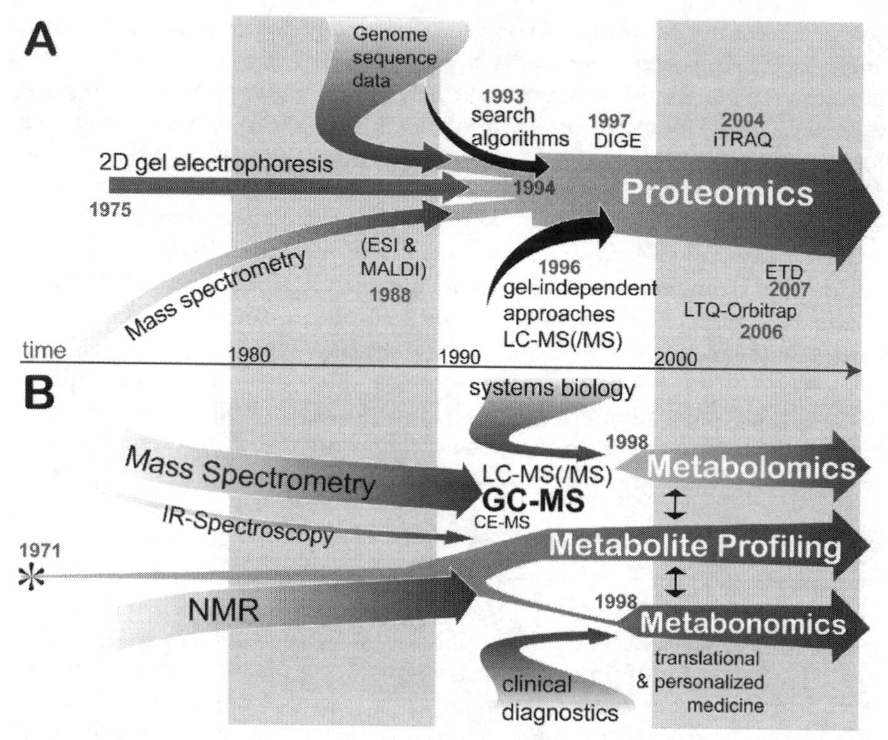

Figure 11-1 Timeline for the development of proteomics and metabolomics. Numerous disciplines were brought together in the formation of both proteomics (A) and metabolomics (B). The development of proteomics depended on several key advances in MS (the invention of ESI and MALDI for protein ionization in 1988), genomics (accumulation of a critical mass of amino acid sequence information for a given organism) and data analysis (implementation of MS database search algorithms), before high-throughput protein identification could be accomplished. Metabolomics, in contrast, depended on developments in individual methodologies and thus emerged from many different laboratories using a wide variety of approaches. This is part of the reason for the diverse nomenclature in the field. Panel A is based on a figure from Patterson and Aebersold (2003).

11.2.1.1 MS and MS/MS Protein Identification

While it is possible to analyze intact proteins by MS, proteomic approaches that examine intact proteins are still largely under development (Han et al. 2008) and would be extremely challenging with samples derived from 2DE gels. Most successful proteomic strategies instead utilize a protease treatment to cleave proteins into shorter well-defined peptides. Trypsin is, for example, widely employed for this purpose as it cleaves proteins consistently at the carboxy terminal side of arginine and lysine residues. Proteolysis can be performed directly on proteins that are imbedded in excised 2DE gel pieces using a so-called "in-gel digest" protocol. The earliest proteomics protocols used this approach to generate samples of peptides derived from different 2DE protein spots that could be analyzed by MS to provide a list of peptide masses for each sample. These masses are then compared to theoretical digests of all of the known amino acid sequences for proteins derived from the organism in question to find statistically significant matches for protein identification. This approach, called peptide mass fingerprinting, works well only for highly abundant, well-resolved proteins and was quickly improved upon by the application of tandem mass spectrometry (MS/MS) (Patterson and Aebersold 1995). In MS/MS experiments multi stage or ion trap mass analyzers are used to 1) select the intact peptide ion, 2) fragment it, and 3) measure the resulting fragmentation pattern. Fragmentation patterns can be quite informative for identifying the peptide amino acid sequence as bond cleavage with the most common type of fragmentation typically occurring at the peptide bonds to produce an overlapping series of *b*- and *y*- type ions (Fig. 11-2). In rare cases amino acid sequence can be assigned directly to MS/MS spectra on the basis of the patterns, which is called de novo sequencing (Falick et al. 2008). Otherwise, as in peptide mass fingerprinting, amino acid sequence databases can be used with various types of assignment algorithms to statistically assign amino acid sequences to MS/MS peptide fragmentation patterns

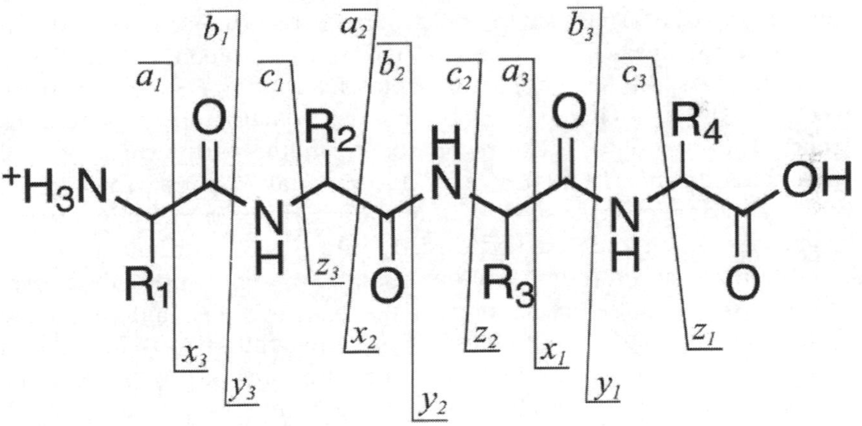

Figure 11-2 Peptide backbone cleavage in tandem mass spectrometry. Tandem mass spectrometry (MS/MS) allows a peptide molecular ion to be isolated and subjected to fragmentation conditions. Cleavage along the peptide back bone can be accomplished using different techniques, all of which provide paired series of ions containing information that can be used to reconstruct the amino acid sequence of the peptide. Typical MS/MS fragmentation is accomplished by collisionally induced dissociation (CID), which breaks weaker bonds preferentially and thus results predominately in *b*- and *y*-ion series. Electron transfer dissociation (ETD) is effective for fragmentation of larger, more highly charged peptides (often use Lys-C instead of trypsin for proteolysis) and results in higher energy cleavages for predominantly *c*- and *z*-ion series.

(Yates et al. 1995). This approach is much more sensitive than fingerprinting and can be used to identify hundreds of proteins from typical 2DE gels of complex protein mixtures. As a result of wide application of MS/MS techniques to 2DE analysis, clear limitations in the resolving power of 2DE have become apparent as most analyzed spots contain multiple overlapping proteins. To this day the vast majority of proteomics is performed using data from two experimental sources: 1) MS and MS/MS data and 2) protein amino acid sequence data that are assembled using statistics (commercial or public domain algorithms include Mascot, Sequest, ProteinPilot, XTandem, and OMSSA) (Kumar and Mann 2009). The quality and assembly of both MS and sequence data are critically important for the proper and confidant interpretation of the statistical output (Huttlin et al. 2007; Wang et al. 2008). A lack of adequate protein sequence information will dramatically affect the confidence of protein identifications. While homology-based search options exist, they tend to be marginally effective and have resulted in high profile errors in cases where they have been improperly used (Asara et al. 2007; Pevzner et al. 2008).

11.2.1.2 Gel-free Proteomics

Significant improvements in instrumentation and amino acid sequence availability since 1994 have led to a dramatic increase in the quality and depth of proteomic analyses. The introduction of gel-free proteomic strategies in the last 10 years, though a departure from 2DE roots, has helped increase throughput by enhancing the performance of chromatographic separations directly interfaced with the mass spectrometer. Multi dimensional protein identification technology or "Mudpit" is an example whereby ion exchange and reverse phase separations are performed in tandem using chromatographic media packed directly into ~ 100 μm diameter fused silica electrospray emitters (Wolters et al. 2001). This approach allows thousands to tens of thousands of peptides to be separated and sequenced quickly using fast and sensitive MS/MS instrumentation. As this approach is performed at the peptide level, complex mixtures of proteins may be proteolyzed directly without the need for intact protein separations, which is a distinct advantage with certain classes of proteins such as integral membrane proteins that (at least until recently, see below) are difficult to resolve by 2DE (Blonder et al. 2006). Although these "shotgun proteomics" approaches provide significantly higher throughput and increased depth of proteomic coverage than 2DE, several disadvantages include the need for complex reconstruction and visualization methods for proteins from peptide measurements, the loss of information regarding sub-populations and partial coverage of protein modifications and the requirement for alternate means of relative protein quantification.

11.2.1.3 Protein Quantification

A major strength of the 2DE-based proteomics approaches is the apparent ease with which one can visually comprehend the data. Comparison of protein spot intensities can provide relative abundance with or without protein identification such that changing species may be specifically selected for follow-up. This apparent simplicity is deceptive as shown in a 2007 study performed by the Association of Biomolecular Resource Facilities (ABRF) in which 2DE-based approaches were found to dramatically under-perform when compared with shotgun sequencing employing either stable isotope labeling approaches or statistical methods (Turck et al. 2007). Isotopic labeling is employed as an internal control in MS that allows one to simultaneously analyze the same chemical species distributed over multiple distinguishable masses (reviewed in Tao and Aebersold 2003). Isotopic labels can be introduced chemically, as with commercial reagents such as iTRAQ (Ross et al. 2004), or into an entire organism, by providing labeled nutrients through normal metabolic processes (Beynon and Pratt 2005). Metabolic stable isotopic labeling has particular value for plant proteomic applications as it can also provide internal control for protein extraction and fractionation steps that are particularly variable in plants due to cellular architecture and high vacuolar protease content (Van Wijk 2001). Potato was, in fact, the first plant to be fully metabolically labeled with ^{15}N using hydroponics (Ippel et al. 2004).

11.2.1.4 Other Advances in Shotgun Proteomics

The commercialization of the LTQ-Orbitrap hybrid mass spectrometer has had a dramatic impact on the depth and confidence of proteomic measurements in the past few years (Makarov et al. 2006). This instrument combines complementary ion trap components: 1) the linear ion trap stage provides high sensitivity and speed in MS/MS data collection and 2) the orbitrap stage provides high resolution, high-mass accuracy measurements for unambiguous assignments. A recent study, for example, cited 4,124 membrane proteins identified at high confidence from rat brain tissues via shotgun sequencing using the LTQ-Orbitrap (Lu et al. 2009). For the purpose of identifying sites of post-translational modifications (PTMs) by MS/MS, a major recent advance has been the development of an accessible alternative peptide fragmentation strategy called electron transfer dissociation (ETD; Fig. 11-2). Weakly bonded PTMs such as phosphorylation and glycosylation are typically removed right away by CID such that the sites of attachment are lost in the fragmentation pattern. ETD, however, preserves these linkages and has become an important tool for mapping PTMs (Swaney et al. 2009). Other advances in shotgun sequencing of PTMs include various modification enrichment strategies such as Fe or Ga immobilized metal affinity chromatography (IMAC), which has been used in conjunction with ^{15}N-metabolic labeling *in planta* to probe signaling pathways involved in biotic stress responses (Benschop et al. 2007; Nuhse et al. 2007).

11.2.1.5 Advances in Gel-based Proteomics

While gel-free approaches have excelled for high complexity proteomic applications, 2DE-based proteomics have benefited from several key advances as well. Fluorescent labeling methods such as Difference Gel Electrophoresis (DIGE; Unlu et al. 1997) have made it possible to perform differential proteomic analysis in a single gel providing a high degree of control over post-labeling technical variability (Riederer 2008). Fluorescent staining reagents provide an alternative method for detecting PTMs at the level of the intact protein (Jacob and Turck 2008; Agrawal and Thelen 2009). Methods have been developed that improve the analysis and resolution of hydrophobic proteins (Rabilloud et al. 2008). Variability in quantification reported in the ABRF study may have several sources including technical error or failure to resolve overlapping protein spots (Turck et al. 2007). A strategy that is increasing in popularity augments 2DE analysis with additional chromatographic separation (geLC-MS), enabling quantification by spectral counting as a means of compensating for confounded protein quantification due to overlapping protein spots (Graham et al. 2007; Yang et al. 2007).

11.2.1.6 Extracting Proteins from Potato Tissues

Two studies specifically test protein extraction methods for different potato tissues for 2DE analysis. The first, compared SDS lysis buffer to a protocol using phenol phase extraction and found no difference in extraction efficiency or spot count, but reported slightly higher spot resolution with the SDS extraction (Delaplace et al. 2006). The second described an effective general extraction protocol for green plant tissues, including potato leaves, but did not appear to provide any basis for comparison (Flengsrud 2008).

11.2.2 Metabolomics

The term "metabolomics" first appears in the literature in the late 1990s, roughly four years after "proteomics", to provide the final segment of the omics cascade (Dettmer et al. 2007). Metabolomics as part of systems biology attempts to identify and measure metabolites in a comprehensive unbiased way (Dixon et al. 2006). This is distinct from metabolite profiling, which has been performed for decades, and which measures specific targeted population or chemically related families of metabolites (Fiehn 2001). These are also distinct from metabonomics, which emerged at approximately the same time as metabolomics, but that seeks to utilize peaks from high throughput characterization of unidentified metabolite populations as statistical input for diagnostic purposes (Lindon et al. 2004). Metabonomics is also often called "metabolic fingerprinting" and despite its fairly recent implementation was proposed in theory by Pauling and coworkers (1971). All three of these approaches for high throughput metabolite characterization are related in that they benefit from technology driven improvements in analysis and frequently share resources.

A much larger diversity of approaches exists for metabolite analysis than proteomics. This to some extent reflects the greater degree of chemical diversity of metabolites compared to proteins, but it is also a function of the way each discipline originated. Figure 11-1 shows a timeline covering the origins of proteomics, metabolomics, metabolic profiling and metabonomics. While proteomics originated following several key innovations, high throughput metabolite analysis did not, and whereas proteomics was initially limited to a fairly small set of laboratories in which key resources were available, metabolite analysis could be performed by many different laboratories using many different techniques, yielding more ideas about what to do with the data. Another key difference is that, while proteomics benefits from the application of genomic information for peptide sequence assignment, metabolomics neither benefits from the presence nor is constrained by the absence of genomic information for compound identification (Sumner et al. 2003). High throughput metabolite analysis is performed using several different separations coupled to mass spectrometry including gas chromatography (GC-MS), liquid chromatography (LC-MS) and capillary electrophoresis (CE-MS) mass spectrometry (Dettmer et al. 2007). Nuclear magnetic resonance (NMR) spectroscopy has been equally important as MS for metabolite analysis; metabonomics was in fact first propelled by the NMR community (Lindon et al. 2004). Other analytical strategies, in addition to infrared (IR) spectroscopy, are occasionally employed, each technique with its own advantages and disadvantages.

For the purposes of this chapter it is worth, at least briefly, describing the status of NMR, GC-MS and LC-MS based metabolite analysis (Sumner et al. 2003). NMR is capable of providing measurements *in situ* and can provide excellent absolute quantitative and structural information. Sensitivity limitations for NMR typically mean that only the top 20 to 50 most abundant metabolites can be observed in a sample. MS techniques are at least several orders of magnitude more sensitive but have some analytical limitations imposed by the various ionization processes.

11.2.2.1 Identifying Metabolites by MS

Identifying metabolites by MS is significantly more difficult than by NMR. For common metabolites that are either volatile or can be made volatile by derivatization, GC-MS coupled to spectral library searching provides a robust and high throughput means for metabolomic identification (Glinski and Weckwerth 2006). Although compromises are always made in the selection of general derivatization strategies, fairly broad applicable chemistries have been developed for this purpose (Fiehn 2008). Fragmentation patterns by LC-MS (MS/MS) instruments can be used for spectral searches, but to date these appear to have a high degree of platform-dependent spectral variability that has limited utility for spectral-library driven metabolite assignment (Dettmer et al. 2007). As a result, in cases where novel or unknown compounds are observed or where LC-MS approaches are required, spectral libraries are of limited value, so one must rely on assignment strategies that are more general. One such approach is to calculate the elemental composition(s) for a spectral feature by using very high accuracy mass measurements (Hegeman et al. 2007; Kind and Fiehn 2007).

11.2.2.2 Quantifying Metabolites by MS

Like many analytical techniques, MS requires standard compounds for absolute quantification. GC-MS produces fairly consistent results and can be amenable to external calibration curves for absolute quantification or "run-to-run" comparisons for relative quantification. LC-MS, on the other hand, is complicated by competitive phenomena during ionization of complex heterogeneous samples called "ion suppression", so that a given signal may change as a function of the concentration of coeluting species, rather that its own concentration (Annesley 2003; Antignac et al. 2005). Stable isotopic labeled compounds have been used as internal standards for quantification in mass spectrometry for several decades largely because of their usefulness in controlling for these and other effects. Isotope dilution, for example, in which introduction of a known quantity of labeled standard compound into a sample is used to quantify the unlabeled compound in the mixture, is standard practice in quantitative MS (Rittenberg and Foster 1940; Tolgyessy et al. 1972) and has been employed recently for metabolite analysis in microbes (Wu et al. 2005). Because of the cost and complexity of these techniques, experimentalists often assume that mass spectral signal intensity is proportional to compound abundance. In practice this approximation degrades as sample complexity and variation increase.

11.2.2.3 Data Analysis

Data for multiple biological replicates of each sample population is collected and processed yielding a data matrix composed of: 1) compound identities, or spectral characteristics (if unidentified), 2) retention times (if chromatographed), and 3) signal intensities. Subsequent statistical analysis of MS data matrices shares many aspects with microarray interpretation (Riva et al. 2005). These matrices are analyzed using multivariate statistical approaches such as principal component analysis (PCA), independent component analysis (ICA), or random forest modeling to identify metabolites that significantly contribute to differentiation of each population (Carpentier et al. 2004).

11.3 Current State of (Untargeted) Potato Proteomics and Metabolomics

Numerous unbiased proteomics and metabolomics studies of potato have been conducted over the past decade. The proteomics studies are predominantly 2DE-based and the metabolomics studies are mostly GC-MS based with some NMR and LC-MS. While contemporary microarray studies are capable of providing many thousands of transcriptional observations, many of the proteomics and metabolomics projects described below provide tens to hundreds of observations of the most abundant proteins and metabolites. This in some ways reflects the added analytical challenges associated with measuring chemically diverse molecules, but should in no way diminish the importance of the observations. Changes in transcription do not always correlate well with protein abundance (Gygi et al. 1999) and are even more removed from metabolite concentrations. The poor correlation of protein and metabolite abundance with gene expression may be understood by considering the multiple possibilities for regulation after transcription. One example is protein degradation, which is partly regulated by hormones in plants (Mockaitis and Estelle 2008). As a result, transcriptional information often does not provide adequate information regarding bona fide changes in plant metabolism or physiology. In one study, transgenic potatoes could not be statistically differentiated from wildtype using large transcriptional datasets, but were easily distinguished through large obvious metabolic perturbations using modest metabolomics datasets (Urbanczyk-Wochniak et al. 2003).

11.3.1 Proteomics

Most of the potato proteomics studies to date focus on some component of potato tubers. While it is widely known that green plant tissues are dominated by the protein RuBisCo, which may constitute as much as 50% of the total protein content, potato tubers contain two classes of proteins: patatins and Kunitz protease inhibitors, which together make up close to 90% of the total protein isolated from cv. Kuras tubers (Bauw et al. 2006). These proteins are thought to serve a dual role as storage and defensive proteins. The presence of hyper abundant proteins limits one's ability to analyze lower abundance proteins in most proteomics methods; RuBisCo is often cited as a major limitation to proteomic studies of green tissues for this reason. Similar problems exist with the hyper abundant proteins of potato tubers. Figure 11.3A shows a 2DE gel of cv. Kuras tuber proteins extracted using phenol over a broad pI range that shows the two dominant classes of proteins. In subsequent analyses with more narrow pI ranges (4–7 and 6–11) the authors were able to detect as many as 600 spots with silver stain, 550 with epicocconone and 250 with Coomassie Brilliant Blue on pI 4–7 gels and 300 spots with silver stain, 280 with epicocconone and only a few well resolved with Coomassie Brilliant Blue for pI range 6–11 (Bauw et al. 2006). From 2DE gels the authors used peptide mass fingerprinting to identify multiple peptides for six of seven known patatin proteins and all 11 known Kunitz protease inhibitors, each with fairly high sequence coverage. Only five other proteins were identified: glyoxylase I, enolase, annexin p34 and two proteins of unknown function. An additional 20 proteins were identified using 1D gels by performing in-gel digests with bands above 50kD. While this study was able to examine the hyper abundant tuber proteins, it was limited in depth of coverage. Several novel insights concerning the phylogenetics of the Kunitz protease inhibitors and observations in large inconsistencies in gene expression and overall protein abundance were made (Bauw et al. 2006).

11.3.1.1 Tuber Skin Proteome

The tuber skin proteome was examined by Barel and Ginzberg using 2DE and peptide mass fingerprinting for protein spot identification (Barel and Ginzberg 2008). This study identified 46 proteins from *S. tuberosum* cv. Desirèe that are more abundant in tuber skin when compared with flesh. The authors point out the value of tuber skin as a model for suberization, a process that occurs in less tractable woody tissues and that is of great interest due to its impact on carbon availability in cellulasic biofuel feed stocks. More than half of the identified proteins can be associated with plant defenses and stress responses and include multiple enzymes involved in the synthesis of monolignol precursors of lignin and several peroxidases thought to be involved in the suberization process. Differential expression was confirmed by RT-PCR.

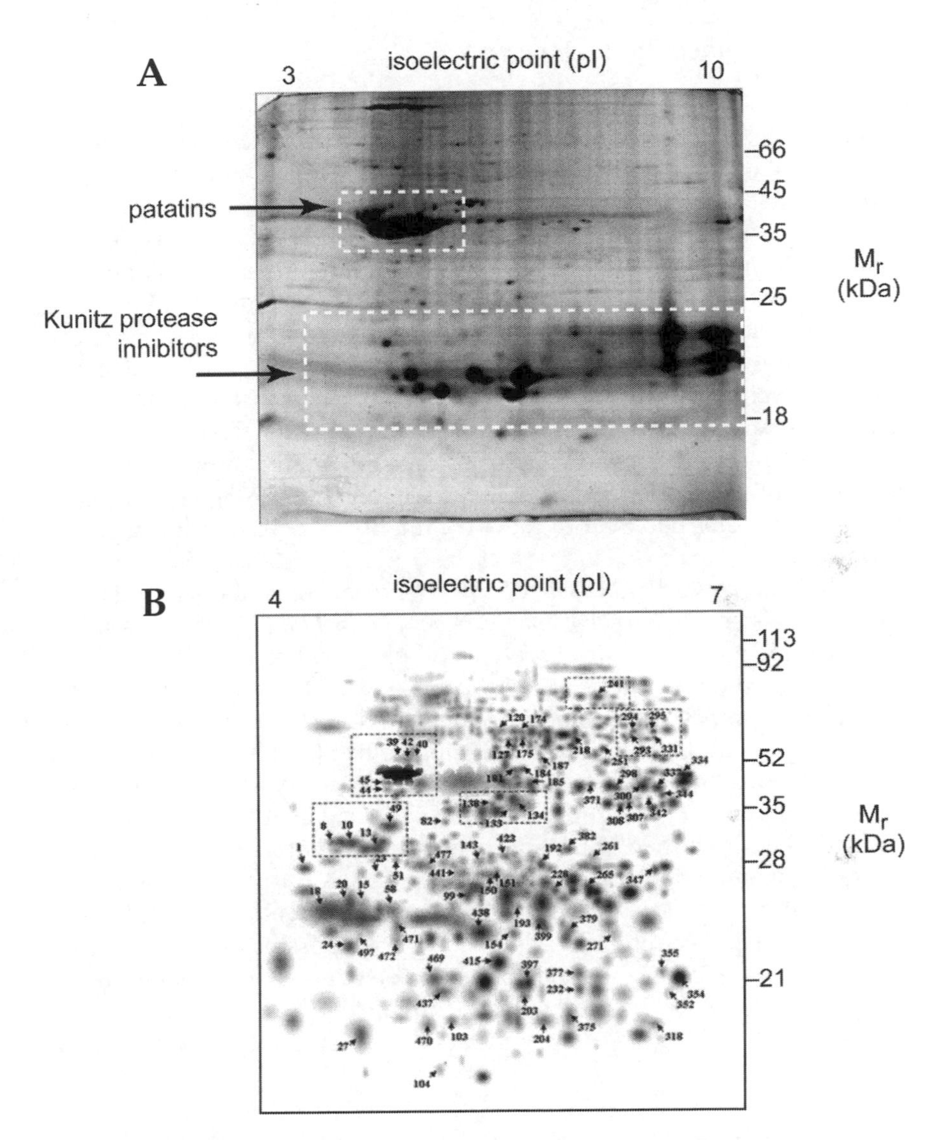

Figure 11-3 2DE of potato tuber proteins. Two examples of potato tuber protein 2DE images are provided from the literature. Panel A was taken from Fig. 1 of Bauw et al. (2006), and modified to show the protein spots associated with hyper abundant patatin and Kunitz protease inhibitors in this broad 3–10 pI range, 12.5% SDS PAGE gel stained with epicocconone. Panel B shows a simulated protein map amalgamation of spots observed over four developmental stages of tuborigenesis taken and modified from Agrawal et al. (2008). The potato variety used for this study has lower levels of patatins than for that used in A. In contrast to A, which was meant to provide a broad overview, larger spot resolution and higher dynamic range in B is accomplished by narrowing the isolectric focusing range so that more proteins in that range can be analyzed and separated to a greater degree.

11.3.1.2 Sub-cellular Tuber Proteomes

The proteomes of potato tuber sub-cellular components have been examined. Stensballe and coworkers isolated amyloplasts from mini- (1–3 g) and micro- (20-30 mg) tubers of *S. tuberosum* cv. Kuras for analysis by 1DE, peptide mass fingerprinting and MS/MS (Stensballe et al. 2008). Twenty-seven and 20 proteins were identified from plastid preparations derived from mini- and microtubers, respectively, by fingerprinting. Following MS/MS characterization, the total number of uniquely identified proteins across both samples increased to 90. Several differences between the mini- and microtubers were noted, although major components of starch synthesis were identical. Functional classes of proteins included starch and sucrose metabolism, the pentose phosphate pathway, glycolysis, amino acid metabolism and plastid chaperonins. While many abundant protein gene sequences encoded expected plastid transit peptides, a significant number of proteins had a motif more commonly associated with targeting to the thylakoid.

Eubel and coworkers isolated mitochondria from tubers and 20 day old etiolated stems of *S. tuberosum* cv. Cilena (Eubel et al. 2004). This study focuses on large mitochondrial protein complexes called respirasomes that function in respiration. Mitochondrial protein extracts were fractionated by 2DE in which the first dimension was blue-native gel electrophoresis, which maintains protein complex integrity, followed by SDS-PAGE in the second dimension, which denatures the complexes. The study identifies five new super complexes in potato previously not observed in other plants. Balmer and coworkers identified 42 putative thioredoxin interacting proteins from potato tuber mitochondria as part of a larger study that also included mitochondria from pea and spinach leaves (Balmer et al. 2004). This study used thioredoxin-activated fluorescence to locate possible redox targets by 2DE. Lastly, 14 phosphorylation sites have been mapped in potato tuber mitochondrial proteins including one in formate dehydrogenase and others that are conserved with similarly mapped sites in other species (Bykova et al. 2003a, b).

11.3.1.3 Tuber Development

Several studies examine tuber development including one by Agrawal and coworkers that compares the proteomes from stolons, and initial, developing and mature tubers of *S. tuberosum* diploid line A16 (Agrawal et al. 2008). The sampled developmental stages were analyzed by 2DE and 219 protein spots were identified in which the intensity varied by at least 2.5 fold in one or more stages. Figure 11-3B shows a 2DE map of all of the combined protein spots from the four samples, focusing on the particularly rich pI range from

4 to 7. Of the 219 developmentally changing spots, 97 were identified by LC-MS/MS. Most of the identified proteins included the abundant patatins, protease inhibitors and primary metabolic enzymes. An earlier study by Lehesranta and coworkers examined 13 developmental stages ranging from stolons to fully mature sprouting tubers from cv. Desirèe (Lehesranta et al. 2006). Samples from different stages were analyzed by 2DE and LC-MS/MS. Around 800 reproducible spots were observed across the samples, 150 of which were shown to change significantly across stages. Of these 150, 59 proteins were identified; an additional 50 unchanging spots were also identified. Hierarchical clustering allowed five well-defined patterns of expression to be described. The complexity of developmental processes and the absence of information concerning regulatory elements like transcription factors or signaling molecules make it impossible to draw many conclusions from these studies at this point.

Recently, an article appeared by Delaplace and coworkers that uses biochemical, physiological and proteomics approaches to examine the molecular changes in post harvest potato tubers (Delaplace et al. 2009). This study used cv. Desirèe and 2DE with DIGE to follow changes in protein after various storage periods, up to 270 days, at 4°C. Several proteome changes were noted including an increase in enzymes of starch catabolism and specific proteolysis of patatins, as well as other changes.

Though some important insights have been achieved, potato proteomics has in general failed to benefit from many of the recent innovations in proteomics methodology. Despite the fact that most of the major publications are fairly recent (2004 or newer), potato proteome coverage is still very shallow. Tuber proteomics presents some special challenges, as was already discussed, but the lack of accessible high quality genomic data of sufficient coverage has, until recently, also contributed to the low coverage. This fact was eluded to by Bauw and coworkers in the stated under-representation of genes or transcripts encoding proteins larger than 50 kDa despite having a library of (at the time) ~ 180,000 ESTs from cv. Kuras (Bauw et al. 2006). As genomic resources continue to improve and higher throughput/sensitivity technologies become more accessible, the information quantity and quality of these studies will increase dramatically.

11.3.2 Metabolomics

Unlike proteomics, potato metabolomics has been near the forefront of technical innovation and has produced some early, relatively high profile publications and other 'firsts' often besting examples from early model genomic plant species (Roessner et al. 2000; Ippel et al. 2004; Weckwerth et al. 2004). This difference in success of these approaches is at least partly due to the fact that metabolomics does not require genomic information as an input for data assignment as does proteomics. Metabolomics methods were also being developed in many more and diverse laboratories than were those of proteomics, which required a nexus of 2DE, specialized mass spectrometry instrumentation and data assignment using genome database search algorithms. Other factors include potato's economic importance, and the accessibility of large amounts of its tissue for biochemical study (especially when compared to the minuscule *Arabidopsis thaliana*).

11.3.2.1 Tuber Metabolome Changes with Growth Medium

One of the most widely and successfully used general derivatization strategies for GC-MS metabolomics studies was first applied by Roessner and coworkers to compare the potato tuber metabolomes of soil and in vitro grown cv. Desirèe (Roessner et al. 2000). The procedure uses a combination of derivitization with methoxylamine and N-methyl-N-(trimethylsilyl) trifluoro acetamide (MSTFA) to make volatile a broad functional range of polar metabolites including sugars, sugar alcohols, dimeric and trimeric saccharides, amines, amino acids, and organic acids. In this study more than 150 volatile metabolites were observed and 77 were identified in a 60 minute GC-MS analysis by comparison of electron impact ionization (EI) spectra with standard compound reference libraries. The protocol is fast and reproducible despite multiple derivatization steps, and has become one of the best methods for identifying dozens of metabolites with directly comparable signal intensities for relative quantification. Large and significant metabolic differences were observed between soil and in vitro grown tubers. An additional component of the study examined the consequences of over expression of a yeast invertase transgene and antisense repression of native ADP-glucose pyrophosphorylase. Several predictable metabolic changes were noted with the addition of invertase, as was an increase in several disaccharides including maltose, isomaltose and trehalose. These latter results were surprising and highlight the importance of broad unbiased sampling.

11.3.2.2 Metabolomics and Phenotyping

Weckwerth and coworkers propose using differential metabolic networks to expose molecular phenotypes where no macroscopic phenotypes are observable (Weckwerth et al. 2004). This study (also in cv. Desirèe) compares wildtype and an antisense repressed sucrose synthase II transgenic line, which has no obvious growth phenotype. More than 1,000 compounds were observed in leaf extracts and 500 were observed in tuber extracts using a GC-time of flight (TOF) MS. Although only 100 or so compounds provided reasonable (> 89% match factor) spectral library matches, all of the compounds were used for the differential analysis. This study incorporates some of the power of a metabolic fingerprinting approach for modeling a biological system. Significant differences in metabolite correlations were observed between the two varieties. Enot and coworkers provide a discussion of the interpretation of similar metabolomic data mining techniques, especially the random forest model approach, and include an example from potato and several from *Arabidopsis* (Enot et al. 2006).

11.3.2.3 Metabolome vs. Transcriptome: Correlation and Predictive Power

Urbanczyk-Wochniak and coworkers describe in two separate papers correlations between metabolic profile and transcriptional data for systems biology. Both papers correlate data from cv. Desirèe collected essentially using the GC-MS strategy of Roessner et al. (2000) discussed above. The first study followed changes in gene transcription and metabolite abundance over diurnal cycles. Some 56 metabolites and 832 transcripts were found to change significantly as a function of diurnal cycle. However, very little correlation was found in the profiles of metabolites and transcripts and mostly only in central metabolic pathways. PCA revealed that metabolite levels change throughout the diurnal cycle with tight temporal regulation (Urbanczyk-Wochniak et al. 2003). The subsequent study examines the correlative predictive power of sets of transcriptional and metabolic fingerprinting data to identify different plant populations, which for this experiment were wildtype cv. Desirèe and a transgenic line, each harvested at various developmental stages. Here metabolic fingerprinting was found to have higher resolution for distinguishing different plant populations than transcript analysis (Urbanczyk-Wochniak et al. 2005).

11.3.2.4 Assessing Metabolic Consequences of Transgene Insertions

Metabolomics and metabolic fingerprinting have become increasingly popular for investigating expected and unexpected changes resulting from transgenes. Defernez and Colquhoun discuss details of experimental design and control experiments for ^{1}H NMR-based metabolite fingerprinting that includes examples and protocols for sample preparation from freeze dried potato tuber tissue (Defernez and Colquhoun 2003). Junker and coworkers use the same basic GC-MS methodology to compare transgenic populations of cv. Desirèe transformed with yeast invertase targeted to the cytosol, the apoplast or the vacuole (Junker et al. 2006). Tuber growth and metabolism were significantly different for the apoplastic and cytosolic transgenes, but only minor changes were observed with vacuolar invertase expression. Another study examines the consequences of 14-3-3 protein repression by comparing panels of enzyme activities and metabolite profiles in cv. Desirèe (Szopa 2002; Szopa et al. 2001). Large changes were observed in metabolites of the TCA cycle suggesting a role for 14-3-3 proteins in its regulation. We will examine the role of metabolomics in food and environmental safety in a subsequent section.

11.3.2.5 Metabolic Flux

Metabolic flux measurements provide additional information about the net flow and partitioning of matter through metabolism that is not obtainable through the analysis of steady state concentrations. Roessne-Tunali and coworkers have begun to use potato tuber disks transferred into medium containing ^{13}C-glucose to begin to probe metabolic flux in potatoes (Roessner-Tunali et al. 2004). Whole potato plants were the first vascular plant to be raised fully labeled with ^{15}N by hydroponics, providing an additional strategy for probing flux analysis in nitrogen containing compounds (Ippel et al. 2004).

11.4 Potato Protein and Metabolite Quality Traits

As the fourth most consumed food crop species globally, there is significant demand for crop improvements for both plant traits relating to agronomic and hardiness properties and tuber qualities affecting nutrition, anti-nutrition, toxicity, consumer preference and post-harvest physiology. Applications of genomics to these goals are discussed in other sections of this volume and are reviewed elsewhere (Duguay et al. 2006). Davies has reviewed some of his efforts with potato metabolic fingerprinting in this and related areas in the VIth International Solanaceae Conference proceedings (Davies 2007). In this section, the use and potential application of metabolite fingerprinting and profiling to potato quality traits will be discussed starting with tuber properties and then examining plant stress responses.

11.4.1 Tuber Quality Traits–by Fingerprinting

While several studies have utilized metabolic fingerprinting as a means of differentiating between groups of potato plants differing by treatment, developmental stage or genetic background, Beckmann and coworkers have taken the approach a step further to identify metabolites that contribute to diagnostic differentiation (Beckmann et al. 2007). Using a random forest model, the authors rank MS features of directly infused extracts (a method called flow infusion electrospray ionization MS or FIE-MS) by predictive significance in differentiating cultivars Desirèe, Agria, Granola, Linda and Solara. The ranking of > 1,000 ions identified ~ 30 ions that provided full differential diagnostic significance with anywhere from nine to 24 of these signals providing cultivar identity with a significance threshold of 0.003 for each variety. The identities of these 30 compounds were confirmed by GC-MS and included raffinose, tyrosine, phenylalanine, aspartate, gluconate, methionine, leucine/isoleucine, γ-aminobuteric acid, quinate and chlorogenate, among others. Patterns of intensities of these ions in each cultivar were then correlated with quality traits including: 1) blackening after cooking; 2) taste; 3) French fry suitability; and 4) frying color. For example, high levels of tyrosine and isoleucine/leucine were associated with desirable flavor profiles through previously characterized Strecker degradation to volatile aldehydes that may contribute an almond or toasted/sweet odor component to the aroma of boiled potatoes. While the method seems a bit circular, given adequately diverse populations of cultivars with well-characterized quality traits, this approach could be quite effective for identifying metabolites associated with specific macroscopic properties. The approach would also benefit from the application of more generalist fingerprinting strategies than direct infusion, which is usually accompanied by dramatic matrix effects that decrease the numbers of observable species and confound attempts to estimate abundance differences (Antignac et al. 2005).

A novel protein fingerprinting approach for identifying polymorphisms in proteins extracted from potato tubers that bypasses the typical dependency of proteomics on high quality genomic sequence data is described by Hoehenwarter and coworkers (2008). This approach starts out like a shotgun proteomics experiment with protein extraction, proteolysis and high throughput MS/MS analysis. Rather than assigning peptide IDs by comparison to genomic information, high mass accuracy peptide masses measured on an Orbitrap mass spectrometer are used as input for PCA and ICA to identify peptides that contribute strongly to the statistical separation of tested populations. In this way differences in protein expression, sequence polymorphisms, and protein modifications can be identified. Once the discriminating peptide populations are identified, sequence-guided and de novo assignment approaches can be applied to identify those key peptides. These can be correlated with specific cultivar quality profiles to generate lead protein markers for quality traits as for metabolites (Beckmann et al. 2007).

11.4.2 Tuber Quality Traits–by Metabolic Profiling

This section details examples of targeted metabolite analysis that provide important information regarding tuber quality. Very little work has been done specifically with protein profiling for tuber quality traits, although the hyper abundant patatins and Kunitz protease inhibitors are likely sources for antigenic species for food allergies and for other properties.

These proteins can be monitored by 1DE if needed and the anti-nutritional properties of the protease inhibitors are largely negated by cooking. A major topic of this section concerns potato steroid alkaloid glycosides which are toxic to humans, important for pest resistance and impact flavor profiles.

Steroid alkaloid glycosides (SAGs) are toxic secondary metabolites found in the green tissues of potato plants and, in lower levels, in tubers (Barceloux 2009; Maga 1980). SAG production is induced by a variety of environmental factors including most biotic and abiotic stresses and light. While their levels are typically low in tubers their synthesis can be stimulated by infections or improper storage, especially in the tuber epidermis (Friedman 2006). While applying metabolic engineering approaches to modify anthocyanin production, Sobiecki and coworkers noted a positive correlation between SAG and anthocyanin abundance that led them to further investigate this phenomenon (Stobiecki et al. 2003). SAGs are large (> 850 Da) polar metabolites (Fig. 11-4) that are not easily volatilized or analyzed by GC-MS. Consequently, HPLC-MS with ESI was used to separate and identify SAGs. While ESI is considered a "soft ionization" strategy and results in far fewer bond cleavages than EI and other types

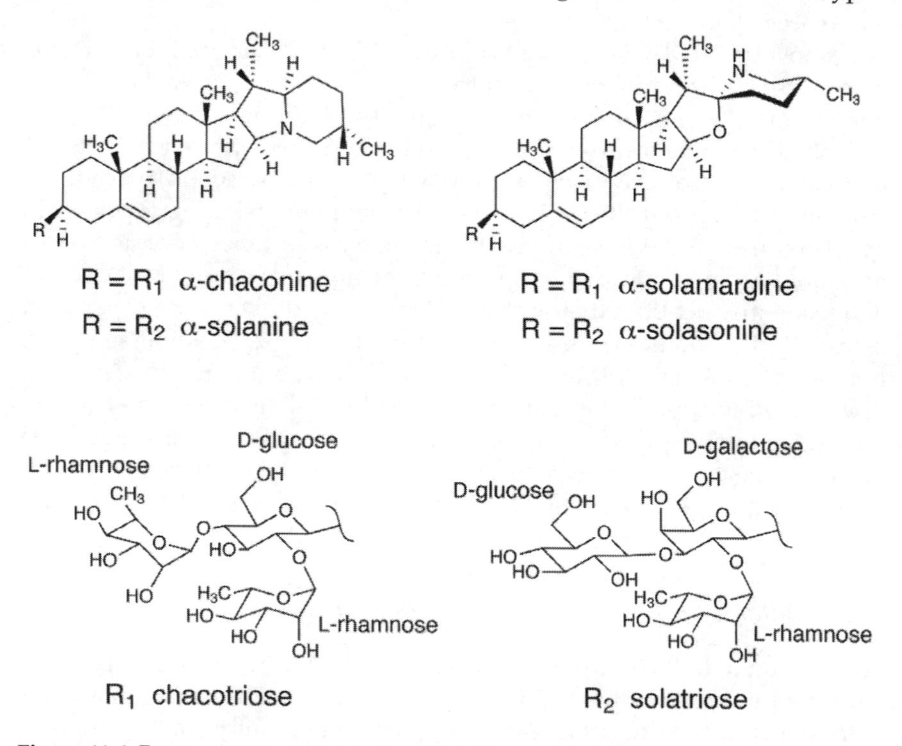

Figure 11-4 Potato steroid alkaloid glycosides. The molecular structures of the four most abundant steroid alkaloid glycosides are shown.

of high energy ionization, weak bonds such as esters, phosphoesters and glycosidic linkages are often cleaved to some degree during ESI. As a result, the SAGs break apart into a series of peaks resulting from the sequential loss of sugars from the triose moiety.

Other groups have recently used this phenomenon to provide optimized LC-MS methods for measuring free SAGs in soil (Jensen et al. 2008). Anthocyanins in this study were characterized using HPLC and UV spectroscopy for detection. With an extensive panel of flavonoid pathway enzyme genes overexpressed and repressed in transgenic lines, including modulation of different isoforms of chalcone synthase (CHS), chalcone isomerase (CHI) and dihydroflavonol reductase (DFR), the best positive correlation with SAG abundance was with DFR. Linkage between SAG and flavonoid biosynthesis is likely at the level of regulation as both types of molecules are upregulated during stress. This study is important because it highlights the unforeseen consequences of metabolic engineering with potentially harmful consequences and it illustrates the need for complementary analytical platforms combining LC and GC techniques as many of these important compounds are inaccessible to GC-MS.

LC techniques have been somewhat underutilized and continue to supply access to novel compound classes such as the recently discovered polyamines in potato, kukoamine A (N^1-, N^{12}-bis(dihydrocaffeoyl) spermine), N^1-, N^4-, N^{12}-tris(dihydrocaffeoyl)spermine, and N^1-, N^4-, N^6-tris(dihydrocaffeoyl)spermidine (Parr et al. 2005). A study mentioned earlier from the same group also saw a small increase in SAGs due to 14-3-3 over expression, which similarly points to a regulatory effect (Szopa 2002). Another metabolic engineering project attempted to decrease SAG abundance by antisense repression of key biosynthetic genes (McCue et al. 2005). This attempt ultimately resulted in no significant change in SAG abundance due to unforeseen compensatory effects of feedback regulation.

Sobiecki and coworker's SAG study (Sobiecki et al. 2003) demonstrated the importance of developing robust and sensitive methods for SAG analysis and to further evaluate the natural variation in SAG levels to assess new transgenic potato varieties. Other studies have shown that new potato varieties should be screened for SAGs as at least two classically bred lines ("Lenape" and "Magnum Bonum") have been shown to have elevated SAG levels (Hellenas and Branzell 1997). Zywicki and coworkers compared C_{18} reverse phase (RP) and hydrophilic interaction chromatography (HILIC) based LC-MS/MS (triple quadrupole multiple reaction monitoring) methods for high throughput SAG analysis (Zywicki et al. 2005). The authors found that the C_{18}RP and HILIC LC methods were comparable in initial optimization of chromatographic separation, but that the HILIC medium was less robust over multiple runs and suffered from loss of ideal peak shape after hundreds of analyses while the RP did not. The authors tested the variation in SAG content of a set of six genetically modified varieties including several that were designed to produce inulins (natural sweeteners from *Helianthus tuberosus*), comparing those levels with the natural variation in six classically bred cultivars. All of the varieties had higher levels of α-chaconine than α-solanine or other low abundance SAGs. The levels of SAGs for each of the genetically modified varieties was indistinguishable from their parent "Desirèe" line, but had more than two times as much α-chaconine as did cv. Linda. Other classically bred lines showed α-chaconine levels similar to the GM lines, thus supporting a claim of substantial equivalence with regard to SAGs.

11.4.3 Biotic and Abiotic Stress

Metabolomics or proteomics approaches to the analysis of stress responses can be valuable for diagnosis of early stages of stress, for understanding the molecular basis of stress and to help devise treatments or resistance mechanisms to mitigate the consequences of the stress condition. The review of dicot plant proteomics for the study of plant-environment interactions by Agrawal et al. (2005) may be useful as a basis for comparison, although it contains minimal references to potato. A significant proteomic analysis of potato salt stress was recently published by Aghaei and coworkers (2008). This study uses 2DE differential comparison of the salt sensitive cv. Concord and a relatively salt tolerant cv. Kennebec. In this study, 305 and 322 proteins spots were detected in these two strains, respectively, and 47 were differentially expressed between strains, as determined by 2DE analysis of shoot tissue with and without exposure to 90 mM NaCl. Of the 47 differentially expressed proteins, 39 were identified by a combination of N-terminal sequencing and Cleveland mapping. The study was accompanied by limited metabolic profiling of proline and total soluble sugar concentrations, both of which are known to increase in response to osmotic stress in many plant species. Proline and soluble sugars increased in response to salt treatment as expected, as did several previously characterized stress proteins such as HSP, calreticulin and an osmotin-like protein. Some proteins linked to proline production also increased, while glutamine synthase and RuBisCo were down regulated. An extensive proteomic analysis of wound healing in tubers was conducted by Chaves et al. (2009). In this 2DE-based study, 182 protein spots were examined from gels of protein extracts taken over eight days following wounding with ~ 90% protein identification success. PCA of proteomic data revealed two distinct time phases in the proteome wound response from wounding to day two, and from four to eight days following wounding (Chaves et al. 2009).

The study of biotic stress in potato adds additional requirements for genomics resources with proteomics considerations. Some methods have focused on identification of unknown pathogens such as viruses (Cooper et al. 2003). Others have examined aspects of the biology of various potato pests including aphids (Francis et al. 2006; Nguyen et al. 2009) and pathogenic bacteria (Coulthurst et al. 2005; Mattinen et al. 2007).

Metabolite analyses of biotic stress typically look at potato stress responses such as induction of wound-induced suberization (Yang and Bernards 2007) or the production of stress-induced secondary metabolites, including phytoalexins (Beier 1990). Compounds produced during fungal invasion of potato have been identified and tested for biological activity. The molecular structures of three potato-derived antifungal phytoalexins are shown in Fig. 11-5. These are among 10 sesquiterpines identified by Stoessl and coworkers: rishitin, lubimin, hydroxylubimin, phytuberin, phytuberol, solavetivone, 15-dihydrolubimin, isolubimin, 10-epilubimin and 15-dihydro-10-epilubimin, all of which are produced after invasion by fungal pathogen *Monilina fructicola* (Stoessl et al. 1978). Katsui and coworkers identified two sesquiterpenoid phytoalexins, oxyglutinosone and epioxylubimin, from potato infected by *Phytophthera infestans* (Katsui et al. 1978). In order to develop tests for potato tuber diseases in stored tubers, Lui and coworkers examined volatile compounds released by potato tubers three to six days following inoculation with *Erwinia caratovora* ssp. *caratovora* and ssp. *atroseptica* and *Fusarium sambucinum* (Lui et al. 2005). The headspace above the stored inoculated tubers was sampled and analyzed by GC-MS. Of more than 1,000 detectable compounds, 81 were observed consistently and 58 could be associated with one or more infections. Multiple compounds specific to each infection were identified: acetic acid ethenyl ester was specific for *E. caratovora* ssp. *caratovora;* cyclohexene, diazene and methoxy-(1,1-dimethyl-2-hydroxyethyl)-amine were unique

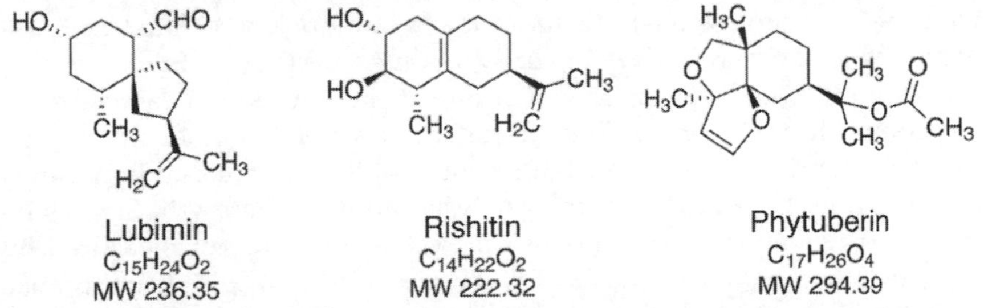

Lubimin
$C_{15}H_{24}O_2$
MW 236.35

Rishitin
$C_{14}H_{22}O_2$
MW 222.32

Phytuberin
$C_{17}H_{26}O_4$
MW 294.39

Figure 11-5 Potato phytoalexins. Molecular structures for three examples of fungus inducible sesquiterpenoid phytoalexins are shown. All three compounds have been shown to possess anti-fungal activity and are presumably produced by the plant as defensive compounds. They can also be used as markers for fungal infections.

to *E. caratovora* ssp. *atroseptica;* and 2,5-norbornadiene and styrene were unique to *Fusarium sambucinum* infections. Large quantities of acetone were present above tubers with both *Erwinia* infections but not in controls or *Fusarium* infections.

11.5 Evaluation of Potato Tuber Nutrition, Safety and Genetic Modifications

Having the capacity to measure large numbers of chemical components in the edible portions of crop plants naturally lends itself to the evaluation of nutrition and food safety concerns. Potato has been the subject of both significant research and discourse in this area. Transcriptional analysis is nearly comprehensive and has been applied to questions regarding food safety with regard to different potato propagation practices (van Dijk et al. 2009). Multiple layers of regulation separate transcription from metabolic activity and metabolite concentrations so that transcriptional analyses do not provide as much resolution for differentiation of populations as does metabolite analysis (Urbanczyk-Wochniak et al. 2005). One of the first obstacles in assessing chemical changes that may relate to food quality and safety is understanding the degree of normal variation for a given crop species. Investigations of normal variation have been carried out for proteins using 2DE-based fingerprinting (Lehesranta et al. 2005) and for metabolites using GC-MS fingerprinting (Roessner et al. 2001; Dobson et al. 2008; Dobson et al. 2010). Normal variation for specific compounds can be used to evaluate the consequences of genetic modifications or changes in agricultural practice by providing a basis for comparison that reflects diversity among current consumer accepted products.

Once natural variation has been evaluated how does one conscientiously make a determination of acceptable similarity? The concept of "substantial equivalence" has been introduced into the scientific literature (possibly from the legal professions) as a tool for assessing risk associated with the introduction of new (mostly genetically modified, GM) crops. Simply stated, a new food should be considered safe if it has the same characteristics and composition as conventional food. The concept is problematic from a scientific standpoint as it is virtually impossible to prove that two things are the same. Depending on the analysis method, there will always be characteristics and components of the system that are not measured by any given technique. We have already seen one example of the phrase "substantial equivalence" used with regard to SAG analysis in GM and conventionally bred potatoes (Zywicki et al. 2005). This example is important as it shows how substantial equivalence is used to describe similarity based on a small profile of exceptionally important toxic compounds. We already know from conventional breeding efforts that these compounds can be dramatically elevated and need to be tested in new varieties. This shows how it is important to perform smaller targeted assays for compounds that are of high potential impact (such as SAGs) and not just provide broad-based profiles of the most abundant metabolites for comparison. While we know what compounds to measure in the case of SAGs, it is hard to imagine what other compounds might need to be assayed to provide adequate support for substantial equivalence, especially in cases of environmental impact in which important interactions have not yet been determined. Several groups have presumably dealt with these issues, demonstrating what they believe is substantial equivalence between various genetically modified and conventional potatoes. A recent study by Khalf and coworkers claims to have demonstrated substantial equivalence for both key proteins and metabolites in transgenic potatoes expressing tomato Kunitz protease inhibitors (Khalf et al. 2010). Catchpole and coworkers performed an extensive study of their inulin-producing GM tubers using GC-MS and LC-MS (and MS/MS) to demonstrate substantial equivalence (Catchpole et al. 2005). A commentary on Catchpole et al. was published by Colquhoun and coworkers (2006). Deferenez and coworkers explore the extent of metabolite variation in 40 GM and control potato lines using NMR and HPLC with a spectrophotometric UV detector (Defernez et al. 2004). Lastly, Badri and coworkers provide a good example of an unexpected protein targeting outcome while attempting to express bovine aprotinin in potato (Badri et al. 2009).

11.6 Conclusions

Progress in potato proteomics has been slowed by the lack of available high quality genome sequence information. Continued efforts to expand genomic sequence coverage can only improve the quality and quantity of potato proteomics information. Metabolomics potato studies have to some extent compensated for the technical barriers to proteomic characterization and have led the field in several areas. The application of metabolomics and metabolic fingerprinting to food safety assessment of genetically modified potatoes has been a particular focus of the research community and presents some interesting unresolved questions.

References

Aghaei K, Ehsanpour AA, Komatsu S (2008) Proteome analysis of potato under salt stress. J Proteome Res 7: 4858–4868.

Agrawal GK, Thelen JJ (2009) A high-resolution two dimensional Gel- and Pro-Q DPS-based proteomics workflow for phosphoprotein identification and quantitative profiling. Meth Mol Biol 527: 3–19, ix.

Agrawal GK, Yonekura M, Iwahashi Y, Iwahashi H, Rakwal R (2005) System, trends and perspectives of proteomics in dicot plants. J Chromatogr, B: Anal Technol Biomed Life Sci 815: 137–145.

Agrawal L, Chakraborty S, Jaiswal DK, Gupta S, Datta A, Chakraborty N (2008) Comparative proteomics of tuber induction, development and maturation reveal the complexity of tuberization process in potato (*Solanum tuberosum* L.). J Proteome Res 7: 3803–3817.

Annesley TM (2003) Ion suppression in mass spectrometry. Clin Chem 49: 1041–1044.

Antignac J-P, de Wasch K, Monteau F, De Brabander H, Andre F, Le Bizec B (2005) The ion suppression phenomenon in liquid chromatography-mass spectrometry and its consequences in the field of residue analysis. Anal Chim Acta 529: 129–136.

Asara JM, Schweitzer MH, Freimark LM, Phillips M, Cantley LC (2007) Protein sequences from mastodon and *Tyrannosaurus rex* revealed by mass spectrometry. Science 316: 280–285.

Badri MA, Rivard D, Coenen K, Michaud D (2009) Unintended molecular interactions in transgenic plants expressing clinically useful proteins: the case of bovine aprotinin traveling the potato leaf cell secretory pathway. Proteomics 9: 746–756.

Balmer Y, Vensel WH, Tanaka CK, Hurkman WJ, Gelhaye E, Rouhier N, Jacquot J-P, Manieri W, Schuermann P, Droux M, Buchanan BB (2004) Thioredoxin links redox to the regulation of fundamental processes of plant mitochondria. Proc Natl Acad Sci USA 101: 2642–2647.

Barceloux DG (2009) Potatoes, tomatoes, and solanine toxicity (*Solanum tuberosum* L., *Solanum lycopersicum* L.). Dis Mon 55: 391–402.

Barel G, Ginzberg I (2008) Potato skin proteome is enriched with plant defence components. J Exp Bot 59: 3347–3357.

Bauw G, Nielsen HV, Emmersen J, Nielsen KL, Joergensen M, Welinder KG (2006) Patatins, Kunitz protease inhibitors and other major proteins in tuber of potato cv. Kuras. FEBS J 273: 3569–3584.

Beckmann M, Enot DP, Overy DP, Draper J (2007) Representation, comparison, and interpretation of metabolome fingerprint data for total composition analysis and quality trait investigation in potato cultivars. J Agri Food Chem 55: 3444–3451.

Beier RC (1990) Natural pesticides and bioactive components in foods. Rev Environ Contam Toxicol 113: 47–137.

Benschop JJ, Mohammed S, O'Flaherty M, Heck AJ, Slijper M, Menke FL (2007) Quantitative phosphoproteomics of early elicitor signaling in Arabidopsis. Mol Cell Proteom 6: 1198–1214.

Beynon RJ, Pratt JM (2005) Metabolic labeling of proteins for proteomics. Mol Cell Proteom 4: 857–872.

Blonder J, Chan KC, Issaq HJ, Veenstra TD (2006) Identification of membrane proteins from mammalian cell/tissue using methanol-facilitated solubilization and tryptic digestion coupled with 2D-LC-MS/MS. Nat Protoc 1: 2784–2790.

Bykova NV, Egsgaard H, Moller IM (2003a) Identification of 14 new phosphoproteins involved in important plant mitochondrial processes. FEBS Lett 540: 141–146.

Bykova NV, Stensballe A, Egsgaard H, Jensen ON, Moller IM (2003b) Phosphorylation of formate dehydrogenase in potato tuber mitochondria. J Biol Chem 278: 26021–26030.

Carpentier AS, Riva A, Tisseur P, Didier G, Henaut A (2004) The operons, a criterion to compare the reliability of transcriptome analysis tools: ICA is more reliable than ANOVA, PLS and PCA. Comput Biol Chem 28: 3–10.

Catchpole GS, Beckmann M, Enot DP, Mondhe M, Zywicki B, Taylor J, Hardy N, Smith A, King RD, Kell DB, Fiehn O, Draper J (2005) Hierarchical metabolomics demonstrates substantial compositional similarity between genetically modified and conventional potato crops. Proc Natl Acad Sci USA 102: 14458–14462.

Chaves I, Pinheiro C, Paiva JA, Planchon S, Sergeant K, Renaut J, Graça JA, Costa G, Coelho AV, Ricardo CP (2009) Proteomic evaluation of wound-healing processes in potato (*Solanum tuberosum* L.) tuber tissue. Proteomics 9: 4154–75.

Colquhoun IJ, Le Gall G, Elliott KA, Mellon FA, Michael AJ (2006) Shall I compare thee to a GM potato? Trends Genet 22: 525–528.

Cooper B, Eckert D, Andon NL, Yates JR, Haynes PA (2003) Investigative proteomics: identification of an unknown plant virus from infected plants using mass spectrometry. J Am Soc Mass Spectrom 14: 736–741.

Coulthurst SJ, Lilley KS, Salmond GPC (2005) Genetic and proteomic analysis of the role of luxS in the enteric phytopathogen, *Erwinia carotovora*. Mol Plant Pathol 7: 31–45.

Davies HV (2007) Metabolomics: applications in functional biodiversity analysis in potato. Acta Hort 745: 471–483.

Defernez M, Colquhoun IJ (2003) Factors affecting the robustness of metabolite fingerprinting using 1H NMR spectra. Phytochemistry (Elsevier) 62: 1009–1017.

Defernez M, Gunning YM, Parr AJ, Shepherd LVT, Davies HV, Colquhoun IJ (2004) NMR and HPLC-UV profiling of potatoes with genetic modifications to metabolic pathways. J Agri Food Chem 52: 6075–6085.

Delaplace P, van der Wal F, Dierick J-F, Cordewener JHG, Fauconnier M-L, du Jardin P, America AHP (2006) Potato tuber proteomics: comparison of two complementary extraction methods designed for two-dimensional electrophoresis of acidic proteins. Proteomics 6: 6494–6497.

Delaplace P, Fauconnier M-L, Sergeant K, Dierick J-F, Oufir M, van der Wal F, America AHP, Renaut J, Hausman J-F, du Jardin P (2009) Potato (*Solanum tuberosum* L.) tuber ageing induces changes in the proteome and antioxidants associated with the sprouting pattern. J Exp Bot 60: 1273–1288.

Dettmer K, Aronov PA, Hammock BD (2007) Mass spectrometry-based metabolomics. Mass Spectrom Rev 26: 51–78.

Dixon RA, Gang DR, Charlton AJ, Fiehn O, Kuiper HA, Reynolds TL, Tjeerdema RS, Jeffery EH, German JB, Ridley WP, Seiber JN (2006) Applications of metabolomics in agriculture. J Agri Food Chem 54: 8984–8994.

Dobson G, Shepherd T, Verrall SR, Conner S, McNicol JW, Ramsay G, Shepherd LVT, Davies HV, Stewart D (2008) Phytochemical diversity in tubers of potato cultivars and landraces using a GC-MS metabolomics approach. J Agri Food Chem 56: 10280–10291.

Dobson G, Shepherd T, Verrall SR, Griffiths WD, Ramsay G, McNicol JW, Davies HV, Stewart D (2010) A metabolomics study of cultivated potato (*Solanum tuberosum*) groups Andigena, Phureja, Stenotomum, and Tuberosum using gas chromatography-mass spectrometry. J Agri Food Chem 58: 1214–1223.

Duguay JL, Li X-Q, Regan S (2006) The potential for genomics in potato improvement. J Therm Sci Technol (Tokyo, Jpn) 1.

Enot DP, Beckmann M, Overy D, Draper J (2006) Predicting interpretability of metabolome models based on behavior, putative identity, and biological relevance of explanatory signals. Proc Natl Acad Sci USA 103: 14865–14870.

Eubel H, Heinemeyer J, Braun H-P (2004) Identification and characterization of respirasomes in potato mitochondria. Plant Physiol 134: 1450–1459.

Falick AM, Kowalak JA, Lane WS, Phinney BS, Turck CW, Weintraub ST, West KA, Neubert TA (2008) ABRF-PRG05: de novo peptide sequence determination. J Biomol Technol 19: 251–257.

Fenn JB, Mann M, Meng CK, Wong SF, Whitehouse CM (1989) Electrospray ionization for mass spectrometry of large biomolecules. Science 246: 64–71.

Fiehn O (2001) Combining genomics, metabolome analysis, and biochemical modelling to understand metabolic networks. Comp Funct Genom 2: 155–168.

Fiehn O (2008) Extending the breadth of metabolite profiling by gas chromatography coupled to mass spectrometry. Trends Anal Chem 27: 261–269.

Flengsrud R (2008) Protein extraction from green plant tissue. Meth Mol Biol 425: 149–152.

Francis F, Gerkens P, Harmel N, Mazzucchelli G, De Pauw E, Haubruge E (2006) Proteomics in *Myzus persicae*: effect of aphid host plant switch. Insect Biochem Mol Biol 36: 219–227.

Friedman M (2006) Potato glycoalkaloids and metabolites: roles in the plant and in the diet. J Agri Food Chem 54: 8655–8681.

Glinski M, Weckwerth W (2006) The role of mass spectrometry in plant systems biology. Mass Spectrom Rev 25: 173–214.

Graham RL, Sharma MK, Ternan NG, Weatherly DB, Tarleton RL, McMullan G (2007) A semi-quantitative GeLC-MS analysis of temporal proteome expression in the emerging nosocomial pathogen *Ochrobactrum anthropi*. Genome Biol 8: R110.

Gygi SP, Rochon Y, Franza BR, Aebersold R (1999) Correlation between protein and mRNA abundance in yeast. Mol Cell Biol 19: 1720–1730.

Han X, Aslanian A, Yates JR, 3rd (2008) Mass spectrometry for proteomics. Curr Opin Chem Biol 12: 483–490.

Hegeman AD, Schulte CF, Cui Q, Lewis IA, Huttlin EL, Eghbalnia H, Harms AC, Ulrich EL, Markley JL, Sussman MR (2007) Stable isotope assisted assignment of elemental compositions for metabolomics. Anal Chem 79: 6912–6921.

Hellenas KE, Branzell C (1997) Liquid chromatographic determination of the glycoalkaloids alpha-solanine and alpha-chaconine in potato tubers: NMKL Interlaboratory Study. Nordic Committee on Food Analysis. J AOAC Int 80: 549–554.

Henzel WJ, Billeci TM, Stults JT, Wong SC, Grimley C, Watanabe C (1993) Identifying proteins from two-dimensional gels by molecular mass searching of peptide fragments in protein sequence databases. Proc Natl Acad Sci USA 90: 5011–5015.

Hoehenwarter W, van Dongen JT, Wienkoop S, Steinfath M, Hummel J, Erban A, Sulpice R, Regierer B, Kopka J, Geigenberger P, Weckwerth W (2008) A rapid approach for phenotype-screening and database independent detection of cSNP/protein polymorphism using mass accuracy precursor alignment. Proteomics 8: 4214–4225.

Hollywood K, Brison DR, Goodacre R (2006) Metabolomics: current technologies and future trends. Proteomics 6: 4716–4723.

Huttlin EL, Hegeman AD, Harms AC, Sussman MR (2007) Prediction of error associated with false-positive rate determination for peptide identification in large-scale proteomics experiments using a combined reverse and forward peptide sequence database strategy. J Proteome Res 6: 392–398.

Ippel JH, Pouvreau L, Kroef T, Gruppen H, Versteeg G, van den Putten P, Struik PC, van Mierlo CPM (2004) In vivo uniform 15N-isotope labeling of plants: Using the greenhouse for structural proteomics. Proteomics 4: 226–234.

Jacob AM, Turck CW (2008) Detection of post-translational modifications by fluorescent staining of two-dimensional gels. Meth Mol Biol 446: 21–32.

James P, Quadroni M, Carafoli E, Gonnet G (1993) Protein identification by mass profile fingerprinting. Biochem Biophys Res Comm 195: 58–64.

Jensen PH, Juhler RK, Nielsen NJ, Hansen TH, Strobel BW, Jacobsen OS, Nielsen J, Hansen HCB (2008) Potato glycoalkaloids in soil-optimizing liquid chromatography-time-of-flight mass spectrometry for quantitative studies. J Chromatogr, A 1182: 65–71.

Junker BH, Wuttke R, Nunes-Nesi A, Steinhauser D, Schauer N, Buessis D, Willmitzer L, Fernie AR (2006) Enhancing vacuolar sucrose cleavage within the developing potato tuber has only minor effects on metabolism. Plant Cell Physiol 47: 277–289.

Karas M, Hillenkamp F (1988) Laser desorption ionization of proteins with molecular masses exceeding 10,000 daltons. Anal Chem 60: 2299–2301.

Katsui N, Yagihashi F, Murai A, Masamune T (1978) Studies on the phytoalexins. XVIII. Structure of oxyglutinosone and epioxylubimin, stress metabolites from diseased potato tubers. Chem Lett 1205–1206.

Kell DB (2006) Systems biology, metabolic modelling and metabolomics in drug discovery and development. Drug Discov Today 11: 1085–1092.

Khalf M, Goulet C, Vorster J, Brunelle F, Anguenot R, Fliss I, Michaud D (2010) Tubers from potato lines expressing a tomato Kunitz protease inhibitor are substantially equivalent to parental and transgenic controls. Plant Biotechnol J 8: 155–69.

Kind T, Fiehn O (2007) Seven Golden Rules for heuristic filtering of molecular formulas obtained by accurate mass spectrometry. BMC Bioinformat 8: 105.

Kumar C, Mann M (2009) Bioinformatics analysis of mass spectrometry-based proteomics data sets. FEBS Lett 583: 1703–1712.

Lehesranta SJ, Davies HV, Shepherd LVT, Nunan N, McNicol JW, Auriola S, Koistinen KM, Suomalainen S, Kokko HI, Kaerenlampi SO (2005) Comparison of tuber proteomes of potato varieties, landraces, and genetically modified lines. Plant Physiol 138: 1690–1699.

Lehesranta SJ, Davies HV, Shepherd LVT, Koistinen KM, Massat N, Nunan N, McNicol JW, Karenlampi SO (2006) Proteomic analysis of the potato tuber life cycle. Proteomics 6: 6042–6052.

Lindon JC, Holmes E, Bollard ME, Stanley EG, Nicholson JK (2004) Metabonomics technologies and their applications in physiological monitoring, drug safety assessment and disease diagnosis. Biomarkers 9: 1–31.

Lu A, Wisniewski JR, Mann M (2009) Comparative proteomic profiling of membrane proteins in rat cerebellum, spinal cord, and sciatic nerve. J Proteome Res 8: 2418–2425.

Lui LH, Vikram A, Abu-Nada Y, Kushalappa AC, Raghavan GSV, Al-Mughrabi K (2005) Volatile metabolic profiling for discrimination of potato tubers inoculated with dry and soft rot pathogens. Am J Potato Res 82: 1–8.

Maga JA (1980) Potato glycoalkaloids. Crit Rev Food Sci Nutr 12: 371–405.

Makarov A, Denisov E, Lange O, Horning S (2006) Dynamic range of mass accuracy in LTQ Orbitrap hybrid mass spectrometer. J Am Soc Mass Spectrom 17: 977–982.

Mann M, Hojrup P, Roepstorff P (1993) Use of mass spectrometric molecular weight information to identify proteins in sequence databases. Biol Mass Spectrom 22: 338–345.

Mattinen L, Nissinen R, Riipi T, Kalkkinen N, Pirhonen M (2007) Host-extract induced changes in the secretome of the plant pathogenic bacterium *Pectobacterium atrosepticum*. Proteomics 7: 3527–3537.

McCue KF, Shepherd LVT, Rockhold DR, Allen PV, Davies HV, Belknap WR (2005) Modification of potato alkaloids: A lesson in applied metabolomics. Abstr of Papers, 229th ACS National Meeting, San Diego, CA, USA, March 13–17, 2005 AGRO-096.

Mockaitis K, Estelle M (2008) Auxin receptors and plant development: a new signaling paradigm. Annu Rev Cell Dev Biol 24: 55–80.

Nguyen TTA, Michaud D, Cloutier C (2009) A proteomic analysis of the aphid *Macrosiphum euphorbiae* under heat and radiation stress. Insect Biochem Mol Biol 39: 20–30.

Nuhse TS, Bottrill AR, Jones AM, Peck SC (2007) Quantitative phosphoproteomic analysis of plasma membrane proteins reveals regulatory mechanisms of plant innate immune responses. Plant J 51: 931–940.

Pappin DJ, Hojrup P, Bleasby AJ (1993) Rapid identification of proteins by peptide-mass fingerprinting. Curr Biol 3: 327–332.

Parr AJ, Mellon FA, Colquhoun IJ, Davies HV (2005) Dihydrocaffeoyl polyamines (kukoamine and allies) in potato (*Solanum tuberosum*) tubers detected during metabolite profiling. J Agri Food Chem 53: 5461–5466.

Patterson SD, Aebersold R (1995) Mass spectrometric approaches for the identification of gel-separated proteins. Electrophoresis 16: 1791–1814.

Patterson SD, Aebersold RH (2003) Proteomics: the first decade and beyond. Nat Genet 33(Suppl): 311–323.

Pauling L, Robinson AB, Teranishi R, Cary P (1971) Quantitative analysis of urine vapor and breath by gas-liquid partition chromatography. Proc Natl Acad Sci USA 68: 2374–2376.

Pevzner PA, Kim S, Ng J (2008) Comment on "Protein sequences from mastodon and *Tyrannosaurus rex* revealed by mass spectrometry". Science 321: 1040.

Rabilloud T, Chevallet M, Luche S, Lelong C (2008) Fully denaturing two-dimensional electrophoresis of membrane proteins: a critical update. Proteomics 8: 3965–3973.

Riederer BM (2008) Non-covalent and covalent protein labeling in two-dimensional gel electrophoresis. J Proteom 71: 231–244.

Rittenberg D, Foster G (1940) A new procedure for quantitative analysis by isotope dilution, with application to the determination of amino acids and fatty acids. J Biol Chem 133: 737–744.

Riva A, Carpentier AS, Torresani B, Henaut A (2005) Comments on selected fundamental aspects of microarray analysis. Comput Biol Chem 29: 319–336.

Roessner U, Wagner C, Kopka J, Trethewey RN, Willmitzer L (2000) Simultaneous analysis of metabolites in potato tuber by gas chromatography-mass spectrometry. Plant J 23: 131–142.

Roessner U, Willmitzer L, Fernie AR (2001) High-resolution metabolic phenotyping of genetically and environmentally diverse potato tuber systems. Identification of phenocopies. Plant Physiol 127: 749–764.

Roessner-Tunali U, Liu J, Leisse A, Balbo I, Perez-Melis A, Willmitzer L, Fernie AR (2004) Kinetics of labelling of organic and amino acids in potato tubers by gas chromatography-mass spectrometry following incubation in 13C labelled isotopes. Plant J 39: 668–679.

Ross PL, Huang YN, Marchese JN, Williamson B, Parker K, Hattan S, Khainovski N, Pillai S, Dey S, Daniels S, Purkayastha S, Juhasz P, Martin S, Bartlet-Jones M, He F, Jacobson A, Pappin DJ (2004) Multiplexed protein quantitation in *Saccharomyces cerevisiae* using amine-reactive isobaric tagging reagents. Mol Cell Proteom 3: 1154–1169.

Stensballe A, Hald S, Bauw G, Blennow A, Welinder KG (2008) The amyloplast proteome of potato tuber. FEBS J 275: 1723–1741.

Stobiecki M, Matysiak-Kata I, Franski R, Skala J, Szopa J (2003) Monitoring changes in anthocyanin and steroid alkaloid glycoside content in lines of transgenic potato plants using liquid chromatography/mass spectrometry. Phytochemistry (Elsevier) 62: 959–969.

Stoessl A, Stothers JB, Ward EWB (1978) Carbon-13 NMR studies, part 76. Post infectional inhibitors from plants, part XXIX. Biosynthetic studies of stress metabolites from potatoes: incorporation of sodium acetate-13C2 into 10 sesquiterpenes. Can J Chem 56: 645–653.

Sumner LW, Mendes P, Dixon RA (2003) Plant metabolomics: large-scale phytochemistry in the functional genomics era. Phytochemistry 62: 817–836.

Swaney DL, Wenger CD, Thomson JA, Coon JJ (2009) Human embryonic stem cell phosphoproteome revealed by electron transfer dissociation tandem mass spectrometry. Proc Natl Acad Sci USA 106: 995–1000.

Szopa J (2002) Transgenic 14-3-3 isoforms in plants: the metabolite profiling of repressed 14-3-3 protein synthesis in transgenic potato plants. Biochem Soc Trans 30: 405–410.

Szopa J, Wrobel M, Matysiak-Kata I, Swiedrych A (2001) The metabolic profile of the 14-3-3 repressed transgenic potato tubers. Plant Sci (Shannon, Irel) 161: 1075–1082.

Tanaka K, Ido, Y, Akita, S, Yoshida, Y, Yoshida, T (1987) Detection of high mass molecules by laser desorption time-of-flight mass spectrometry. In: HX-t, LMatsuda (ed) 2nd Japan-China Joint Symp Mass Spectrom Osaka, Japan, pp 185–188 .

Tao WA, Aebersold R (2003) Advances in quantitative proteomics via stable isotope tagging and mass spectrometry. Curr Opin Biotechnol 14: 110–118.

Tolgyessy J, Braun T, Kyrs M (1972) Isotope Dilution Analysis. 1st edn. Pergamon Press, Oxford, UK.

Turck CW, Falick AM, Kowalak JA, Lane WS, Lilley KS, Phinney BS, Weintraub ST, Witkowska HE, Yates NA (2007) The Association of Biomolecular Resource Facilities Proteomics Research Group 2006 study: relative protein quantitation. Mol Cell Proteom 6: 1291–1298.

Unlu M, Morgan ME, Minden JS (1997) Difference gel electrophoresis: a single gel method for detecting changes in protein extracts. Electrophoresis 18: 2071–2077.

Urbanczyk-Wochniak E, Luedemann A, Kopka J, Selbig J, Roessner-Tunali U, Willmitzer L, Fernie AR (2003) Parallel analysis of transcript and metabolic profiles: a new approach in systems biology. EMBO Rep 4: 989–993.

Urbanczyk-Wochniak E, Baxter C, Kolbe A, Kopka J, Sweetlove LJ, Fernie AR (2005) Profiling of diurnal patterns of metabolite and transcript abundance in potato (*Solanum tuberosum*) leaves. Planta 221: 891–903.

van Dijk JP, Cankar K, Scheffer SJ, Beenen HG, Shepherd LV, Stewart D, Davies HV, Wilkockson SJ, Leifert C, Gruden K, Kok EJ (2009) Transcriptome analysis of potato tubers—effects of different agricultural practices. J Agri Food Chem 57: 1612–1623.

Van Wijk KJ (2001) Challenges and prospects of plant proteomics. Plant physiol 126: 501–508.

Wang G, Wu WW, Zhang Z, Masilamani S, Shen RF (2008) Decoy methods for assessing false positives and false discovery rates in shotgun proteomics. Anal Chem 81(1): 146–159.

Weckwerth W, Loureiro ME, Wenzel K, Fiehn O (2004) Differential metabolic networks unravel the effects of silent plant phenotypes. Proc Natl Acad Sci USA 101: 7809–7814.

Wolters DA, Washburn MP, Yates JR, 3rd (2001) An automated multidimensional protein identification technology for shotgun proteomics. Anal Chem 73: 5683–5690.

Wu L, Mashego MR, van Dam JC, Proell AM, Vinke JL, Ras C, van Winden WA, van Gulik WM, Heijnen JJ (2005) Quantitative analysis of the microbial metabolome by isotope dilution mass spectrometry using uniformly 13C-labeled cell extracts as internal standards. Anal Biochem 336: 164–171.

Yang W-L, Bernards MA (2007) Metabolite profiling of potato (*Solanum tuberosum* L.) tubers during wound-induced suberization. Metabolomics 3: 147–159.

Yang Y, Thannhauser TW, Li L, Zhang S (2007) Development of an integrated approach for evaluation of 2-D gel image analysis: impact of multiple proteins in single spots on comparative proteomics in conventional 2-D gel/MALDI workflow. Electrophoresis 28: 2080–2094.

Yates JR, 3rd, Speicher S, Griffin PR, Hunkapiller T (1993) Peptide mass maps: a highly informative approach to protein identification. Anal Biochem 214: 397–408.

Yates JR, 3rd, Eng JK, McCormack AL, Schieltz D (1995) Method to correlate tandem mass spectra of modified peptides to amino acid sequences in the protein database. Anal Chem 67: 1426–1436.

Zywicki B, Catchpole G, Draper J, Fiehn O (2005) Comparison of rapid liquid chromatography-electrospray ionization-tandem mass spectrometry methods for determination of glycoalkaloids in transgenic field-grown potatoes. Anal Biochem 336: 178–186.

12

Future Challenges and Prospects

Toni Wendt and *Ewen Mullins**

ABSTRACT

As we progress through the technology era, the application of genomics to potato research has revolutionized our ability to respond to those a/biotic stresses that continue to threaten the commercial viability of the crop. The single greatest challenge to potato's dominance as a stable food crop is the prospect of a changing climate and with it a change in disease/pest occurrences. The issue is further compounded with increasing oil prices and legislative hurdles. Encouragingly, output from genomics-based studies has demonstrated the potential to address several of these issues especially when combined with dynamic management regimes. Conversely, one of the more intractable challenges remains non-scientific in design but targets the applied output of many genomics programs. Legislative impediments (within the European Union) and societal concern (within and outside the EU) against transgenic varieties impedes the future development of potato. So while we can identify and develop novel genetic resources to address many of the macro-challenges facing the global potato industry, to succeed going forward we must complement our genomics-based research with an expansion of the public's knowledge and understanding of the challenges and possible solutions that potato genomics can provide in support of the potato.

Keywords: Challenges, future, GM, prospects, transformation

12.1 Introduction

Once "humble", now a "hero"; the theme of the recent World Potato Congress was resolute (WPC Inc 2009). As the world's population advances to 9 billion by 2050 (UN Secretariat 2007), the potato is expected to play a

Plant Biotechnology Unit, Teagasc Crops Research Centre, Oak Park, Carlow, Ireland.
*Corresponding author: *ewen.mullins@teagasc.ie*

pivotal role in underpinning global food reserves, with particular relevance to less developed regions whose population is projected to rise by 32% (5.4 billion to 7.9 billion) over the next 40 years (WPC Inc 2009). Promoted by the UN's Food and Agriculture Organization as the "food of the future" through 2008 (The International Year of the Potato), the potato is already an integral component of the global food supply with production reaching a record 325 million tons in 2007 (FAO 2008). In contrast to the volatility of global cereal prices, the potato is not a globally traded commodity, meaning potato prices are typically determined at a local/regional level, making the potato a recommended food security crop that can help low-income farmers and vulnerable consumers ride out extreme events in world food supply and demand (FAO 2008). But while the adaptability of the potato is its greatest asset (from China to India to Europe to the Americas) the crop faces significant challenges into the near future if it is to realize the aspirations of so many.

12.2 Challenges

Climate change and legislative constraints combined with depleting petroleum reserves (i.e., "peak oil") represent the most significant challenges to the sustainable growth of global potato production. Publication of the 4th Assessment Report of the Intergovernmental Panel on Climate Change has indicated the likelihood of a significant change to regional climates due to global warming (IPCC 2007). As to be expected the severity of climate fluctuations will be latitude dependent, with global warming likely to lead to changes in the time of planting and a shift of the location of potato production at higher latitudes. In many of these regions, changes in potato yield are likely to be relatively small, and sometimes positive. Indeed, year to year variability in regional climates will require crop management systems to become more flexible to ensure maximum yields. At lower latitudes, shifting planting time or location is less feasible and in these regions global warming could have a strong negative effect on potato production (Hijmans 2003).

It can be expected that shifting climates will inevitably lead to the migration of diseases/pests to new regions and/or the evolution of new disease strains. Exerting effective control over *Phytophthora infestans* has always been a challenge for potato growers but more so in the last 10 years, which has led to increased fungicide use (Hinds 2008). At a European level, the issue is compounded with the emergence of a new metalaxyl resistant A2 isolate ("Blue 13"), which has commanded a dominant position in European *Phytophthora* populations. The rapid prevalence of Blue 13 is of serious concern due to its ability to over winter and infect early, its aggressiveness at low temperatures and its shorter latent period (Lees et al. 2008). While the Hungarian-derived *S. tuberosum* var. Sárpo Mira has demonstrated foliage resistance to Blue 13 in the field (White and Shaw 2008), the absence of resistance in European main crop varieties (e.g., *S. tuberosum* var. Bintje, Desiree, Agria, Spunta, Maris Piper) is likely to lead growers to increase their rate of fungicide application per season.

Aside from increasing the potential for pathogen evolution, increasing the number of fungicide applications erodes the grower's profit margin through additional fuel, labor and chemical costs. Indeed, surveyed potato growers in the US have stated fuel costs as the greatest challenge to maintaining a competitive enterprise (Grusenmeyer 2008). Although, the price of oil has decreased from the heights of mid-2008, the common consensus among economists is that the long-term trend for oil prices is up. As with all sectors of society, such a scenario will challenge the profitability of existing potato management regimes, which typically require fungicide applications every 7–10 days through the growing season.

Introducing combinations of resistance (*R*) genes conferring broad-spectrum resistance to *P. infestans* from wild *Solanum* species into cultivated potatoes through genetic modification (comprehensively detailed in Chapter 8) demonstrates the potential of delivering potato lines with durable field resistance (Song et al. 2003; van der Vossen et al. 2003, 2005; Liu and Halterman 2006; Park et al. 2009). Unfortunately, owing to the present patent landscape the development of modified potato (and all crops) for commercialization is impractical for those institutions/agencies that are not partnered with the patent holders, who according to a recent review of plant transformation procedures have placed *"a stranglehold on transformation technology"* (Nottenburg and Rodriguez 2007). This only serves to impede the improvement of the potato by breeders and scientists who are tasked with responding to regional pressures as they arise.

But even for those few that are in the position of being able to develop genetically modified (GM) potato lines for commercialization, the greatest challenge at present is non-scientific in nature. A case in point has been the length of time it took for the Amflora™ starch potato to be processed through the EU authorization pipeline. First notified to Swedish authorities in 2003, Amflora was developed by BASF Plant Science for industrial starch production. Yet, while the European Food Safety Authority repeatedly concluded that the GM line posed no risk to human, animal health and/or the environment (EFSA 2006), due to the political obstacles that exist for GM technology cultivation was not authorized until March 2nd, 2010. For the future development of potato, the consequence of such inaction is real as the same company has developed a GM blight resistant line of *S. tuberosum* var. Agria (termed Fortuna™), which is claimed to be resistant to *P. infestans* Blue 13 (Farmers Guardian 2009). While the durability of the material against continuously evolving *P. infestans* populations is something that warrants independent study, the initial economic benefit of growing such a crop would be significant as would be its impact on reducing pesticide loads on the environment.

Of course, the legislative/political obstacles challenging adoption of GM technology across the EU are in contrast with the science-based approach adopted by non-EU countries. Regrettably, this could have a greater impact in the long term, particularly for those nations who wish to export seed into Europe. While potato imports into Europe are traditionally small (due to plant health certification requirements) the current situation is likely to result in the development of "asynchronous authorizations"; where the EU authorization process for GM lines lags behind the equivalent approval mechanisms of non-EU exporting countries. As such, they are unable to export into the EU market place because of the EU-wide policy of zero tolerance for those GM lines not already approved by the EU. As demonstrated in the food and feed market, the issue of asynchronous authorizations can have significant repercussions on global trade (Stein and Rodríguez-Cerezo 2009).

Undeniably, there is a growing consensus of the necessity to farm in a more sustainable manner so as to minimize the environmental impact of traditional crop regimes and negate the potential contamination of food stocks with pesticide residues. But new European legislation (to replace Pesticide Directive 91/414/EEC) governing the application of pesticides has now come into force and presents a considerable challenge to European potato growers in the future. In an attempt to encourage the pesticide industry to develop more benign chemistries, the imposition of this regulation represents a fundamental change from science-based risk assessment, to hazard-based regulatory cut-off criteria. This is in contrast with the risk assessment approach endorsed by the World Trade Organization. At a time when growers seek viable alternatives to underpin their resistance management strategies, the implementation of this regulation in full will only serve to further erode management options for disease control (Bielza et al. 2008).

12.3 Prospects

So taking into consideration the above listed issues, what is the outlook for the potato and its associated industries? In short the potato is well positioned to directly/indirectly address the majority of the listed issues. The support for such optimism is described in the preceding chapters. With the integration of genomics-derived output into breeding programs traditional bottlenecks in conventional breeding are being removed, allowing breeders to accurately select for specific traits earlier in their programs (see Chapter 3). Undeniably, the application of genomics to potato research has revolutionized the development of potato; equipping it with novel traits to counter a/biotic stresses and post-harvest issues, establishing it as a principle crop for specific non-food purposes and fortunately providing it with the tools to counter future challenges. However, advancements in potato research must be matched equally with an informed understanding by the non-scientific community of the potential of each achievement. This is critical if the full potential of genomics-derived research is to materialize and we are to see future challenges addressed.

12.3.1 In Response to Climate Change

For global potato production, the possible impacts of climate change have been assessed (Hijmans 2003). Assuming that the average global temperature will increase by between 1.4°C and 5.8°C over the period 1990 to 2100 (Houghton et al. 2001), global potato yield could decrease by 18% to 32% without adaptation and by 9% to 18% with adaptation (Hijmans 2003). In simulations reported by Tubiello and Ewert (2002) the role of adaptation proved critical in reducing (by up to 50%) projected losses in potato production in the Pacific Northwest. Using the POTATOS and NPOTATO models, Wolf (2002) applied climate change scenarios to European potato production, concluding that moderate yield increases could be returned in northern Europe (Finland and Denmark). Similar output could be attained for southern Europe through the cultivation of an earlier crop variety coupled with advanced planting dates (Wolf 2002). Here in Ireland, such an approach is already being adopted by some producers who have shifted to sowing earlier varieties to offset the higher than average rainfall through late summer and autumn, which has in recent times seen harvesting postponed until the following spring.

On the contrary, for water deficit conditions the industry cannot rely solely on management-based solutions. The potential of applying transcriptional profiling to field grown Andean potato germplasm to identify drought tolerance traits has been demonstrated with the protein phosphatase 2C gene identified as being positively associated with yield maintenance under drought (Schafleitner et al. 2007). Of interest was the fact that the Andigena landrace "Sullu" was reported to possess a lower stress load during drought as a result of lower levels of ROS accumulation, greater mitochondrial activity and active chloroplast defenses, as opposed to the protective roles of compatible solutes such as hexoses and complex sugars (Vasquez-Robinet et al. 2008). In contrast and via a transgenic approach, Stiller et al. (2008) have demonstrated the potential of expressing the *trehalose-6-phosphate synthase* (*TPS1*) gene from *Saccharomyces cerevisiae* in potato. Transcription of the *TPS1* gene was found to maintain a maximal CO_2 assimilation rate for six to nine days during drought treatment as opposed to decreasing rapidly after three days in wild type plants (Stiller et al. 2008). Although, the overproduction of TPS in transgenic plants can result in growth aberrations, Karim et al. (2007) have described strategies to circumvent this issue by altering promoter functionality (Karim et al. 2007).

12.3.2 In Response to Increasing Management Costs

The biggest expense for the majority of potato growers is the necessity for continuous fungicide applications to control *P. infestans*, which causes annual losses to the European potato sector alone of over M€900 (Haverkort et al. 2007). The introduction of novel sources of resistance that remove this requirement will clearly impact positively on a grower's profit margins via reduced traffic through the potato crop. Factoring in the anticipated reduction in fuel, labor, machinery depreciation and fungicide costs savings of up to 200€/ha/year can be made (dependent on regional circumstances) through the cultivation of a durable blight resistant potato crop (Flannery et al. 2005).

Pyramiding *R* genes conferring broad-spectrum resistance from wild *Solanum* species into cultivated potatoes is a worthwhile strategy to achieve durable resistance (see Chapter 8). As potato lines containing the first suite of such *R* genes are readied for commercialization towards 2015, the discovery and functional profiling of additional broad-spectrum *R* genes continues through map-based cloning (Chapter 8) and "effector genomics". The latter of which has led to the rapid cloning of *S. stoloniferum Rpi-sto1* and *S. papita Rpi-pta1*, which are functionally equivalent to *S. bulbocastanum Rpi-blb1* (Van der Vossen et al. 2003) and promises to accelerate the engineering of late blight resistance in potato varieties (Vleeshouwers et al. 2008).

Similarly, *Bt* potato was generated in the late 1990s in response to significant insect damage on the US potato crop. Predicted to cut pesticides by up to 80% (Monsanto 1999), its introduction was anticipated by growers who had seen pesticide efficacy decrease in line with evolving insect resistance. In spite of a clear need for the crop and its potential to reduce management costs, the product was discontinued after initial uptake due to consumer resistance. More recently, its effectiveness against eastern European populations of the Colorado Beetle (*Leptinotarsa decemlineata*) was reported (Kalushkov and Batchvarova 2005). The application of genomics-based techniques against *P. infestans* and *L. decemlineata* in this manner demonstrates the potential of the discipline to deliver, in a relatively short period of time, an applied output for growers that can minimize their management costs. This highlights the prospects for the potato industry to tackle a major challenge, assuming consumer antipathy to the technology can be addressed.

12.3.3 Adopting Crop Modification Technologies

In the last decade transformation technology has advanced to a point where few technical limitations remain. An extensive range of monocotyledonous and dicotyledonous crops has been successfully transformed and reports of the stable transformation of new species are described regularly. Although the issue of genotype dependency remains with large variations in transformation efficiencies reported between potato varieties (0.3%–25%) (Herres et al. 2002; Petti 2007), the genetic transformation of potato is now common practice with ATMT being the principle technique adopted (Mullins et al. 2006). As described previously, a further limitation is the degree of patent complexity covering ATMT, with all aspects of *Agrobacterium*-mediated transformation licensed, from the treatment of the target tissue before the actual transformation to specific inoculation times of the plant material with the bacteria (Nottenburg and Rodriguez 2007).

In an attempt to bypass the challenge of having to negotiate with patent holders, Broothaerts et al. (2005) described the functionality and availability of their Transbacter technology through the BiOS licensing system of CAMBIA (CAMBIA 2009). They showed that the non-*Agrobacterium* strains from the Rhizobia family (*Sinorhizobium meliloti*, *Rhizobium* spp. NGR234, *Mesorhizobium loti*) have the ability to transform tobacco (*Nicotiana tabaccum*), rice (*Oryza sativa*) and *Arabidopsis thaliana* while bypassing all existing patent claims (Broothaerts et al. 2005; Nottenburg and Rodriguez 2007). We have demonstrated the ability of this technology to genetically transform potato (*S. tuberosum* var. Maris Peer) after modifying a standard ATMT transformation protocol to facilitate gene delivery. Using this optimized protocol the transformation frequency (calculated as % of shoots equipped with root systems with the ability to grow in rooting media supplemented with 25 μg/ml hygromycin) of the rhizobia strains was calculated at 4.72%, 5.85% and 1.86% for *S. meliloti*, *R.* sp. NGR234 and *M. loti* respectively, compared to 47.6% for the *A. tumefaciens* control (Wendt et al. in press). While significantly lower than conventional ATMT this finding overcomes an existing challenge for researchers and provides an alternative mode of gene transfer to the potato research community.

12.3.4 In Response to Legislative Issues

European policy towards the cultivation of GM crops is driven by Directive 2001/18/EC. Approvals under 2001/18/EC are granted once dossiers demonstrate the absence of "harm" and that the GM variety is substantially equivalent to a conventional counterpart. But proving "substantial equivalence" as per 2001/18 is akin to asking *"how long is a piece of string"* and has encouraged the level of ambiguity and misinformation that is now intrinsic to the EU "debate" on GM crops. Attempts have been made to offset the issue using cisgenesis [also termed "intragenesis" (Nielsen 2003)] with reports suggesting that cisgenic-derived crops are adequately equivalent to traditionally bred plants (Schouten et al. 2006), which should in effect exempt cisgenic-derived material from EU Directive 2001/18/EC (Jacobsen and Schouten 2008; Rommens et al. 2007). Others argue that it is but an approach to loosen the restrictions on the regulation of genetically modified crops (de Cock Buning et al. 2006; Giddings 2006; Schubert and Williams 2006), concluding that it is unlikely to generate a true cisgenic crop since microbial elements during the *Agrobacterium*-based transformation will remain in the plant genome (Petti et al. 2009).

This could be resolved though through the work of Conner et al. (2007) who have taken the cisgenic approach further, showing that in addition to the gene(s) of interest all regulatory and structural components of a standard transformation vector can be isolated from the potato genome, leading to a "P-vector". Based on this research, it would be possible to create a vector derived solely from potato sequence, which in turn could be used to deliver major *R* genes from other *Solanum* spp. The development of this material would then be in a position to challenge its inclusion in to the Directive 2001/18/EC. In fact, as genomics has provided us with the opportunity to develop such material, the process highlights the urgent need for policymakers to update/refine existing legislation to reflect these advancements. The necessity to redraw the boundaries of current legislation is all the more obvious following the successful application of zinc finger nuclease-based transformation in tobacco (Townsend et al. 2009), which negates the random nature of existing transformation systems with targeted gene disruption now possible via homologous recombination or non-homologous end joining.

Pesticides are fundamental to the way potatoes are currently grown as they provide a cost-effective mechanism to control major weeds, pests and diseases. The efficacy of this approach is now under pressure due to changes in European legislation (revision of 91/414/EEC) and the implementation of the Water Framework Directive. Although the new EU pesticide regulation is unlikely to be enacted through individual state legislatures until 2012, the impact on potato growing could be significant with estimated losses in production of between 22–45% (Clarke et al. 2008) through the likely prohibition of mancozeb and chlorothalonil fungicides along with glufosinate, linuron and pendimethalin-based herbicides. Ironically, these issues could be addressed through the development of engineered varieties (e.g., glyphosate tolerant potato, nematode resistant and blight tolerant potato) but legislative impediments to this approach ensure this will not be achieved in the time left before the new directive is adopted.

12.3.5 Novel Applications

Principally a stable food crop, the introgression of non-food traits into potato germplasm has expanded the remit of the crop into the synthesis of pharmaceutical and industrial products. The first plant-derived pharmaceutical (PDP, human serum albumin) produced in potato, was reported in 1990 (Sijmons et al. 1990). By the end of 2003, a total of 186 applications had been made for the release of PDP plants in the US, Canada and the EU (Sauter and Husing 2005). Although in the US priority is given to maize for PDP production and in Europe it is with tobacco, the potato has increased in popularity due to the potential of the potato tuber as a bioreactor and the considerably lower health risk by avoiding potential pathogen contaminations (Giddings et al. 2000; Ma et al. 2003; Commandeur et al. 2003; Goldstein and Thomas 2004). Whereas animal, fungal, and bacterial productions systems are more expensive due to the prerequisite purification and specialized labor costs. On the contrary potatoes are easy to grow and maintain and the start-up costs are minimal since existing infrastructure for cultivation, harvesting and storage can be used immediately (Giddings et al. 2000; Sparrow et al. 2007). Considering downstream applications, additional advantages include the stability of the PDP inside the tuber (Sparrow et al. 2007) and the accessibility of an already significant market. As such, it is not surprising that many potato-based PDPs are already in clinical trail or in the process of development (Table 12-1).

Ensuring the efficient segregation of food and non-food produce will be critical to alleviate public concerns and is already an issue with policy-makers (EFSA 2009). Fortunately, the phenotypic diversity of potato varieties provides a simple solution to complement standard coexistence measures. A search of the European Cultivated Potato Database (*www.europotato.org*) describes 84 varieties with blue tuber skin color. Of these, varieties such as *S. tuberosum* var. Blue Potato and Blue Catriona maintain the traditional white tuber flesh color, ensuring no technical complications from the high levels of tissue anthocyanidin in the other "blue" varieties. Utilizing one of these genotypes, which contrasts with the traditional white-to-yellow/red tuber skin colors, presents an ideal phenotypic marker with which to prevent seed-mediated gene flow from PDP-producing potatoes into food stocks. Although, the clonal nature of potato is believed to negate the consequence of a pollen-mediated gene flow event, studies have shown that fallen berries can provide a viable F_1 population in a managed field environment (Petti et al. 2007). Again, the issue can be resolved by capitalizing on the diversity of the potato: *S. tuberosum* var. Shetland and Edzell Blue rarely produce flowers, negating the potential for pollen-mediated gene flow, but still possess white tuber flesh color and a blue tuber skin finish.

Table 12-1 Demonstration of *Solanum tuberosum* as a biofactory for pharmaceutical products.

Author	Variety	Active substance/gene product	Application	Class
(Sijmons et al. 1990) (Farran et al. 2002)	Désirée Kennebec	Recombinant human serum albumin	Stabilizing agent and carrier for drug delivery	Recombinant biopharmaceutical
(Tacket et al. 1998) (Mason et al. 1998)	Frito-Lay 1607	*E. coli* heat-labile enterotoxin LT-B[a]	Diarrhoea	Recombinant vaccine
(Mason et al. 1996) (Tacket et al. 2000)	Frito-Lay 1607	Norwalk virus capsid protein[a]	Norwalk virus	Recombinant vaccine
(Ma et al. 1997)	Désirée	Glutamic acid decarboxylase (mouse GAD67)	Autoantigen in diabetes	Recombinant vaccine
(Arakawa et al. 1998)	Bintje	Cholera toxin B subunit—insulin fusion protein	Autoimmune diabetes	Recombinant vaccine
(Richter et al. 2000)	N/A	Hepatitis B surface antigen (HBsAg S-protein)[b]	Hepatitis B virus vaccine	Recombinant antigen
(Artsaenko et al. 1998)	Désirée	Single-chain Fv antibodies	Clinical immunology	Recombinant antibody
(Yu and Langridge 2003)	Bintje	Rotavirus capsid protein VP6	Viral gastroenteritis	Recombinant antigen
(Chong et al. 2000)	Bintje	Human lactoferrin (hLF)	Antimicrobial, immune regulatory agent	Recombinant biopharmaceutical
(Schünmann et al. 2002)	Stellar	Antibody fusion protein (SimpliRed™ HIV diagnostic reagent)	Detection of HIV-1 antibodies in human blood	Diagnostic reagent
(Chakraborty et al. 2000)	A16	Seed albumin gene (*AmA*1)	Amino acid deficiency	Therapeutic agent
(Yu and Langridge 2001)	Bintje	Cholera toxin B and A2 subunits fused to rotavirus enterotoxin and enterotoxigenic *E.coli* fimbrial antigen	Antigen for enteric diseases	Recombinant antigen

[a]clinical trial phase 1 (completed).
[b]clinical trial phase 2.

Similar to PDP development, the potato offers substantial economic potential in the production of industrial products compared to conventional microbial based processes. Several industrial applications have already been reported (Table 12-2) and several more are currently under development (Table 12-2). The most prominent of those are applications surrounding the optimization of starch composition (Roeper 2002), which led to the development of the Amflora™ potato. Through a gene silencing approach, the expression of the major starch producing enzyme (granule starch bound synthase) has been modified so that starch synthesis in the engineered variety is focused solely on amylopectin, as opposed to amylose. Other industrial applications for potato include the production of biodegradable plastic (Conrad 2005; Scheller and Conrad 2005; Mooibroek et al. 2007) that can be used as a substitute for polyacrylate (Mooibroek et al. 2007). A viable substitute for petroleum-based products, the potato-derived product is produced from the waste products of potatoes processed for starch and/or protein extraction.

12.4 Conclusion

There is little doubt that potato production, as with all aspects of global agriculture, will be confronted with significant challenges in the near future. The warming of the climate will produce complex consequences to species numbers and distributions across all of the globe's biodiversity (Pimm 2009). Modeling has provided some insight into the possible impact of climate change across potato production zones and in contrast to other major crops, potato is arguably one of the better positioned crops to respond with its diversity of landraces, the robustness of potato management systems coupled with the application of genomics to dissect the genetic and molecular pathways of potato physiology. The latter has uncovered durable sources of resistance to potatoes nemesis, *P. infestans*, and the introduction of these novel *R* genes into commercial varieties presents an opportunity to reduce the environmental impact of potato farming while increasing margins for the grower. Although the potential to expand potato productivity is region specific, global potato production is set to increase with China alone predicted to be producing up to 81 MT by 2010 (Wang and Zhang 2004). For other regions efforts will be solely focused on preserving existing tuber yield for food production into the future. The increasing function of potato demonstrates its versatility for non-food applications such as the synthesis of therapeutic and/or industrial products and its potential as a worthwhile derivative of bioethanol (Gerbens-Leenes and Hoekstra 2009). So as the development of potato as a crop continues, in no short part due to the output from genomics-based studies, it is hopeful to think that there is little that open, collaborative multi-disciplinary research cannot

Table 12-2 Demonstration of *Solanum tuberosum* as a biofactory for industrial related products.

Author	Variety	Active substance/gene product	Application	Class
(Kok-Jacon et al. 2005)	Kardal, *amf* mutant	Dextran (dextransucrase *DsrS*)	Chromatographic media, biodegradable hydrogels, medical applications (antithrombotic)	Biopolymer
(Kok-Jacon et al. 2007)	Kardal	Alternan (alternansucrase *Asr*)	Functional foods (probiotics), potential alternative to gum arabic	Biopolymer
(Hellwege et al. 2000)	Désirée	Inulin (1-*SST*, 1-*FFT*)	Production of fat-reduced foods, ethanol production	Biopolymer
(Crowell et al. 2008)	Spunta	Vitamin E (At-HPPD, At-HPT)	Enhanced nutritional value, cosmetics (antioxidant)	Vitamin
(Scheller et al. 2001)	Solaria, Desi	Synthetic spidroins	Production of spider silk for various applications	---
(Sugama 1997)	N/A	Oxidized potato starch polymers	Potato starch films for coatings	Biopolymer
(Mooibroek et al. 2007) (Neumann et al. 2005)	Désirée Albatros	Cyanophycin synthetase (*cph*A)[a]	Biodegradable substitute for polycarboxylates (biodegradable plastic)	Biopolymer
(Visser et al. 1991)	PD007 *amf* mutant	Inhibition of granuale-bound starch synthase (amylose free potato starch)	Food industry, paper manufacturing	Biopolymer
(Schwall et al. 2000) (Edwards et al. 2002)	*amf* mutant Désirée	High-amylose starches	Nutrition, paper manufacturing	Biopolymer
(Börnke et al. 2002)	Solara	Palatinose (sucerose isomerase *pal*I).	Sugar substitute	Biopolymer
(Ji et al. 2004)	*amf* mutant	Modified starch granule size	Biodegradable plastic films	Biopolymer

[a]field trials.

achieve in addressing the meta-challenges listed earlier. Yet, it is the non-scientific obstacles that will require the most attention going forward. It is discouraging to note that present anxieties against modified potato varieties are based on perceived risks while at the same time the definitive benefits of a product remain ignored. This serves no purpose but to undermine the development of the crop and ultimately the sector. At a time when global agriculture is set to face unprecedented environmental pressures, it is clear that societal concerns must be addressed in a coherent manner in order to underpin the continued development of the potato.

Acknowledgements

The research of EM and TW is supported through the Teagasc Core Research Fund. TW is also supported by the Teagasc Walsh Fellowship Fund.

References

Arakawa T, Yu J, Chong DKX, Hough J, Engen PC, Langridge WHR (1998) A plant-based cholera toxin B subunit-insulin fusion protein protences against the development of autoimmune diabetes. Nat Biotechnol 16: 934.

Artsaenko O, Kettig B, Fiedler U, Conrad U, During K (1998) Potato tubers as a biofactory for recombinant antibodies. Mol Breed 4: 313–319.

Bielza P, Denholm I, Ioannidis P, Sterk G, Leadbeater A, Leonard P, Jorgensen LN (2008) Declaration of Ljubljana. The Impact of Declining European Pescticide Portfolio on Resistance Management. Outlooks Pest Manag, Dec, pp 246–248.

Börnke F, Hajirezaei M, Sonnewald U (2002) Potato tubers as bioreactors for palatinose production. J Biotechnol 96: 119–124.

Broothaerts W, Mitchell HJ, Weir B, Kaines S, Smith LMA , Yang W, Mayer JE, Rodriguez CR, Jefferson RA (2005) Gene transfer to plants by diverse bacteria. Nature 433: 629–633.

CAMBIA (2009) *http: //www.cambia.org/daisy/cambia/home.html* (Cited 24 July 2009).

Chakraborty S, Chakraborty N, Datta A (2000) Increased nutritive value of transgenic potato by expressing a nonallergenic seed albumin gene from *Amaranthus hypochondriacus*. Pro Natl Acad Sci USA 97.

Chong DKX, Chong DKX, Langridge WHR (2000) Expression of full-length bioactive antimicrobial human lactoferrin in potato plants. Transgen Res 9: 71–78.

Clarke J, Gladders P, Green K, Lole M, Ritchie F, Twining S, Wynn S (2008) Evaluation of the impact on UK agriculture of the proposal for a regulation of the European Parliament and of the Council concerning the placing of plant protection products on the market: *http: //www.adas.co.uk/media_files/ Publications/ecep_ppp_loss_impacts_final_exec_summary_4_june.pdf*

Commandeur U, Twyman RM, Fischer R (2003) The biosafety of molecular farming in plants. AgBiotechNet 5: 110–119.

Conner AJ, Barrel PJ, Baldwin SJ, Lokerse AS, Cooper PA, Erasmuson AK, Nap JA, Jacobs JME (2007) Intragenic vectors for gene transfer without foreign DNA. Euphytica 154: 341–353.

Conrad U (2005) Polymers from plants to develop biodegradable plastics. Trends Plant Sci 10: 511–512.

Crowell E, McGrath J, Douches D (2008) Accumulation of vitamin E in potato (*Solanum tuberosum*) tubers. Transgen Res 17: 205–217.

de Cock Buning T, Lammerts van Bueren ET, Haring MA, de Vriend HC, Struik PC (2006) 'Cisgenic' as a product designation. Nat Biotech 24: 1329–1331.

Edwards A, Vincken J-P, Suurs LCJM, Visser RGF, Zeeman S, Smith A, Martin C (2002) Discrete forms of amylose are synthesized by isoforms of GBSSI in pea. Plant Cell 14: 1767–1785.

EFSA (2006) Opinion of the Scientific Panel on Genetically Modified Organisms on a request from the Commission related to the notification (Reference C/SE/96/3501) for the placing on the market of genetically modified potato EH92-527-1 with altered starch composition, for cultivation and production of starch, under Part C of Directive 2001/18/EC from BASF Plant Science. EFSA J 323: 1–20.

EFSA (2009) Scientific Opinion on Guidance for the risk assessment of genetically modified plants used for non-food or non-feed purposes. EFSA J l: 1164.

FAO (2008) International Year of the Potato. *http: //wwwpotato2008org/en/aboutiyp/indexhtml*

Farmers Guardian (2009) Politicians delaying a blight-resistant potato. *http:// wwwfarmersguardiancom/storyasp?sectioncode=1&storycode=27421* (Cited 9th July 2009).

Farran I, Sánchez-Serrano J, Medina J, Prieto J, Mingo-Castel A (2002) Targeted expression of human serum albumin to potato tubers. Transgen Res 11: 337–346.

Flannery M-L, Thorne F, Kelly P, Mullins E (2005) An economic cost-benefit analysis of GM crop cultivation: an Irish case study. AgBioforum—J Agrobiotechnol Manag Econ 7: 1.

Gerbens-Leenes W A, Hoekstra Y (2009) The water footprint of bioenergy. Proc Natl Acad Sci USA 106: 10219–10223.

Giddings G, Allison G, Brooks D, Carter A (2000) Transgenic plants as factories for biopharmaceuticals. Nat Biotechnol 18: 1151–1155.

Giddings VL (2006) 'Cisgenic' as a product designation. Nat Biotech 24: 1329–1329.

Goldstein DA, Thomas JA (2004) Biopharmaceuticals derived from genetically modified plants. QJM, An Int J Med 97: 705–716.

Grusenmeyer D (2008) Potato Industry Challenges & Opportunities. Farm Viability Institute, Syracuse, NY, USA: *http: //www.nyfvi.org/documents/40.pdf*

Haverkort A, Boonekamp P, Hutten R, Jacobsen E, Lotz L, Kessel G, Visser R, van der Vossen E (2008) Societal costs of late blight in potato and prospects of durable resistance through cisgenic modification. Potato Res 51: 47–57.

Hellwege EM, Czapla S, Jahnke A, Willmitzer L, Heyer AG (2000) Transgenic potato (*Solanum tuberosum*) tubers synthesize the full spectrum of inulin molecules naturally occuring in globe artichoke (*Cynara scolymus*) roots. Pro Natl Acad Sci USA 97: 8699–8704.

Herres P, Schippers-Rozenboom M, Jacobsen E, Visser R (2002) Transformation of a large number of potato varieties: genotype-dependent variation in efficiency and somaclonal variability. Euphytica 124: 13–22.

Hijmans R (2003) The effect of climate change on global potato production. Am J Potato Res 80: 271–280.

Hinds H (2008) Influence of Recent Climate Change on Late Blight Risk in the UK. Euroblight Workshop, Hamar, Norway.

Houghton J, Ding Y, Griggs D, Noguer M, Van Der Linden P, Xiaosu D, Maskell K, Johnson C (2001) Climate Change 2001. The Scientific Basis. Contribution of Working Group I to the Third Assessment Report of the Intergovernmental Panel on Climate Change (IPCC). Cambridge Univ Press, Cambridge, UK.

IPCC (2007) Climate Change 2007: Synthesis Report. Contribution of Working Groups I, II and III to the Fourth Assessment Report of the Intergovernmental Panel on Climate Change. In: R Pachauri, A Reisinger (eds) IPCC, Geneva, Switzerland, p 104.

Jacobsen E, Schouten H (2008) Cisgenesis, a New Tool for Traditional Plant Breeding, Should be Exempted from the Regulation on Genetically Modified Organisms in a Step by Step Approach. Potato Research 51: 75–88.

Ji Q, Oomen RJFJ, Vincken J-P, Bolam DN, Gilbert HJ, Suurs LCJM, Visser RGF (2004) Reduction of starch granule size by expression of an engineered tandem starch-binding domain in potato plants. Plant Biotechnol J 2: 251–260.

Kalushkov P, Batchvarova R (2005) Effectiveness of Bt NewLeaf potato to control Leptinotarsa decemlineata in Bulgaria. Biotechnol Biotechnol Equip 19: 28–34.

Karim S, Aronsson H, Ericson H, Pirhonen M, Leyman B, Welin B, Mäntylä E, Palva E, Van Dijck P, Holmström K-O (2007) Improved drought tolerance without undesired side effects in transgenic plants producing trehalose. Plant Mol Biol 64: 371–386.

Kok-Jacon GA, Vincken J-P, Suurs LCJM, Wang D, Liu S, Visser RGF (2005) Production of dextran in transgenic potato plants. Transgen Res 14: 385–395.

Kok-Jacon G, Vincken J-P, Suurs L, Wang D, Liu S, Visser R (2007) Expression of alternansucrase in potato plants. Biotechnol Lett 29: 1135–1142.

Lees A, Cooke D, Carnegie S, Stewart JM, Sullivan L, Williams N (2008) *P. infestans* populations changes: implications. Euroblight Workshop, Hamar, Norway.

Liu Z, Halterman D (2006) Identification and characterization of RB-orthologous genes from the late blight resistance wild potato species *Solanum verrucosum*. Physiol Mol Plant Pathol 69: 230–239.

Ma JK-C, Drake PMW, Christou P (2003) The production of recombinant pharmaceutical proteins in plants. Nat Rev Genet 4: 794–805.

Ma S-W, Zhao D-L, Yin Z-Q, Mukherjee R,Singh B,Qin H-Y, Stiller CR, Jevnikar AM (1997) Transgenic plants expressing autoantigens fed to mice to induce oral immune tolerance. Nat Med 3: 793–796.

Mason HS, Ball JM, Shi J-J, Jiang X, Estes MK, Arnetzen CJ (1996) Expression of Norwalk virus capsid protein in transgenic tobacco and potato and its oral immunogenity in mice. Pro Natl Acad Sci USA 93: 5335–5340.

Mason HS, Hag TA, Clements JD, Arntzen CJ (1998) Edible vaccine protects mice against *Eschlerichia coli* heat-labile enterotoxin(Lt): potatoes expressing a synthetic LT-B. Vaccine 16: 1336–1343.

Monsanto (1999) NewLeaf Plus Potato. *http: //wwwmonsantocouk/primer/newleafhtml*

Mooibroek H, Oosterhuis N, Giuseppin M, Toonen M, Franssen H, Scott E, Sanders J, Steinbüchel A (2007) Assessment of technological options and economical feasibility for cyanophycin biopolymer and high-value amino acid production. Appl Microbiol Biotechnol 77: 257–267.

Mullins E, Milbourne D, Petti C, Doyle-Prestwich BM, Meade C (2006) Potato in the age of biotechnology. Trends Plant Sci 11: 254–260.

Neumann K, Stephan DP, Ziegler K, Huhns M, Broer I, Lockau W, Pistorius EK (2005) Production of cyanophycin, a suitable source for the biodegradable polymer polyaspartate, in transgenic plants. Plant Biotechnol J 3: 249–258.

Nielsen KM (2003) Transgenic organisms-time for conceptual diversification? Nat Biotech 21: 227–228.

Nottenburg C, Rodriguez CR (eds) (2007) *Agrobacterium*-mediated Gene Transfer: A Lawyer's Perspective. Springer, New York, USA.

Park TH, Vleeshouwers VGAA, Jacobsen E, Vossen Evd, Visser RGF (2009) Molecular breeding for resistance to *Phytophthora infestans* (Mont.) de Bary in potato (*Solanum tuberosum* L.): a perspective of cisgenesis. Plant Breed 128: 109–117.

Petti C (2007) Elucidating the propensity to genetically transform *Solanum tuberosum* L. and investigating the consequences for subsequent risk assessment studies. Biology. National Univ of Ireland, Maynooth, Maynooth, Ireland, p 384.

Petti C, Meade C, Downes M, Mullins E (2007) Facilitating co-existence by tracking gene dispersal in conventional potato systems with microsatellite markers. Environ Biosaf Res 18.

Petti C, Wendt T, Meade C, Mullins E (2009) Evidence of genotype dependency within *Agrobacterium tumefaciens* in relation to the integration of vector backbone sequence in transgenic *Phytophthora infestans*-tolerant potato. J Biosci Bioeng 107: 301–306.

Pimm SL (2009) Climate disruption and biodiversity Curr Biol 19: R595–R601.

Richter LJ, Thanavala Y, Arntzen CJ, Mason HS (2000) Production of hepatitis B surface antigen in transgenic plants for oral immunization. Nat Biotechnol 18: 1167–1171.

Roeper H (2002) Renewable raw materials in Europe—Industrial utilisation of starch and sugar. Starch/Stärke 54: 89–99.

Rommens CM, Haring MA, Swords K,Davies HV, Belknap WR (2007) The intragenic approach as a new extension to traditional plant breeding. Trends in Pl Sci 12: 397–403.

Sauter A, Hüsing B (2005) Green genetic engineering—transgenic plants of the 2nd and 3rd generation. TAB (Büro für Technikfolgen-Abschätzung beim Deutschen Bundestag) Report 104.

Schafleitner R, Gutierrez Rosales RO, Gaudin A, Alvarado Aliaga CA, Martinez GN, Tincopa Marca LR, Bolivar LA, Delgado FM, Simon R, Bonierbale M (2007) Capturing candidate drought tolerance traits in two native Andean potato clones by transcription profiling of field grown plants under water stress. Plant Physiol Biochem 45: 673–690.

Scheller J, Conrad U (2005) Plant-based material, protein and biodegradable plastic. Curr Opin Plant Biol 8: 188–196.

Scheller J, Guhrs K-H, Grosse F, Conrad U (2001) Production of spider silk proteins in tobacco and potato. Nat Biotechnol 19: 573–577.

Schouten HJ, Krens FA, Jacobsen E (2006) Cisgenic plants are similar to traditionally bred plants: international regulations for genetically modified organisms should be altered to exempt cisgenesis. EMBO reports 7: 750–753.

Schubert D, Williams D (2006) 'Cisgenic' as a product designation. Nat Biotech 24: 1327–1329.

Schünmann PHD, Coia G, Waterhouse PM (2002) Biopharming the SimpliRED™ HIV diagnostic reagent in barley, potato and tobacco. Mol Breed 9: 113–121.

Schwall GP, Safford R, J.Westcott R, Jeffcoat R, Tayal A, Shi Y-C, Gidley MJ, Jobling SA (2000) Production of a very-hygh-amylose potato starch by inhibition of SBE A and B. Nat Biotechnol 18: 551–554.

Sijmons PC, Dekker BM, Schrammeijer B, Verwoerd TC, Van den Elzen PJM, Hoekema A (1990) Production of correctly processed human serum albumin in transgenic plants. Biotechnology 8: 217–221.

Song J, Bradeen JM, Naess SK, Raasch JA, Wielgus SM, Haberlach GT, Liu J, Kuang H, Austin-Phillips S, Buell CR, Helgeson JP, Jiang J (2003) Gene RB cloned from *Solanum bulbocastanum* confers broad spectrum resistance to potato late blight. Proc. Natl Acad Sci USA 100: 9128–9133.

Sparrow P, Irwin J, Dale P, Twyman R, Ma J (2007) Pharma-Planta: road testing the developing regulatory guidelines for plant-made pharmaceuticals. Transgen Res 16: 147–161.

Stein AJ, Rodríguez-Cerezo E (2009) The global pipeline of new GM crops. Implications of asynchronous approval for international trade. Joint Research Centre of the Institute of Prospective Technological Studies, Seville, Spain. p 114.

Stiller I, Dulai S, Kondrák M, Tarnai R, Szabó L, Toldi O, Bánfalvi Z (2008) Effects of drought on water content and photosynthetic parameters in potato plants expressing the trehalose-6-phosphate synthase gene of *Saccharomyces cerevisiae*. Planta 227: 299–308.

Sugama T (1997) Oxidized potato-starch films as primer coatings of aluminium. J Materials Sci 32: 3995–4003.

Tacket CO, Mason HS, Losonsky G, Clements JD, Levine MM, Arntzen CJ (1998) Immunogenicity in humans of a recombinant bacterial-antigen delivered in transgenic potato. Nat Med 4: 607–609.

Tacket CO, Mason HS, Losonsky G, Estes MK, Levine MM, Arntzen CJ (2000) Human immune responses to a novel Norwalk virus vaccine delivered in transgenic potatoes. J Infect Dis 182: 302–305.

Townsend JA, Wright DA, Winfrey RJ, Fu F, Maeder ML, Joung JK, Voytas DF (2009) High-frequency modification of plant genes using engineered zinc-finger nucleases. Nature 459: 442–445.

Tubiello FN, Ewert F (2002) Simulating the effects of elevated CO2 on crops: approaches and applications for climate change. Eur. J. Agron. 18: 57–61.

UN Secretariat (2007) World Population Prospects, The 2006 Revision. The Department of Economic and Social Affairs, United Nations, New York, USA.

Van Der Vossen E, Sikkema A, te Lintel Hekkert B, Gros J, Stevens P, Muskens M, Wouters D, Pereira A, Stiekema W, Allefs S (2003) An ancient R gene from the wild potato species *Solanum bulbocastanum* confers broad-spectrum resistance to *Phytophthora infestans* in cultivated potato and tomato. Plant J 36: 867–882.

van der Vossen E, Gros J, Sikkema A, Muskens M, Wouters D, Wolters P, Pereira A, Allefs S (2005) The *Rpi-blb2* gene from *Solanum bulbocastanum* is an *Mi-1* homolog conferring broad-spectrum late blight resistance in potato. Plant J 44: 208–222.

Vasquez-Robinet C, Mane SP, Ulanov AV, Watkinson JI, Stromberg VK, De Koeyer D, Schafleitner R, Willmot DB, Bonierbale M, Bohnert HJ, Grene R (2008) Physiological and molecular adaptations to drought in Andean potato genotypes. J Exp Bot 59: 2109–2123.

Visser RGF, Somhorst I, Kuipers GJ, Ruys NJ, Feenstra WJ, Jacobsen E (1991) Inhibition of the expression of the gene for granule-bound starch synthase in potato by antisense constructs. Mol Gen Genet 225: 289–296.

Vleeshouwers VG, Rietman H, Krenek P, Champouret N, Young C, Oh S-K, Wang M, Bouwmeester K, Vosman B, Visser R, Jacobsen E, Govers F, Kamoun S, Van der Vossen EA (2008) Effector denomics accelerates discovery and functional profiling of potato disease resistance and *Phytophthora infestans* avirulence genes. PloS ONE 3: e2875.

Wang Q, Zhang W (2004) China's potato industry and potential impacts on the global market. Am J Potato Res 81: 101–109.

Wendt T, Doohan F, Wincklemann D, Mullins E (in press) Gene transfer into *Solanum tuberosum* via *Rhizobium* spp. Transgenic Res.

White S, Shaw D (2008) Resistance of Sárpo clones to the new, aggressive strain, of *Phytophthora infestans*, Blue 13. Euroblight workshop, Hamar, Norway.

Wolf J (2002) Comparison of two potato simulation models under climate change. II. Application of climate change scenarios. Clim Res 21: 187–198.

WPC Inc (2009) 7th World Potato Congress: *http: //wwwpotatocongressorg*

Yu J, Langridge WHR (2001) A plant-based multicomponent vaccine protects mice from enteric diseases. Nat Biotechnol 19: 548–552.

Yu J, Langridge W (2003) Expression of rotavirus capsid protein VP6 in transgenic potato and its oral immunogenicity in mice. Transgen Res 12: 163–169.

Index

B

C

D

E

F

G

H

I

J

K

T

U

V

W

X

Y

Z

Color Plate Section

Chapter 1

Figure 1-1 Potato morphology. (a) The flower of primitive wild potato species classified as superseries *Stellata* lack a fused corolla, giving the flower a star-shaped or stellate appearance. (b) In contrast, the five petals of the flower of the cultivated potato (*Solanum tuberosum* ssp. *tuberosum*) and other species classified as superseries *Rotata* are fused, giving the flower a wheel-shaped or rotate appearance. Potato flowers range in color from white to pinkish to purple. (c) The potato tuber is an underground stem modified for carbohydrate storage. Potato tubers come in a variety of shapes, sizes, and colors.

Chapter 3

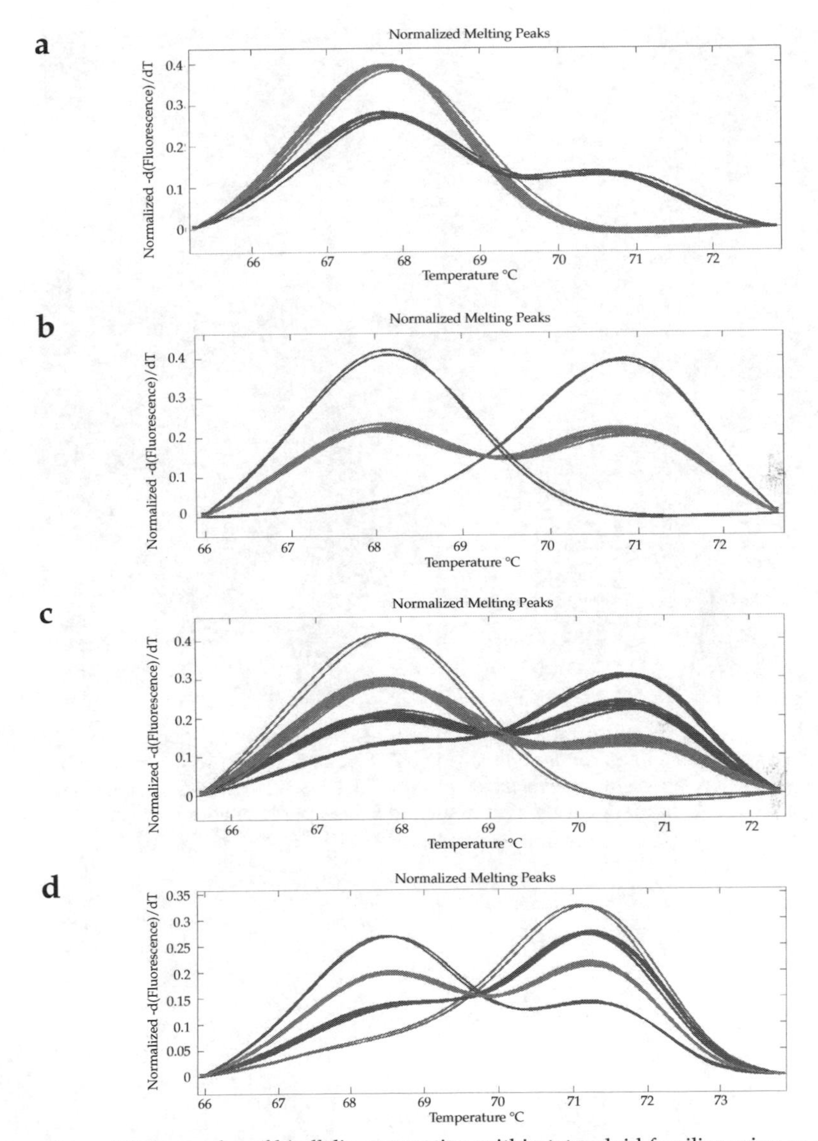

Figure 3-1 Examples of bi-allelic segregation within tetraploid families using an unlabeled probe HRM assay for the *dfr* locus, described in De Koeyer et al. (2010). The peak at 71°C represents the perfect match to the dominant R allele producing red pigment and the peak at 67.5°C represents a mismatch or non-red (r) pigment allele. The relative heights of the two peaks indicate the genotype at this locus. **A.** rrrr (*grey* samples) x Rrrr (*red* samples) produce rrrr and Rrrr progeny in a 1:1 ratio; **B.** RRRR (*red*) x rrrr (*blue*) produce all RRrr (*grey*) progeny; **C.** RRrr (*red*) x Rrrr (*grey*) produce rrrr (*green*), Rrrr, RRrr, and RRRr (*blue*) progeny; **D.** RRRR (*green*) x Rrrr (*blue*) produce RRrr (*grey*), and RRRr (*red*) progeny.

Chapter 5

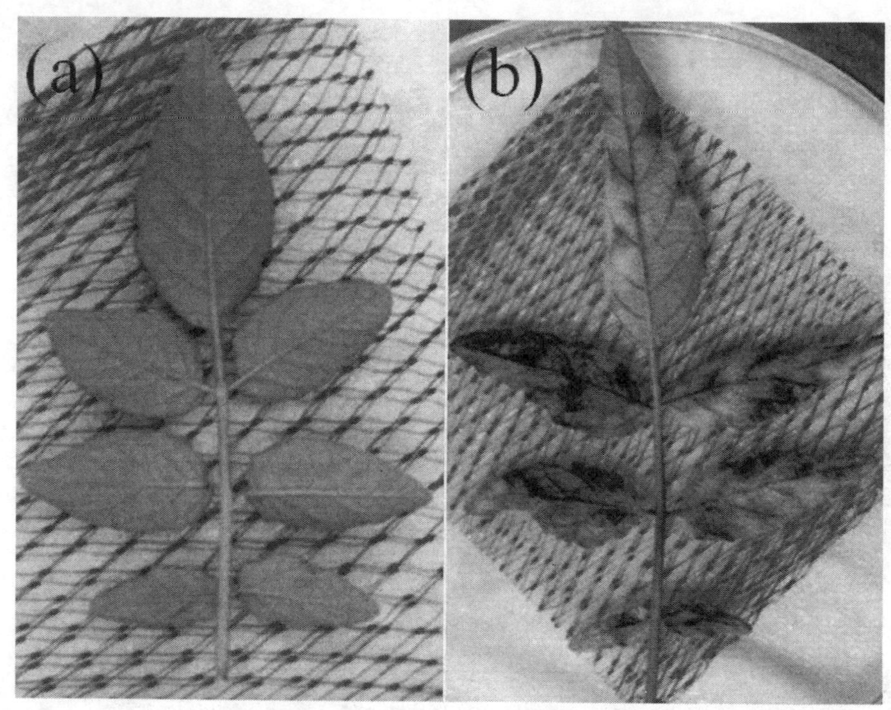

Figure 5-1 Backcross 1 progeny from a cross between resistant clone *S. pinnatisectum* (PI# 253214) and susceptible clone *S. cardiophyllum* (PI# 347759), backcrossed to the susceptible parent, 6 days post inoculation with *P. infestans* isolate MSU96 (US-8, A2 mating type), incubated at 18°C under 12 hour light/dark cycles, **a** resistant and **b** susceptible phenotypes.

Chapter 7

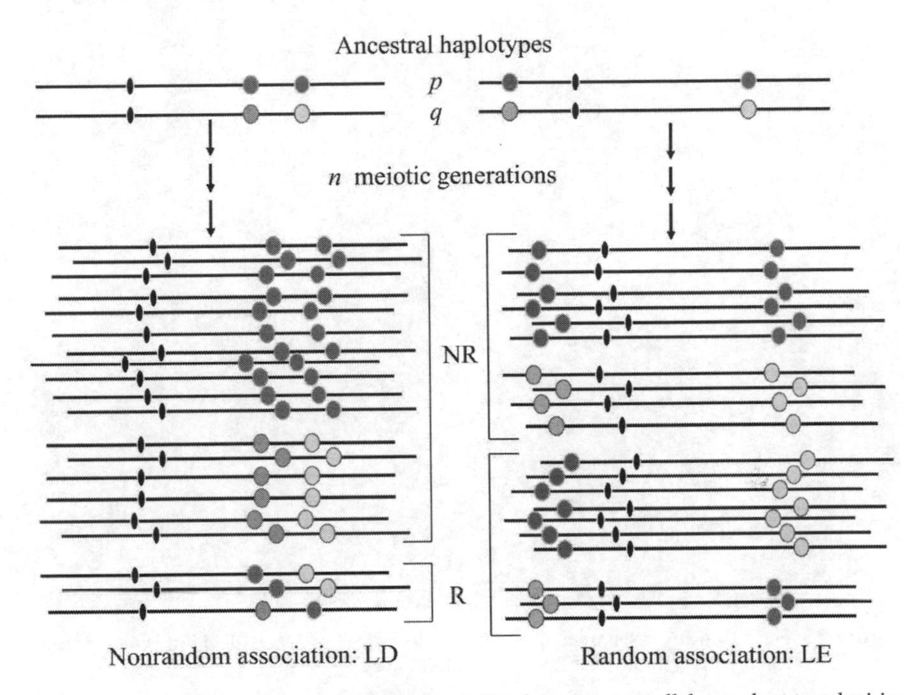

Figure 7-1 Nonrandom versus random association between two alleles each at two loci in populations of chromosomes descending from two ancestral haplotypes. Loci are symbolized by spheres and alleles by different colored spheres. On the left, the loci are closely linked. After n meiotic generations, the majority of the descendant chromosomes carry the same, non-recombinant (NR) allele configurations (haplotypes) as the ancestral chromosomes with haplotype frequencies p and q. The alleles are in linkage disequilibrium (LD). On the right, the loci are more distantly linked or unlinked. After n meiotic generations, non-recombinant (NR) and recombinant (R) haplotypes are equally distributed in the descendant chromosomes. The alleles are in linkage equilibrium (LE).

Figure 7.3 Field trial for resistance to late blight (courtesy of Dr. H.-R. Hofferbert, Böhm-Nordkartoffel Agrarproduktion GbR, Ebstorf, Germany).

Chapter 8

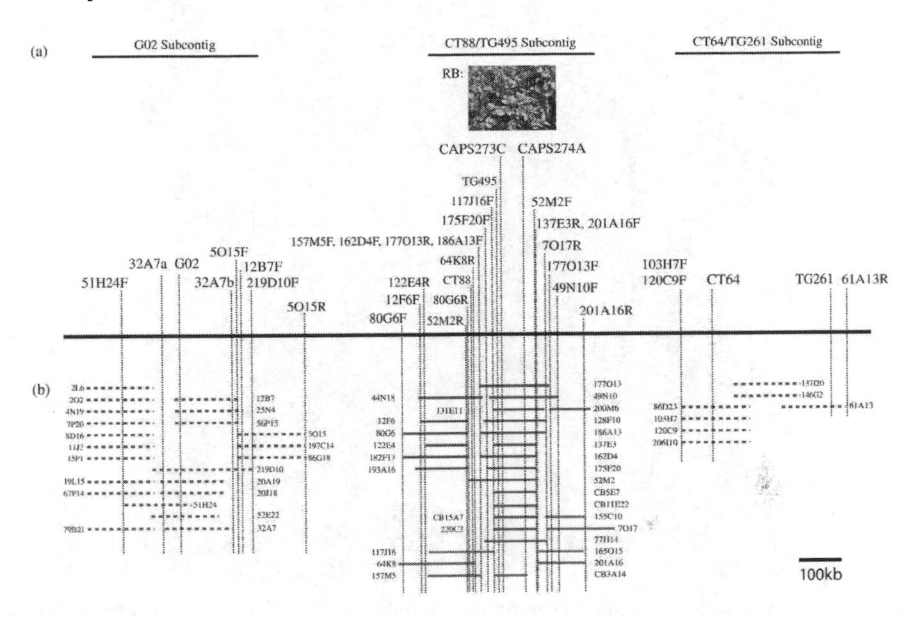

Figure 8-1 A representation of chromosome walking efforts at the *S. bulbocastanum* late blight resistance locus *RB*. (a) An integrated physical map of the *S. bulbocastanum* chromosome VIII in the *RB* vicinity. Genetically mapped markers include RFLP (CT64, CT88, TG261, TG495), RAPD (G02), and BAC-derived markers. RFLP markers were selected to initiate chromosome walking efforts; BAC markers are the result of iterative steps in the walk. Marker locations are based on combined genetic and physical data for the region. All measurements between markers reflect actual or calculated physical distances. Subcontig groupings are indicated. Late blight resistance (*RB*) is mapped to a region flanked by BAC-derived markers CAPS273C and CAPS274A. (b) BAC contigs. All BAC clones were isolated from the *S. bulbocastanum* BAC library developed by Song et al. (2000). BAC names indicate plate of origin and register location within a plate. Individual BAC clones are derived from the *RB* (resistant) homolog (*green*), the *rb* (susceptible) homolog (*blue*), or are of unknown homolog origin (*red*). BACs represented by *dashed lines* are of unknown insert size and their placement is tentative. Insert sizes have been estimated for BACs represented by *solid lines;* these BACs are drawn to scale (note 100-kb marker). [Reprinted with kind permission from Springer Science+Business Media: Bradeen et al. (2003) Concomitant reiterative BAC walking and fine genetic mapping enable physical map development for the broad-spectrum late blight resistance region, *RB*. *Mol Genet Genom* 269:603-611 (Fig. 2).].

Figure 8-3 The transgene *RB* imparts to potato agriculturally meaningful levels of resistance to foliar late blight under field conditions in the absence of fungicides. Photograph of a late blight nursery in Minnesota in 2005. This picture was taken approximately four weeks after inoculation of the field with *Phytophthora infestans* US8. Transgenic plants (green) remain healthy and relatively disease free compared to non-transgenic potato cultivar NorChip (rows of dead, infected plants). Higher copy numbers of the potato *RB* transgene correspond to enhanced transcript and late blight resistance levels. [Reprinted with permission from Bradeen et al. (2009) Mol Plant-Microbe Interact 22: 437–446].

Chapter 9

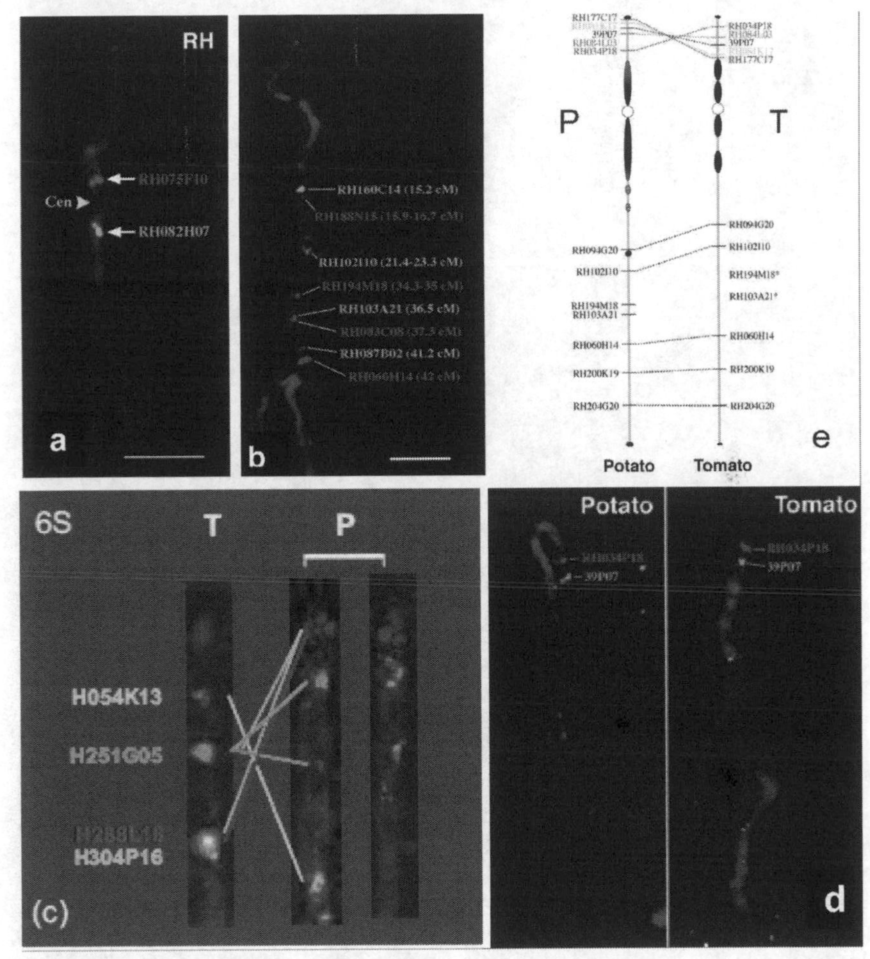

Figure 9-1 FISH mapping of AFLP marker-anchored BAC clones on potato pachytene chromosome VI.

(Panels a-b and d-e – reprinted with permission from Iovene et al. 2008; panel c – reprinted with permission from Tang et al. 2008). (a) Determination of the genetic position of the centromere of potato chromosome VI of genotype RH by FISH mapping of BACs RH075F10 and RH082H07. Scale bar = 5 µm. (b) FISH mapping of eight BACs located in the euchromatic region on the long arm of the chromosome. Scale bar = 5 µm. (c) Cross-species FISH of tomato BACs on the short arm of pachytene chromosome VI of tomato (T) and potato (P) showed inverted order between the homeologues. BAC H251G05 (green) produced a large and small focus on the potato chromosome, suggesting a breakpoint in this BAC and a putative chromosomal rearrangement (Tang et al. 2008). (d) Cross-species FISH of potato BACs on the short arm of pachytene chromosome VI of potato (P) and tomato (T). (e) The comparative chromosomal positions of potato BACs on potato (P) and tomato (T) pachytene chromosome VI. Reproduced with permission of Genetics Society of America, Genetics Editorial Office.

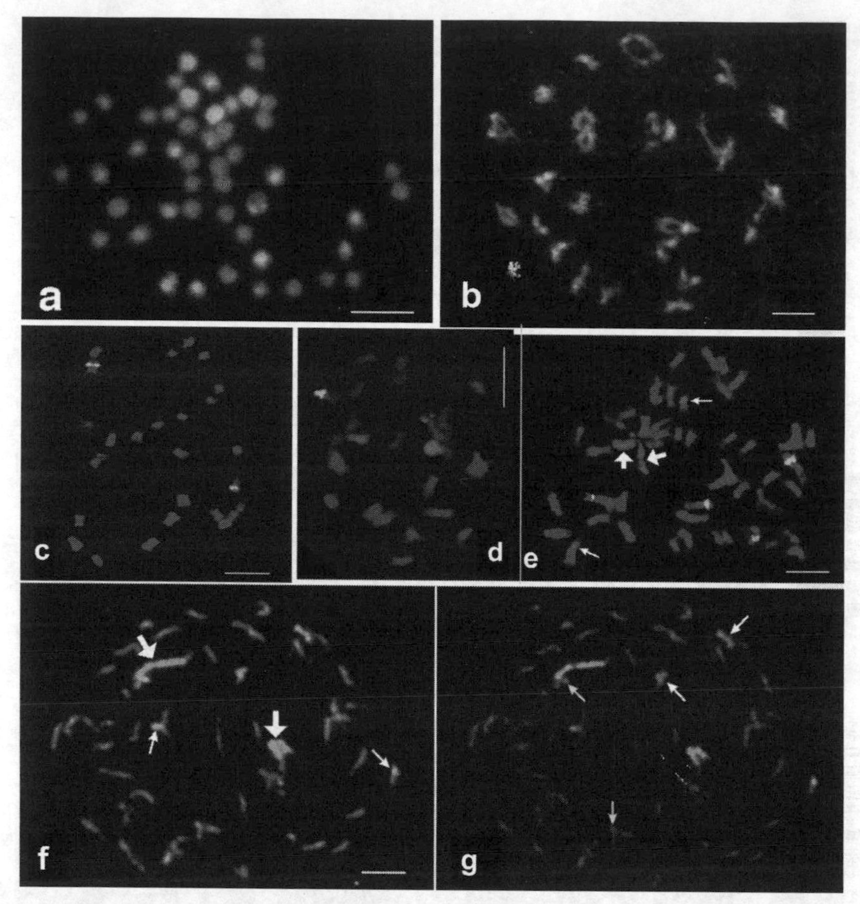

Figure 9-2 GISH analysis of series *Longipedicellata* tetraploid species *Solanum stoloniferum* ($2n = 4x = 48$) reprinted with permission from Pendinen et al. 2008. (a) Somatic chromosomes of *S. stoloniferum* probed with labeled DNA from its putative diploid ($2n = 2x = 24$) progenitor species - *S. verrucosum* (red) and *S. jamesii* (green). (b) GISH analysis of chromosomal pairing in diakinesis of *S. stoloniferum*. Pairing between *A* genome chromosomes (red, detected by labeled DNA of *S. verrucosum*) and *B* genome chromosomes (green, detected by labeled DNA of *S. jamesii*) were not observed.

(c-e) FISH mapping of 45S rDNA (red) and 5S rDNA (green) in (e) tetraploid *S. stoloniferum* and its putative diploid progenitor species (c) *S. verrucosum* and (d) *S. jamesii*. In *S. stoloniferum*, 45S rDNA hybridization sites were observed (large and small arrows). (f) Somatic chromosomes of *S. stoloniferum* probed with labeled genomic DNA from *S. verrucosum* (red) and *S. andreanum* (green). A large fragment (big arrows) and a small fragment (small arrows) from two pairs of *S. stoloniferum* chromosomes showed bright green color. Color differentiation was not observed on the rest of the chromosomes. (g) The same metaphase cell as in (f) was hybridized with 45S ribosomal RNA gene probe (red). The two large white arrows and two small yellow arrows FISH sites are not located on the chromosomes with color differentiation in GISH analysis. Scale bars = 5 µm. © 2008 NRC Canada or its licensors. Reproduced with permission.